Food Safety

07343901
614.3 ROB
ROBERTS
811840

£22.05 U

BATH ... OF HIGH...

DISCARD
B.C.H.E. – LIBRARY
00111058

BATH COLLEGE OF HIGHER EDUCATION
NEWTON PARK LIBRARY
DISCARD
Stock No. 811840
Class No. 614.3

Food Safety

Edited by

HOWARD R. ROBERTS, PH.D.

Vice President, Science and Technology
National Soft Drink Association

A WILEY-INTERSCIENCE PUBLICATION

John Wiley & Sons
New York · Chichester · Brisbane · Toronto

Copyright © 1981 by John Wiley & Sons, Inc.

All rights reserved. Published simultaneously in Canada.

Reproduction or translation of any part of this work beyond that permitted by Sections 107 or 108 of the 1976 United States Copyright Act without the permission of the copyright owner is unlawful. Requests for permission or further information should be addressed to the Permissions Department, John Wiley & Sons, Inc.

Library of Congress Cataloging in Publication Data:

Food Safety.

"A Wiley Interscience Publication."
Includes bibliographies and indexes.
1. Food adulteration and inspection. 2. Food contamination. 3. Food additives. I. Roberts, Howard Richard, 1932–

TX531.F65 363.1'79 80-25335
ISBN 0-471-06458-0

Printed in the United States of America

10 9 8 7 6 5 4 3 2 1

Contributors

Sylvia M. Charbonneau, Ph.D.
Health Protection Branch
Health and Welfare Canada
Ottawa, Canada

John Doull, M.D., Ph.D.
Department of Pharmacology
University of Kansas Medical Center
Kansas City, Kansas

Elmer H. Marth, Ph.D.
Department of Food Science and
the Food Research Institute
University of Wisconsin
Madison, Wisconsin

Ian C. Munro, Ph.D.
Health Protection Branch
Health and Welfare Canada
Ottawa, Canada

Albert E. Pohland, Ph.D.
Bureau of Foods
Food and Drug Administration
Washington, D.C.

Howard R. Roberts, Ph.D.
National Soft Drink Association
Washington, D.C.

Joseph V. Rodricks, Ph.D.
Clements Associates
Washington, D.C

Vala Jean Stults, R.D., Ph.D.
Nutrition Consultant
Washington, D.C.

Preface

Food safety is one of the most often debated subjects of today, but usually the debate focuses on only individual pieces of the total puzzle. A particular food additive or all food additives are condemned or defended, environmental contaminants are deplored or ignored, dietary goals and guidelines are extolled or criticized, but there have been relatively few attempts to systematically evaluate the entire puzzle. That is the purpose of this book. It places the major areas of potential hazard associated with the food supply in perspective by considering the relative risks they present to consumers. Each area of potential hazard is systematically examined relative to its nature, extent, and importance.

Starting with an overview of the major areas impacting on food safety, the text material is organized to take the reader through foodborne hazards of microbial origin, nutritional hazards, environmental contaminants, food hazards of natural origin, and food additives. The text concludes with a discussion of food safety and toxicology, including a critique of current food safety law. Each chapter is a self-contained treatment of its particular area but is written to fit with other chapters to convey a total perspective on food safety. Such a perspective is becoming of increasing importance not only to specialists in the food and nutrition areas and to policymakers, but also to each of us as consumers.

This book is designed to serve both as a textbook and a ready reference work for students and professionals in food science and technology, the biological sciences, and health care and for anyone with more than a passing interest in food safety. No particular background is assumed for the reader other than interest in food safety. Chapter authors have addressed their material to the general reader not necessarily trained in that particular subject area. The attempt has been made to provide sufficient depth to satisfy the technically trained reader without confusing the general reader. In each case references to a broad range of literature have been provided so that those interested may pursue the subject further.

HOWARD R. ROBERTS

Washington, D.C.
January 1981

Acknowledgments

I express my appreciation for the extensive time and effort devoted by my coauthors and to all of our colleagues in government, industry, and academia who inspired ideas, furnished reference material, and provided encouragement during the preparation of this text. In particular, I am especially grateful to my mentor and friend, Dr. Virgil O. Wodicka, to whom the basic credit for the concept of this book must be given.

Many of my former colleagues in the Bureau of Foods, Food and Drug Administration are due a special debt of gratitude for their assistance in identifying background material and in reviewing draft manuscripts. These include Mrs. Maryln K. Cordle, Dr. Allan L. Forbes, Dr. M. R. S. Fox, Dr. Victor P. Frantalli, Dr. Richard M. Jacobs, Dr. A. C. Kolbye, Jr., Dr. Corbin I. Miles, Dr. Y. K. Park, and Mr. Richard J. Ronk. Special thanks for assistance must also go to Dr. Richard L. Hall of McCormick and Company and Dr. Kenneth D. Fisher of the Federation of American Societies for Experimental Biology. The patient effort of Margie Coakley in typing and retyping much of the manuscript is also gratefully acknowledged.

A contribution from the College of Agricultural and Life Sciences, University of Wisconsin—Madison relative to the preparation of Chapter 2 is also acknowledged.

Contents

Food Safety

CHAPTER 1

Food Safety in Perspective

H. R. ROBERTS

Newspaper and magazine articles, television and radio reports, scientific and technical literature, government activity, and public opinion polls all bear witness to a current preoccupation with the nature and the safety of our food supply. We are constantly being told that certain components of our food and what and how much we are eating are jeopardizing our health. For example, in recent years a wide variety of popular books, as well as government reports, have focused on such issues as the controversy over food additives, the role of diet in children's hyperactivity and in heart disease and cancer, and the importance of vitamins (e.g., 1–11). Best-seller lists are seldom without one or more entries representing the latest diet [e.g., the Scarsdale (12) and Pritkin (13) diets]. Even Dr. David Reuben, who explained everything we wanted to know about sex, quickly turned his attention to nutrition (14). Congress, not often in the habit of avoiding areas of popular attention, has repeatedly entered the fray. In addition to addressing the saccharin and nitrite issues, congressional activity has ranged from the recommendation of national dietary goals (15)—inspiring debate in the nutrition community (16)—to a general examination of food safety policy (17). Then, too, scientists, once content to subject preliminary findings to peer review and further research, in some cases now seek popular notice of their unsubstantiated conclusions—finding ready allies to that end in the media and adding to the concern about our food supply (18).

Our preoccupation with food safety has grown along with and has been reenforced by fairly significant changes in the nature of our food supply, especially those that have occurred over the last 20 years or so. In that period the shift of the population to urban areas has continued, accompanied by a concentration and a mechanization of agricultural production. Today, less than 5% of our population produces not only the food consumed by our generally well-fed citizens, but also significant surpluses used in other parts of the world. This highly efficient production system, together with equally efficient processing and distribution systems, has not only provided the convenience and variety of foods demanded by current lifestyles, but has also generated a wide variety of protest.

Of the nearly $200 billion worth of food consumed annually, about 55% is subjected to some processing before distribution. This level of food processing is largely responsible for the variety of food items—on the order of 5000 to 10,000 in today's supermarkets—and for the convenience and the uniformity of quality associated with current lifestyles. Another facet of today's lifestyles is reflected in the expenditure of over one-third of our food dollars on prepared foods purchased and/or consumed away from the home. Trends seem to indicate further increases both in the proportion of processed foods and in away-from-home purchase and consumption. Another trend is the increasing personal interest in nutrition. There are probably few families in the United States in which at some time at least one member is not dieting to gain weight, to lose weight, or to avoid "additives" or other dietary components (e.g., vegetarianism).

On the one hand, the trends in the nature of our food supply have been encouraged by consumer acceptance of the results. At the same time, however, there is some dissatisfaction with highly processed foods and a yearning for the "good old days" when foods were prepared "from scratch." [R. L. Hall, in an article entitled "Faith, Fad, Fear and Food" (19), discusses the assertion that the craze for health foods is a substitute for conventional religion.]

FOOD SAFETY IN HISTORY

Although current concern about food safety is a relatively recent phenomenon, safety problems in the food supply go back to the origins of *Homo sapiens*. Indeed, the first toxicologist was probably the prehistoric person who deduced that unknown plants and fruits were safe to eat if they were consumed without obvious harm by animals. (It is interesting to note that millions of years later we are still making food safety decisions on the basis of effects observed in animals, except that now we are infinitely more sophisticated about looking for those effects.) Observation of the animals left something to be desired, however. As Tannahill (20) has noted, neanderthal people, dependent on hunting, fishing, and plant collecting for their food, were subject to vitamin deficiencies, seasonal malnutrition, plant poisons, and contaminated food.

Lacking any means of preservation, prehistoric people apparently consumed meat left over from the hunt with little, if any, concern for decomposition or its effects on mortality. Later, the value of freezing and drying—climate permitting—and smoking was recognized for preserving meat and fish. Still later the use of salt became a common preservation technique for these products but for other products proper preservation remained a problem for thousands of years.

In addition to the early preservation techniques, another precursor of modern food processing, cooking, was probably in existence in neolithic times. The earliest food additives, salt and smoke, were in use in this era, and others were added when fermentation was discovered and the first simple breads, beers, and wines came into being. Despite improvement in variety, taste, and palatability, though, nutrition and food contamination problems remained.

Convenience foods also go back to prehistoric times, at least in terms of dried foods. Dried snake was a favorite of Chinese travelers more than 2000 years ago. The Tibetan favorite was likely their powdered tea and yak butter mixture (20). Thousands of years later, in the time of Columbus, dried beef and pork and hardtack were the staples for sea voyagers, for whom, among various nutrition problems of that day, scurvy was especially a problem. The introduction of lemons (providing vitamin C) ameliorated the problem and incidentally was one of the first instances of the nutrient fortification of the diet.

Much of the history of humans from the point of view of food safety involved the problems of spoiled and contaminated foods. Thus it is not unexpected that spices would play a major role in the food supply. Spices were among the first food additives used and were important not only to enhance the taste of foods, but also to disguise the taste of spoiled foods. The spice trade was an important factor in world history at least from 1450 B.C. in Egypt throughout the era of the Roman Empire and into the middle ages (20).

With the advent of the industrial revolution, food safety problems were multiplied by increasing urbanization and the age-old problems of nutrition and food spoilage and contamination. The deliberate adulteration of food also increased significantly in this period. Adulteration ranged from the relatively harmless practice of mixing floor sweepings with pepper, ash leaves with tea, and mahogany sawdust with coffee to the addition of copper to pickles, picrotoxin to beer, prussic acid to wine, and lead and copper salts to sweets. Such adulteration became a public issue and probably provided some historical basis for the currently unfavorable, but popular perception of food additives (21).

One of the most significant events in the history of food safety was the initial development of the canning process by Nicolas Appert in 1809. Appert, vying for a prize of 12,000 francs offered by the French government, succeeded in preserving various foods in stoppered glass bottles heated in boiling water. The microbiological explanation of why this approach worked was not to be known until many years later, but Appert's basic approach is that used today in commercial canning (22). The food safety problems involved in improper heat processing of foods will form an important part of our subsequent discussion.

FEDERAL FOOD LAW

As the twentieth century dawned, several states had "pure food" laws, but it took the unstinting efforts of Dr. Harvey Wiley and his "poison squad" to secure passage of the first federal law in 1906 (23, 24). The pure food and drug law of 1906 reflected the simple belief of that time that food was either pure or adulterated. Although amended several times, the changes most pertinent to our food safety discussion occurred in 1958. The 1958 amendments not only shifted the burden of demonstrating the safety of food additives to industry, but also contained certain features that have contributed to today's concern about food safety. Most important is the so-called Delaney Clause, which forbids the use of any food additive shown to induce cancer in humans or animals. Another feature was the "grandfathering" of additives in use at the time through the exemption for generally recognized as safe (GRAS) substances. Subsequently, the banning of cyclamates brought a presidential directive to review the safety of the GRAS substances. The GRAS review initiated in 1970, in part and in turn, has led to the Federal Food and Drug Administration's (FDA's) cyclic review of all direct and indirect additives, now in process. Revisions to the Federal Food, Drug, and Cosmetic (FFD&C) Act have also tightened regulatory control of the safety of color additives (in 1960) and drugs used in food-producing animals (in 1962 and 1968). Interestingly, the trend toward stricter legislative control of food safety was interrupted in 1976 by the vitamin amendments to the act. These amendments were directed at limiting FDA control over composition, labeling, and use of vitamin supplements (25). The current FFD&C Act (26) is in many ways a loosely integrated collection of independent approaches to the various aspects of food safety and is frequently cited in the following chapters of this book.

In the generation since the 1958 Food Additives Amendments, science has made a quantum leap forward, with analytical chemistry increasing the sensitivity of detection from tenths of a percent to parts per billion and, in some cases, parts per trillion. This increase in the sensitivity of detection has produced evidence of trace contaminants in food that were hitherto unsuspected and has raised the fundamental question of the significance of such minute amounts to food safety. Toxicology, too, has advanced with evermore exquisite searching for adverse effects in animal studies and the addition of the new branches of genetics toxicology and teratology. With these advances, science has augmented the inherent concern about food safety and consistently reproves the theorem that absolute safety of the food supply is impossible.

THE RISKS FROM FOOD HAZARDS

All of this brings us to the basic question of this book: "Is our food supply safe?" Strictly speaking, the answer must be "no," since safety implies the absence of hazard or risk (27, 28). Much like all other aspects of life, absolute food safety is impossible for there is virtually no component of our food supply that is without some risk to some part of the population. What are those risks, then? That question has a complicated set of answers that can best be provided by an overall examination of the several areas of hazards associated with our food supply and the relative risks that they present to human health. [For an excellent discussion of the scientific assessment of these risks, the 1978 report of the Food Safety Council (28) is recommended.]

The various types of hazard associated with the food supply can be grouped in several different ways. For any such grouping, the evaluation of risk involves three major criteria: severity, incidence, and onset. *Severity* indicates the type of effect involved, ranging from mild and temporary discomfort through more serious but reversible effects, to irreversible effects, including death. *Incidence* refers to the number of cases or rate of occurrence of a given effect. *Onset* denotes the time of occurrence of the effect after exposure to a hazard and can range from immediate to long term (29).

Quantification of the three risk criteria presents different degrees of difficulty. In certain cases direct human observation is possible, but in most cases there is only anecdotal or indirect evidence based on human experience, epidemiology studies, or laboratory studies with animals or other test systems. Nonetheless, a relative ranking of risks can be made for the various areas of food safety, and a practical perspective can be gained on the total food safety problem by examining each resulting area.

Applying the risk criteria of severity, incidence and onset and grouping food safety hazards into five convenient classes result in the following ranking from greatest to least risk: (*a*) foodborne hazards of microbial origin, (*b*) nutritional hazards, (*c*) environmental contaminant hazards, (*d*) food hazards of natural origin, and (*e*) food and color additive hazards.

Although the public conception of food risks is generally in reverse order, or as Hall (30) has said, perverse order, to this ranking it has been presented and discussed a number of times by Wodicka (31), Hall (32), and Schmidt (33) and is consistent with the thinking of most food technologists. For example, the conclusions of the Marabou Symposium on Food and Cancer (34) was the following:

> Of the potential sources of harm in foods the largest by far are, first, microbiological contamination and, next, nutritional imbalance. Risks from environmental contamination are about 1,000 times less and risks from pesticide residues and food additives can be estimated as about a further 100 times smaller again. Naturally occurring compounds in food are far more likely to cause toxicity than intentional food additives.

The particular classification and ranking just noted formed the basis for an American Institute of Nutrition symposium on food safety in 1978 (29, 35–38), from which the idea for this book originated.

The rationale for this approach to characterizing food safety is presented for these five areas in order of their risk in Chapters 2 through 6, respectively. Briefly, though, to set the stage, the five chosen risk categories encompass the following considerations.

Foodborne Hazards of Microbial Origin

It is well known that foods can serve as vehicles of many pathogenic and toxigenic agents of disease. Foodborne disease agents are characterized by their diversity. Some produce their effects through toxic metabolites resulting from growth of the organisms in the food prior to ingestion (e.g., staphylococcal food poisoning and botulism). Others produce adverse effects through ingestion of living organisms (e.g., *Salmonella*). Still others require the ingestion of large numbers of living organisms that sporulate in the digestive tract and release a toxin (e.g., *Clostridium perfringens* poisoning). The source of these hazards may be on the farm; during food processing; or, more likely, during food service preparation or preparation in the home. The importance of this area prompted a special report in 1975 on microbial hazards in foods by the National Academy of Sciences (39).

The severity of microbial effects ranges from temporary discomfort and rather prompt recovery to the acutely toxic effects of botulism, which, depending on the speed of diagnosis and treatment, can have a very high mortality rate. [Botulinum toxin is the most powerful poison known to humans, with the oral lethal dose estimated to be between 0.1 and 1.0 μg (40)]. Even the normally less dangerous foodborne diseases can have very severe effects, perhaps contributing to death in the case of infants, the elderly, and debilitated persons. Incidence of foodborne diseases as reported to the Federal Center for Disease Control in Atlanta amounts to only a few thousand cases annually, yet the total number of cases may be on the order of 10,000,000 annually (41). The time of onset for the foodborne diseases is fairly immediate. For example, in the case of botulism, symptoms begin to appear within 12 to 36 hours after ingestion of food containing the toxin. Staphylococcal food poisoning generally appears within 1 to 6 hours after the food is ingested, and salmonellosis occurs within 12 to 18 hours after the organisms are ingested.

Nutritional Hazards

The hazards in the food supply associated with nutritional factors can be thought of in terms of both deficiencies and excesses in nutrient intakes. Such nutritional deficiency diseases as scurvy, pellagra, rickets, beriberi, and goiter illustrate the hazards associated with nutrient deficiencies. Some of these were prevalent in the United States early in the twentieth century until they were largely irradicated through improved dietary intake and the nutrient fortification of staple foods. Some still exist in underdeveloped parts of the world. At the other end of the spectrum the toxicity of excess doses of the fat-soluble vitamins in particular is well known (42), as is that of some of the trace elements (43). In addition to the deficit and excess ends of the spectrum, there are various concerns associated with poor nutrition. For example, increased susceptibility to infectious diseases, to foodborne diseases, and to the effects of environmental contaminants arising from malnutrition immediately come to mind. Further, information is gradually evolving in regard to the adverse effects of poor nutritional status at various stages of human development on the etiology of chronic diseases. Of special concern is the increasing prevalence of self-prescribed megavitamin therapy—especially in light of existing knowledge of the toxicity of the fat-soluble vitamins and growing concern about the water-soluble vitamins. Similarly, the Food and Nutrition Board (44) specifically has warned against habitually exceeding the upper intake levels listed for the trace elements because of potential toxicity. Finally, and perhaps most importantly, the continuing appeal of fad diets presents significant potential for hazard, as illustrated by the problems arising from the sole use of certain liquid-protein diet preparations (45).

Thus, although the hazards associated with nutrition are of varying severity, collectively they must be considered of major significance. The incidence of nutritional hazards is difficult to quantitate. Nutritional deficiency diseases and cases associated with the toxicity of vitamins and minerals are of relatively low incidence, but other effects are probably of significant incidence. Generally, effects in this category can be described as occurring in a long-term basis, although acute toxic effects are also possible.

Environmental Contaminant Hazards

Among the plethora of chemicals to which humans are exposed, including those occurring naturally, the most important from a risk point of view are those classified herein as environmental contaminants. Included are the trace elements and organometallic compounds (e.g., arsenic, mercury, cadmium, tin, and lead), as well as a variety of organic substances [e.g., polychlorinated biphenyls (PCBs) and the halogenated hydrocarbon pesticides]. Although this category contains substances of quite diverse chemical structure, there are common characteristics in terms of behavior. The environmental contaminants tend to be stable and thus persistent in the environment, tend to bioaccumulate in the food chain, and can be biotransformed with increased toxicity. Illustrative reviews of the occurrence and the toxicity of the environmental contaminants (e.g., 46–48) indicate the importance of this class of compounds to food safety. Severity ranges over a wide spectrum of effects, depending on the duration and extent of exposure. Some of these substances (e.g., lead, mercury, and PCBs) have been shown to be of special concern in the case of the fetus, infants, and young children because of greater retention as well as susceptibility. The incidence of effects is difficult to quantitate for the environmental contaminants. By and large, incidence can be assessed only indirectly by combining what is known about toxicity with exposure data estimated by monitoring the levels found in food. Fortunately, regulatory controls in terms of permitted levels in foods appear from surveys to be helpful in minimizing human exposure, but unfortunately many consumers assume that since there is government approval, the affected foods are totally without hazard. Consistent with the spectrum of effects associated with these substances, the times of onset can vary widely from acute to long term. As noted, infants and young children are of special concern since effects induced in the short term [e.g., central nervous system (CNS) damage] can be of long-term significance.

Food Hazards of Natural Origin

The tens of thousands or perhaps even hundreds of thousands of compounds naturally present in foods could in part be incorporated with those of microbial origin and in part with the environmental contaminants. However, they are treated separately in this book because of their importance and need for attention in their own right. Only a small fraction of these substances have been identified, but among the known compounds are those with significant acute and chronic toxicity.

If one were to somehow solve all the scientific and public concern problems associated with substances directly or indirectly, deliberately or accidentally added to foods, there would still remain a set of food safety questions of immense magnitude and complexity—that of the natural contaminants. These compounds include a large class of substances occurring in foods of plant origin ranging from the oxalates in spinach, through the glycoalkaloids in the white potato to the mushroom poisons. Also included are the trace elements and the toxicologically important mycotoxins occurring in grains and other foods susceptible to mold growth, such as the aflatoxins, ochratoxin, patulin, zearalenone, and the tricothecene toxins. Other important contaminants of natural origin include the pyrrolizidine alkaloids and paralytic shellfish poison. In addition, a large set of coumpounds produced in foods during storage, processing, or preparation, such as the nitrosamines and polynuclear aromatic hydrocarbons, are included. Two articles by Hall (30, 49) provide an excellent overview of these substances, many of which have also been reviewed by the National Academy of Sciences (46).

These contaminants of natural origin are of importance not only because of direct human consumption, but also because of secondary exposure from the edible by-products of food-producing animals. In terms of severity, this class contains acute toxicants as well as potent carcinogens. The incidence of effects associated with these constituents cannot be well quantitated, but the case of aflatoxins presents one indication of the need for concern and the possible improvements that might result from control measures (50). In accord with the wide range of potential effects, onset can vary from acute to long term.

Food and Color Additive Hazards

If public attention were the criterion, food and color additives would rank highest. However, in terms of risk, they rank below the preceding categories (31–34). This class includes a wide variety of substances—over 2000 direct additives and perhaps on the order of 10,000 indirect additives, although most of the indirect additives probably do not actually end up in finished foods. By our definition of the category, it also includes several hundred drugs used in food-producing animals.

The discrepancy between public attention and inherent hazard is in large part due to a lack of understanding of the definition of the term "additive." Contrary to popular perception, the majority of direct food additives are GRAS substances. These substances, predominantly spices and flavors, include many familiar food ingredients, some of which, like salt and certain spices, have been used for thousands of years. Review of the safety information on certain GRAS substances has revealed that about 90% present no significant hazard with normal human food uses. Most of the remaining, although requiring adddditional testing to meet today's scientific standards, have not been associated with hazards to humans. The other direct additives used in foods have been approved, and their uses have been regulated by the FDA. Indirect additives, such as those used in production, processing, and packaging and that might migrate to food, are numerous but normally occur in foods, if at all, at trace levels—many at parts per billion or less. Examination of severity, incidence, and onset of effects for this very large class of constituents indicates the lowest-ranking hazard.

Regardless of the risk ranking of food additives, it is consistent with sound public policy to periodically review their safety, as well as to review other factors that affect food safety, in light of rapidly advancing science and changing food consumption patterns. This review of food additives, encompassing GRAS substances, direct and indirect food additives, color additives, and drugs used in food-producing animals, is now a standard activity of the Food and Drug Administration (51).

ASSESSING THE SAFETY OF FOOD COMPONENTS

Much of the confusion about the safety of our food supply and the relative risks associated with its various aspects derives from our inability to adequately measure the safety—or, more precisely, the risk—of food components. Whether it is a microbial toxin, an essential vitamin, a natural or other environmental contaminant, or a food additive, we would like to know what risk a given exposure presents (28). In some cases there is risk evidence from direct human experience or from epidemiology studies, but in most cases we are forced to rely on animal or other laboratory tests.

Although animal tests at high dose levels—perhaps a thousand times the equivalent of normal human exposure—are treated with public skepticism, the fear of another thalidomide lurks in the back of our minds. Then, too, if animal tests are to be discounted, what are we to conclude about the many substances whose use is based on an absence of adverse effects in animal tests? Rather than dismissing animal tests (or other laboratory tests) entirely or in contrast accepting their results blindly, they should be considered in perspective. Although the regulatory decision maker or the public policymaker would like yes–no answers from the scientist, such is not always possible; more often than not the scientific answer is "maybe."

There are many staunch defenders as well as critics of the Delaney Clause, which prohibits approval of any food or color additive, "found to induce cancer when ingested by man or animal . . . ." However, the issue is much broader than just the Delaney Clause; the FFD&C Act requires that an additive be "safe" under its conditions of use and the public expects constituents of food to be "safe" regardless of whether they are legally classified as food additives. In this context, "safe" involves not only cancer, but also any other toxic effect. However, having noted the impossibility of assuring absolute safety, the question really concerns the kinds and the levels of potential risk involved. That question, in turn, poses a host of other questions involving the chemical structure of the compound as well as its mutagenic behavior, its metabolism and biological activity, its acute toxicity, its reproductive and teratology effects, and in each case its dose–response characteristics. Finally, there is the question of how to extrapolate from the animal tests to humans. Obtaining sound data in each of these areas and integrating it to form a pattern that permits scientifically valid regulatory decisions is easier said than done. The Food Safety Council has proposed a decision-tree approach incorporating all these facets of toxicology and leading at several stages to a "pass," "fail," or "more testing" decision (28). Of course, a public-policy decision relative to a given food constituent may well weigh other considerations, and that possibility has generated a great deal of discussion relative to the risk–benefit concept (e.g., 52–56).

The preceding problems of safety or risk assessment are discussed in Chapter 7 with the objective of lending additional insight to the overall question of food safety. This final chapter, together with Chapters 2 through 6, should enable the reader to gain an overall perspective on food safety and thus place in proper context the continuing deluge of scientific and media reports on the safety of food constituents. The references in each chapter provide sources for the reader who is interested in pursuing these food safety topics in more depth.

REFERENCES

1. J. S. Turner, *The Chemical Feast: Ralph Nader's Study Group Report on the Food and Drug Administration*, Grossman, New York, 1970.
2. J. Verrett, *Eating May Be Hazardous to Your Health*, Simon and Schuster, New York, 1974.
3. E. M. Whalen and F. J. Stare, *Panic in the Pantry*, Atheneum, New York, 1973.
4. J. Hightower, *Eat Your Heart Out*, Crown, New York, 1975.
5. B. F. Feingold, *Why Your Child Is Hyperactive*, Random House, New York, 1975.
6. E. M. Whalen, *Preventing Cancer*, Norton, New York, 1977.
7. E. M. Whalen and F. J. Stare, *Eat OK, Feel OK: Food Facts and Your Health*, Christopher, North Quincy, Mass., 1978.
8. American Council on Science and Health, *Cancer in the United States: Is There an Epidemic?*, New York, 1978.
9. J. J. Fried, *The Vitamin Conspiracy*, Saturday Review Press, New York, 1975.
10. National Science Foundation, *Chemicals and Health*, U.S. Government Printing Office, Washington, D.C., 1966.
11. Editors, *FDA Consumer*, 13 (3), 2 (1979).
12. H. Tarnower and S. S. Baker, *The Complete Scarsdale Medical Diet*, Rawson, Wade, New York, 1979.
13. N. Pritkin and P. McGrady, Jr., *The Pritkin Program for Diet and Exercise*, Grossit and Dunlap, New York, 1979.

14. D. R. Reuben, *Everything You Always Wanted to Know About Nutrition,* Simon and Schuster, New York, 1978.
15. U.S. Senate Select Committee on Nutrition and Human Needs, *Dietary Goals for the United States,* U.S. Government Printing Office, Washington, D.C., 1977.
16. A. E. Harper, *Am. J. Clin. Nutr.*, 31, 310 (1978).
17. U.S. Senate Committee on Agriculture, Nutrition and Forestry, *Food Safety: Where Are We?*, U.S. Government Printing Office, Washington, D.C., 1979.
18. W. R. Barclay, *J. Am. Med. Assoc.* 242 (7), 657 (1979).
19. R. L. Hall, *Chem. Technol.*, 3 (7), 412 (1973).
20. R. Tannahill, *Food in History,* Stein and Day, New York, 1973.
21. E. Corwin, *FDA Consumer,* 10 (9), 10 (1976).
22. W. C. Frazier, *Food Microbiology,* 2nd ed., McGraw-Hill, New York, 1967.
23. H. W. Wiley, *An Autobiography,* Bobbs-Merrill, Indianapolis, 1930.
24. O. Anderson, *The Health of a Nation; Harvey Wiley and the Fight for Pure Food,* Univ. Chicago Press, 1958.
25. H. Hopkins, *FDA Consumer,* 10 (6), 10 (1976).
26. *Federal Food, Drug, and Cosmetic Act, as Amended,* U.S. Government Printing Office, Washington, D.C., 1979.
27. W. W. Lowrance, *Of Acceptable Risk,* William Kaufmann, Los Altos, California, 1976.
28. Food Safety Council, *Food Cosmet. Toxicol.*, 16, Suppl. 2 (1978).
29. H. R. Roberts, *Fed. Proc.*, 37 (12), 2575 (1978).
30. R. L. Hall, "Naturally Occurring Toxicants and Food Additives: Our Perception and Management of Risks," *Proceedings of Marabou Symposium on Food and Cancer,* Caslon Press, Stockholm, 1978, pp. 6–20.
31. V. O. Wodicka, "Food Safety in 1973," *Proceedings of Flavor and Extract Manufacturers Assoc.*, Washington, D.C., 1973.
32. R. L. Hall, *Nutr. Today,* 8 (4), 20 (1973).
33. A. M. Schmidt, *Address at Food Safety—A Centenary of Progress Symposium,* London, October 1975.
34. A. S. Truswell, N. G. Asp, W. P. T. James, and B. MacMahon, "Conclusions," *Proceedings of Marabou Symposium on Food and Cancer,* Caslon Press, Stockholm, 1978, pp. 112–113.
35. E. M. Foster, *Fed. Proc.*, 37 (12), 2577 (1978).
36. I. C. Munro and S. M. Charbonneau, *Fed. Proc.*, 37 (12), 2582 (1978).
37. J. V. Rodricks, *Fed. Proc.*, 37 (12), 2587 (1978).
38. J. Doull, *Fed. Proc.*, 37 (12), 2594 (1978).
39. National Academy of Sciences, *Prevention of Microbial and Parasitic Hazards Associated with Processed Foods,* Washington, D.C., 1975.
40. Staff Report, *Nutr. Today,* 10 (5, 6), 4 (1975).

41. T. Fodor, C. Reisberg, H. A. Hershey, and H. Berkowitz, *Public Health Rep.*, 85, 1013, (1970).
42. K. C. Hayes and D. M. Hegsted, "Toxicity of the Vitamins," in National Academy of Sciences, *Toxicants Occurring Naturally in Foods*, 2nd ed., Washington, D.C., 1973, pp. 235–253.
43. E. J. Underwood, "Trace Elements," in National Academy of Sciences, *Toxicants Occurring Naturally in Foods,* 2nd ed., Washington, D.C., 1973, pp. 43–87.
44. Food and Nutrition Board, *Recommended Dietary Allowances,* 9th ed., National Academy of Sciences, Washington, D.C., 1980.
45. Food and Drug Administration, *Fed. Reg.*, 43, 60883 (1978).
46. National Academy of Sciences, *Toxicants Occurring Naturally in Foods,* Washington D.C., 1973.
47. World Health Organization, *Environmental Health Criteria—2,* Geneva, 1976.
48. Office of Technology Assessment, *Environmental Contaminants in Food,* U.S. Government Printing Office, Washington D.C., 1979.
49. R. L. Hall, *Nutr. Today,* 12 (6), 1 (1977).
50. Food and Drug Administration, "Assessment of Estimated Risk Resulting from Aflatoxins in Consumer Peanut Products and Other Food Commodities," Washington, D.C., 1978.
51. H. Hopkins, *FDA Consumer,* 11 (5), 8 (1977).
52. P. B. Hutt, *Food Drug Cosmet. Law J.*, 33 (10), 505 (1978).
53. Expert Panel on Food Safety and Nutrition, Institute of Food Technologists, *The Risk/Benefit Concept as Applied to Food,* Chicago, 1978.
54. H. R. Roberts, *Food Technol.*, 32 (8), 59 (1978).
55. H. R. Roberts, *Food Drug Cosmet. Law J.*, 34 (3), 153 (1979).
56. *Principles and Processes for Making Food Safety Decisions,* Report of the Social and Economic Committee of the Food Safety Council, Washington, D.C., 1980.

CHAPTER 2

Foodborne Hazards of Microbial Origin

E. H. MARTH

The presence in food of certain microorganisms or metabolites resulting from their growth can cause various human illnesses that are of two general forms. The illness is commonly called *food poisoning* or *food intoxication* if the responsible microorganism grew in food and produced a toxin that was consumed with the food and then caused the illness. Examples of food intoxication include staphylococcal poisoning and botulism. Another form of illness is food infection, which results when the responsible microorganism is in the food, is consumed with the food, and then causes the physiologic changes that manifest themselves as an illness. Food infections are caused by foodborne viruses, salmonellae, and some other microorganisms.

The primary importance of foodborne diseases in food safety in part arises from the diversity of the diseases involved. Encompassed in this category are diseases of both great severity (e.g., botulism) as well as high incidence (e.g., salmonellosis). This chapter briefly looks at the sources of foodborne illnesses and describes their occurrence. Detailed discussions are presented on staphylococcal food poisoning, botulism, perfringens poisoning, and salmonellosis. Finally, information is summarized on other bacteria of importance in foodborne illness and on viruses. The chapter concludes with a discussion of the control of foodborne illness.

SOURCES OF FOODBORNE ILLNESS

Foodborne illnesses can be caused by bacteria, rickettsia, viruses, molds, and parasites. Bacteria responsible for such illnesses include *Staphylococcus aureus*, *Clostridium botulinum*, *Clostridium perfringens*, *Salmonella* species, *Bacillus cereus*, *Vibrio parahaemolyticus*,

Vibrio cholera, *Shigella* species, *Escherichia coli*, *Brucella* species, *Yersinia enterocolitica*, *Campylobacter* species, and several others. Of these, *S. aureus*, *Salmonella* species, *C. perfringens* (because of the frequency with which these three organisms cause disease), and *C. botulinum* (because of the severity of the disease it causes) are by far the most important.

Of the rickettsia, *Coxiella burnetti* is best known as a cause of foodborne illness. This organism, which sometimes occurs in raw milk and in products made from raw milk, causes Q fever in humans. Because of the organism's heat resistance, the minimum treatment for pasteurization of market milk by the holding method has been set at 145°F for 30 minutes.

An assortment of viruses can also become foodborne and then pose a potential hazard to the consumer. For example, the virus responsible for infectious hepatitis is a well-known cause of foodborne illness.

More than 80 species of molds can produce more than 100 toxic organic compounds (often secondary metabolites) during growth on foods and feeds. These toxins collectively are designated mycotoxins, and their presence in foods represents another form of food intoxication. (Mycotoxins are discussed in Chapter 5.)

Certain parasites can occur in foods and usually in foods of animal origin. Inadequate cooking of the foods can result in infestation of humans by these parasites. Included are *Trichinella spiralis* (a nematode sometimes in pork but that can occur in tissues from other animals), *Gnathostoma spinigerum* (a nematode in freshwater fish), *Angiostrongylus* species (nematodes in mollusks), *Anisakis* species (nematodes in fish such as herring or cod), *Taenia saginata* (beef tapeworm), *Taenia solium* (pork tapeworm), *Diphyllobothrium latum* (fish tapeworm), *Clonorchis sinensis*, *Opisthorchis* species, *Metagonimus yokogawai*, *Heterophyes heterophyces* (the last four are trematodes in fish), *Paragonimus westermani* (the Oriental lung fluke from crabs and crayfish), *Fasciola hepatica* (liver fluke from sheep, cattle, hogs, goats, and deer), *Fasciolopsis buski* (intestinal fluke from contaminated raw plants), *Toxoplasma gondii* (intracellular protozoan parasite from meats or soil), *Entamoeba histolytica*, *Giardia lamblia*, *Dientamoeba fragilis* (the last three are protozoan parasites that can occur in fecally contaminated water or food), *Ascaris lumbricoides*, and *Trichuris trichiura* (these two are helminths that can occur in fecally contaminated water or food) (1).

INCIDENCE OF FOODBORNE ILLNESS

The numbers of reported outbreaks of foodborne illnesses that occurred in the United States during 1975 to 1977 are given in Table 2.1.

TABLE 2.1 Foodborne Illnesses of Microbial Origin, 1975 to 1977

Cause	Outbreaks			Cases			Deaths		
	1975	1976	1977	1975	1976	1977	1975	1976	1977
Bacterial									
A. hinshawii	1	0	1	15	0	13	--[a]	--	--
B. cereus	3	2	0	45	63	0	--	--	--
C. botulinum	14	23	20	19	40	75	2	5	5
C. perfringens	16	6	6	419	509	568	1	0	0
Salmonella	38	28	41	1573	1169	1706	2	3	0
Shigella	3	6	5	413	273	67	0	1	0
S. aureus	45	26	25	2275	930	905	0	0	0
V. cholerae (not 01)	0	0	1	0	0	2	0	0	0
V. parahaemolyticus	2	0	2	222	0	118	--	--	--
Streptococcus (Group D suspect)	1	0	0	50	0	0	--	--	--
Y. enterocolitica	0	1	0	0	286	0	--	--	--
TOTAL	123	92	101	5031	3270	3454	5	9	5

TABLE 2.1 Foodborne Illnesses of Microbial Origin, 1975 to 1977 (Continued)

Cause	Outbreaks			Cases			Deaths		
	1975	1976	1977	1975	1976	1977	1975	1976	1977
Parasitic									
T. spiralis	20	8	14	193	27	87	1	0	0
Anisakidae	1	0	1	1	0	4	--	--	--
D. latum	1	0	0	1	0	0	--	--	--
E. histolytica	0	0	0	0	0	0	--	--	--
TOTAL	22	8	15	195	27	91	1	--	--
Viral									
Hepatitis A	3	2	4	173	37	72	--	--	--
Echo, Type 4	0	1	0	0	80	0	--	--	--
TOTAL	3	3	4	173	117	72	--	--	--

Source. Reference (2)

[a] -- indicates no data reported

This table includes information for diseases of bacterial, parasitic, and viral origin but does not list illnesses caused by chemical contaminants (chemical contaminants are discussed in Chapter 4). The number of reported outbreaks of illnesses caused by bacteria, parasites, and viruses ranged from 92 to 123, 8 to 22, and 3 to 4 per year, respectively. The number of cases (individuals that were ill) ranged from approximately 3500 to 5500 per year.

Most experts agree that these numbers reflect only a small portion of the actual cases of foodborne illness since many go unreported or laboratory work needed for an accurate diagnosis is not done. Some estimates of the number of cases per year in the United States have included 1,000,000 (3), 100,000 to 200,000 (4), and 500,000 to 1,000,000 (5). An expert committee on *Salmonella* of the National Research Council estimated that 2,000,000 cases of salmonellosis occurred annually in the United States (6). More recently, Hauschild and Bryan (7) reviewed available data, did some calculations, and concluded that there were (*a*) 1,400,000 to 3,400,000 cases of foodborne and waterborne disease annually in the United States during 1974 and 1975, (*b*) 1,100,000 to 2,600,000 cases of foodborne and waterborne disease annually in the United States during 1967 to 1976, and (*c*) 740,000 cases of human salmonellosis annually in the United States during 1967 to 1976. Although the true number of outbreaks and cases remains unknown, it is quite likely that several million persons in the United States annually suffer from foodborne illness.

Food Mishandling

Data summarized in Table 2.2 indicate that most mishandling of food occurs in food service establishments and in the home. Mishandling of food resulting in foodborne illness is rare in the food processing industry. Bryan (8) has assembled similar information for foodborne illnesses associated just with meat and poultry in the United States during 1973 to 1977. His data show that when the place of mishandling could be determined, food service establishments accounted for 65% of the outbreaks, homes for 31%, and food processing factories for 4%. It is clearly evident where attention must be directed if we are to minimize the outbreaks and cases of foodborne illness in the United States.

Contributing Factors

For those confirmed outbreaks of food illness in which the factors involved were reported, the resulting data are shown in Table 2.3. These data indicate that the four factors most often contributing to outbreaks of foodborne illness are holding food at the wrong temperature (includes inadequate cooling), inadequate cooking, use of contaminated equipment in handling food, and poor personal hygiene by the food handler. Similar results were found by Bryan (8) in his evaluation of

TABLE 2.2 Places Where Food Was Mishandled to Cause Foodborne Illness, 1977

Cause	Food Processing Establishments	Food Service Establishments	Homes	Unknown
Bacterial				
A. hinshawii	0	0	1	0
C. botulinum	0	1	17	2
C. perfringens	0	6	0	0
Salmonella	4	19	13	5
Shigella	0	3	2	0
Staphylococcus	1	16	6	2
V. cholerae (not 01)	0	0	1	0
V. parahaemolyticus	0	2	0	0
TOTAL	5	47	40	9
Parasitic				
T. spiralis	0	0	14	0
Anisakidae	0	0	0	1
TOTAL	0	0	14	1
Viral				
Hepatitis A	0	4	0	0
TOTAL	0	4	0	0
Total for all types	5	51	54	10

Source. Reference (2)

TABLE 2.3 Factors Contributing to Confirmed Outbreaks of Foodborne Illness, 1977

Cause	Number of Outbreaks		Number of Outbreaks Involving		
	Reported	In Which Factors Reported	Improper Holding Temperature	Inadequate Cooking	Contaminated Equipment
Bacterial					
A. hinshawii	1	1	1	0	0
C. botulinum	20	18	5	15	1
C. perfringens	6	6	6	1	2
Salmonella	41	24	18	8	7
Shigella	5	3	0	0	0
Staphylococcus	25	19	18	3	11
V. cholerae (not 01)	1	1	0	1	0
V. parahaemolyticus	2	2	2	1	0
TOTAL	101	74	50	29	21
Parasitic					
T. spiralis	14	14	0	14	0
Anisakidae	1	1	0	1	0
TOTAL	15	15	0	15	0
Viral					
Hepatitis A	4	4	1	0	1
TOTAL	4	4	1	0	1
Total for all types	120	93	51	44	22

TABLE 2.3 Factors Contributing to Confirmed Outbreaks of Foodborne Illness, 1977 (Continued)

Cause	Number of Outbreaks		Number of Outbreaks Involving		
	Reported	In Which Factors Reported	Food From Unsafe Source	Poor Personal Hygiene	Other
Bacterial					
A. hinshawii	1	1	0	0	0
C. botulinum	20	18	1	0	1
C. perfringens	6	6	0	2	0
Salmonella	41	24	6	7	4
Shigella	5	3	0	3	0
Staphylococcus	25	13	1	11	4
V. cholerae (not 01)	1	1	1	0	0
V. parahaemolyticus	2	2	0	0	0
TOTAL	101	74	9	23	9
Parasitic					
T. spiralis	14	14	1	0	0
Anisakidae	1	1	0	0	0
TOTAL	15	15	1	0	0
Viral					
Hepatitis A	4	4	0	4	0
TOTAL	4	4	0	4	0
Total for all types	120	93	10	27	9

Source. Reference (2)

factors contributing to outbreaks of foodborne illnesses as associated with meat and poultry in the United States during 1968 to 1977. Bryan lists the following factors, in descending order of importance: (*a*) improper cooling of cooked foods, (*b*) foods prepared a day or more before serving, (*c*) inadequate cooking or heat processing, (*d*) infected person handling cooked food, (*e*) inadequate reheating of cooked and chilled foods, (*f*) improper hot storage of cooked foods, (*g*) cross-contamination of cooked foods from raw foods, (*h*) inadequate cleaning of equipment, (*i*) ingestion of raw products, (*j*) use of leftovers, (*k*) improper fermentation, (*l*) improper thawing of cooked foods, (*m*) improper construction of equipment, (*n*) inadequate processing–preparation space, (*o*) abscess on meat, (*p*) feeding animals mercury-treated grain (chemical poisoning), and (*q*) eating meat of animals that were sick or dying at time of slaughter. Of those listed, the first six were most often involved in the outbreaks of foodborne illness.

STAPHYLOCOCCAL FOOD POISONING

About 20 to 40% of the foodborne illness outbreaks that are reported in the United States in one year are accounted for by staphylococcal food poisoning. The illness is caused by one of several enterotoxins that are produced by *Staphylococcus aureus* during its growth in foods that commonly are of animal origin.

Although humans probably have been afflicted with staphylococcal food poisoning for centuries, the first reasonably well documented outbreak of the illness was not reported until 1884 (9). In this case, toxic Cheddar cheese caused illness in more than 300 persons, and when the cheese was consumed experimentally by Vaughn (9), he soon developed symptoms that are characteristic of the illness. Microscopic examination of the cheese also revealed the presence of spherically shaped bacteria. Further reports of this disease seem to be lacking until 1914, when Barber (10) described an outbreak that resulted after consumption of milk produced by a mastitic cow.

The significance of these early observations was not recognized until 1930, when Dack et al. (11) isolated large numbers of a pigment-forming staphylococcus from a Christmas cake that was responsible for an outbreak of food poisoning. Sterile filtrates prepared from broths in which the staphylococcus had grown were consumed by human volunteers who became ill (12). This observation led to the conclusion that staphylococci were responsible for a form of food poisoning (13).

The following discussion deals with symptoms of the disease, the organism, factors that affect the organism, the enterotoxin, and foods commonly involved in staphylococcal intoxications. More detailed discussions of this disease have been provided by Minor and Marth (14–18), Bryan (19), and Bergdoll (20).

Symptoms of the Disease

Characteristic symptoms of this illness have been described by Bergdoll (20). The onset of symptoms commonly occurs within 1 to 6 hours (average of 2 to 3 hours) after the consumption of enterotoxin-containing food. The actual time required to develop symptoms is determined by the amount of toxin consumed (e.g., the amount of toxic food eaten) and the sensitivity of the individual to the enterotoxin.

The most common symptoms are nausea, vomiting, retching, abdominal cramping, and diarrhea. Vomiting can occur without diarrhea, and diarrhea can occur without vomiting. Additional symptoms in severe cases can include headache, muscular cramping, marked prostration, above or below normal temperature, and sometimes a dramatic drop in blood pressure (e.g., from 120/80 to 60/40).

Recovery is usually rapid (e.g., in 1 to 3 days), but the more severe the symptoms, the longer the recovery period. The mortality rate is low, but some deaths among the elderly or the very young have been caused by the disease.

The Organism

Description

The cells of *S. aureus*, from 0.8 to 1.0 μm in diameter, commonly form irregular clusters but can occur singly or in pairs (21). Cells of some strains of the bacterium produce a capsule. *Staphylococcus aureus* is gram-positive except in cultures more than 48 hours old, where gram-negative cocci can appear. The bacterium is nonmotile and does not produce spores. Colonies produced on a suitable agar medium are round, raised, opaque, smooth, and glistening. Pigmentation of the colonies is variable, and the color ranges from deep gold to cream or white.

Metabolism

Staphylococci obtain energy by way of glycolysis, the hexose monophosphate pathway, and the tricarboxylic acid cycle (18). Operation of these cycles is dependent on conditions of growth. Glucose can be metabolized aerobically or anaerobically to form pyruvate.

During anaerobic growth, pyruvate can (*a*) be reduced to lactate by lactate dehydrogenase, (*b*) undergo dismutation in which one molecule of pyruvate is reduced to lactate and the second molecule of pyruvate is oxidized to acetate and carbon dioxide, or (*c*) from α-acetolactic acid by a condensation of acetaldehydediphosphothiamine with pyruvate, followed by decarboxylation of the α-acetolactic acid to form acetylmethylcarbinol. Under some conditions, acetylmethylcarbinol can be reduced to 2, 3-butanediol or oxidized to diacetyl. End-products of anaerobic metabolism are lactate (73 to 94% of final products), acetate (4 to 7%), and traces of pyruvate.

During aerobic growth, pyruvate is oxidized to acetate, which can be further oxidized by entering the tricarboxylic acid cycle through acetyl-CoA (coenzyme A). Primary end-products of aerobic metabolism are acetate, carbon dioxide, and lactate.

Acid is produced from glucose, lactose, maltose, and mannitol during aerobic and anaerobic growth of S. *aureus.* If growth is aerobic rather than anaerobic, the bacterium can use more carbohydrates to obtain carbon and energy. Thus acid is produced from many hexoses, pentoses, disaccharides, and sugar alcohols; however, no acid is produced from arabinose, cellobiose, dextrin, inositol, raffinose, rhamnose, or xylose. Esculin and starch are not hydrolyzed. *Staphylococcus aureus* reduces nitrates and produces ammonia from arginine by arginine dihydrolase. Trace amounts of hydrogen sulfide may be formed from cysteine. Glutamic acid and lysine are not decarboxylated.

Proteases, lipases, phospholipases, lipoproteins, lipases, esterases, and lyases are produced by S. *aureus.* Most strains of the bacterium hydrolyze native animal proteins (e.g., hemoglobin, fibrin, egg white, and casein) and polypeptides (such as gelatin). Lipids, Tweens, Spans, and phospholipoproteins are hydrolyzed with the release of fatty acids. Some strains produce lecithinase C, whereas others synthesize hyaluronic acid, which degrades mucopolysaccharides.

Factors That Affect the Organism

Growth or survival of S. *aureus* is affected by the presence of other bacteria, temperature, acids, salts, and sugars and chemical sanitizers. Each of these factors is discussed in the following paragraphs.

Presence of Other Bacteria

Repression of growth of S. *aureus* has been noted when the bacterium was in the presence of a mixture of psychrotrophic bacteria (22), lactic acid bacteria (23–28), coliforms (29), *Protens vulgaris* (29), *Bacillus cereus* (29), *Serratia marcescens* (29), *Pseudomonas* species (29), and *Achromobacter* species (29). When such tests were made, inhibition of growth was accompanied by inhibition of enterotoxin production (28).

Repression of growth by competing bacteria is affected by (*a*) proportion of S. *aureus* in the mixture, (*b*) pH, (*c*) temperature, (*d*) presence and amount of sodium chloride, and (*e*) presence of various food materials. Although an assortment of bacterial metabolites may contribute to inhibition of S. *aureus*, hydrogen peroxide produced by some lactic acid bacteria is bacteriostatic to S. *aureus*.

Temperature

Staphylococcus aureus can grow at temperatures in the range of 10 to 45°C, with occasional strains able to grow at slightly lower or higher temperatures. The bacterium grows best at 35 to 37°C.

Production of enterotoxin generally is maximal at 35 to 40°C, depending on the strain of *S. aureus.* Less enterotoxin is produced above or below this range of temperature, with the minimum temperature for enterotoxin production being near 16°C and the maximum near 45°C.

Heat can be used to inactivate cells of *S. aureus;* typical *D* values (minutes at a given temperature needed to reduce the population by 90%) are 2.10 at 53°C and 1.75 at 58°C. Exposure of cells to less heat can injure but not kill them. Lesions resulting from sublethal heat treatments include (*a*) damage to the cytoplasmic membrane, (*b*) alteration of the cell's metabolic capabilities, and (*c*) degradation of ribosomal RNA (30).

Staphylococci are rather resistant to inactivation by temperatures below freezing, particularly when compared to such gram-negative bacteria as *E. Coli* and salmonellae. Minor and Marth (31) demonstrated that freezing cells of *S. aureus* in trypticase soy broth followed by storage for 24 hours at −30°C and thawing had little or no effect on viability of the cells.

Acids

Most strains of *S. aureus* grow at pH values between 4.5 and 9.3, with an optimum of 7.0 to 7.5 (20). Factors that affect the pH at which a given strain will grow include, (*a*) type of medium, (*b*) salt concentration (up to about 15%), (*c*) size of inoculum, and (*d*) atmosphere and temperature of incubation. Enterotoxin is produced at pH values less extreme than those that permit growth. One report (32) suggests that enterotoxin is produced at pH values between 5.15 and 9.0. If the substrate contains salt, the pH range permitting entertoxin production becomes narrower.

Staphylococci are relatively sensitive to the presence of specific acids in the environment. The following acids were noted by Nunheimer and Fabian (33) to be germicidal for staphylococci (listed in order of decreasing effectiveness): acetic, citric, lactic, tartaric, and hydrochloric. Survival of 18-hour-old cells of *S. aureus* (10^8/ml) after 24 hours of incubation at 37°C in trypticase soy broth acidified with different acids was determined by Minor and Marth (34). A 99.9% decrease in numbers of viable staphylococci occurred with acetic, lactic, phosphoric, citric, and hydrochloric acid when the pH of the medium was 4.5, 4.4, 4.2, 3.9, and 3.8, respectively. The undissociated acid molecule was responsible for enhanced inactivation of cells by partially dissociated acids since anions of these acids had no effect on cell survival. Staphylococci were most sensitive to the effects of hydrogen ions within 12 to 24 hours during a 120-hour incubation at 37°C. They also were more sensitive at a high incubation temperature (45°C rather than 37°C) and when the initial number of bacteria was small. Exposure of staphylococci to sublethal treatments with acids can result in cell injury that is mild and can be repaired by the cell (31).

Salt and Sugars

Presence of salts and sugars in solution has a direct effect on water activity (a_w) of the medium. Aerobic growth of *S. aureus* occurs in media with a_w values too low for growth of many species of microorganisms (35). The minimal a_w value of *S. aureus* is at or near 0.86; hence staphylococci are highly tolerant to the presence of solutes such as salts and sugars in their growth media.

Nunheimer and Fabian (33) observed that 15 to 20% sodium chloride in broth was inhibitory at the optimum growth temperature of *S. aureus*, whereas a concentration of 20 to 25% was definitely germicidal. Good growth of *S. aureus*, according to Genigeorgis and Sadler (36), occurred at 37°C in brain heart infusion broth with 16% NaCl and at pH 6.9, but no cells survived after 10 days of incubation when the broth was at pH 5.1 and contained 16% salt. According to data by Gojvat and Jackson (37), staphylococci multiplied in broth with 12% NaCl that was held at temperatures between 20 and 35°C.

Iandolo et al. (38) showed that the lag phase and the growth rate of staphylococci increased at 37°C when 4% NaCl was added to trypticase soy broth. When 8% NaCl was present, there was an additional increase in the lag phase but the growth rate was lower than in the control.

Vigorous growth of *S. aureus* was noted by Hucker and Haynes (39) in media that contained up to 50% sucrose, but after 24 hours growth was retarded because acid developed in the medium. Nunheimer and Fabian (33) reported that 50 to 60% sucrose was required to inhibit growth of *S. aureus* and 60 to 70% was needed for germicidal activity. In contrast, 30 to 40% and 40 to 60% glucose concentrations were inhibitory and germicidal, respectively.

Other Chemicals

Minor and Marth (18) summarized data that show *S. aureus* to be inactivated by chlorine (0.8 ppm in 30 seconds at 25°C and pH 7.2, 15 ppm in 5 seconds at pH 8.5 to 8.9, and 15 ppm in 111 seconds at pH 11.2), iodine (5 ppm at 30°C and pH 7), and cetyltrimethylammonium bromide (a quaternary ammonium compound) (50 ppm). In addition, *S. aureus* is inactivated by an assortment of antibiotics and by such chemicals as bromine, *o*-polyphenol, and hexachlorophene. However, such compounds are not suitable for use in food processing.

Staphylococcal Enterotoxins

Five enterotoxins, A, B, C, D, and E, have been identified (40–44). Two forms of enterotoxin C have been purified (45, 46). One type, C_1, has an isoelectric pH of 8.6, and the other, C_2, has an isoelectric pH of 7.0.

The purified enterotoxin molecule is comprised entirely of a single polypeptide chain. Purified preparations of enterotoxins are fluffy, snow-white materials that are hygroscopic and readily soluble in water and salt solutions (45–47). Chemical and physical properties that differentiate the exterotoxins include (*a*) molecular weight, (*b*) isoelectric point, (*c*) extinction value, (*d*) sedimentation and diffusion coefficients, (*e*) reduced viscosity value, and (*f*) partial specific volume. The molecular weights range from 27,800 for enterotoxin A to 34,100 for enterotoxin C_1. The isoelectric point ranges from pH 7.0 for enterotoxins C_2 and E to pH 8.6 for enterotoxins B and C_1 (48).

Enterotoxins are particularly rich in aspartic acid, glumatic acid, and lysine. The total number of amino acid residues per molecule ranges from 239 for enterotoxin B to 296 for enterotoxin C_1, The complete amino acid sequence for the enterotoxin B has been determined (49–51).

Enterotoxins are remarkably resistant to heat, a fact recognized long ago by Davison and Dack (52), who observed that the potency of enterotoxin could be only gradually decreased by prolonged boiling and autoclaving. Schantz et al. (47) showed that the activity of an enterotoxin B preparation was retained even after heating at 60°C for 16 hours. The heat resistance of the individual enterotoxins is dependent primarily on the relative purity of the toxin preparation rather than on the concentration of the toxin or the heating menstruum.

Enterotoxins in their active state are resistant to proteolytic enszymes such as trypsin, chymotrypsin, rennin, and papain (53). Pepsin destroys the activity of enterotoxin at a pH of about 2 (54).

Conditions that are optimal for production of enterotoxin are following (18) (*a*) use of a protein hydrolyzate in the medium, (*b*) aeration of the culture, (*c*) incubation at 37 to 40 °C, and (*d*) a pH value of 6.5 to 7.3. Reduction of the water activity of the medium reduces synthesis of toxin. Maximum production of enterotoxin results when cells are near the ends of the exponential phase of growth.

Foods Involved in Staphylococcal Intoxications

Foods of animal origin, such as dairy products and meats, are most often involved in outbreaks of staphylococcal poisoning. Foods of plant origin are seldom involved, unless the food is a mixture that contains one or more ingredients of animal origin. In that event, the ingredient(s) of animal origin is (are) the most likely cause of the problem.

Dairy Products

Toxigenic staphylococci can enter milk from the mastitic bovine udder, if *S. aureus* causes the mastitis. Other sources could include the exterior of the animal or humans involved in handling milk. Although straphylococci can grow and produce enterotoxin in raw milk under the right conditions, pasteurized milk is a better substrate than raw milk because most of the competing microorgranisms have been eliminated (18). *Staphylococcus aureus* is a poor competitor in the presence of the bacteria likely to be in raw milk (or raw meats).

Outbreaks of staphylococcal poisoning have resulted from such dairy products as nonfat dry milk, butter, and cheese. Use of appropriate hygienic practices, proper control of manufacturing processes, and use of active lactic starter cultures when cultured products are produced will essentially eliminate outbreaks of staphylococcal poisoning associated with these foods.

Meat and Meat Products

Raw meats are contaminated with staphylococci at the time of slaughter or during handling after slaughter. Surveys have shown that from 20 to 100% of raw meats obtained from commercial sources contained staphylococci (18). As with raw milk, staphylococci seldom grow well in raw meats because of the competing microflora.

When the competitive microflora has been partially or completely eliminated, as in semipreserved meat products, growth and enterotoxin production by *S. aureus*, if present, is a distinct possibility, provided other conditions are right. Curing salts do not inhibit *S. aureus* and the pH of fermented meats must be reduced to 4.8 or less for inhibition of the bacterium. Smoking of meats can, through heat, inactivitate *S. aureus*, but smoking of sausages also can afford the bacterium an opportunity to grow if the temperature is right (18). Vacuum packing tends to inhibit growth of *S. aureus* in meat products.

Other Foods

Freshly caught fish are usually free of staphylococci; however, they can become contaminated from humans during icing, filleting, and other handling. Fish can support growth and enterotoxin production by staphylycocci, but this is likely to occur only when the natural microflora of the fish has been partially or completely eliminated (18).

The situation with poultry meat is much like that with other meats. In addition, barbequed poultry can be a problem if heating, storage, and handling are inadequate.

Staphylococci are not likely to enter or grow in uncracked raw eggs. They can, however, enter boiled eggs when such eggs are cooled in staphylococcus-laden water. Since the bacteriostatic properties of raw eggs have been eliminated during boiling, *S. aureus* can grow in the eggs and produce enterotoxin. This has happened and has caused outbreaks of foodborne illness. *Staphylococcus aureus* also can grow in liquid egg products if these products are not handled properly.

If *S. aureus* is present in custards or cream fillings used with pastries, the organism can make the foods toxic if they are handled improperly and the bacterium is allowed to produce enterotoxin. Growth of *S. aureus* in synthetic cream fillings is unlikely unless the composition of the filling is altered when it contacts the pastry and then becomes suitable for proliferation of the staphylococcus.

Delicatessen salads often contain ingredients that are cut, diced, or otherwise processed by hand, and so the presence of *S. aureus* in these foods is possible. As long as the pH and the temperature are properly controlled, these foods are not likely to cause problems.

BOTULISM

Botulism is a serious, often fatal disease caused by the action of a proteinaceous neurotoxin elaborated by *Clostridium botulinum* (55). Botulinus toxin is generally regarded as the most potent poison in the world. The disease occurs in three forms: (*a*) the traditional foodborne intoxication, (*b*) wound botulism, and (*c*) infant botulism (which may result from consumption of spores of the bacterium in a food). To date seven types, A through G, of botulinus toxin are known, based on results of the toxin–antitoxin neutralization test. Human botulism (other than infant botulism) is generally caused by toxins of type A, B or E, whereas botulism in animals generally results from ingestion of type C or D toxin.

Botulism, derived from *botulus*—the Latin word for sausage—has been known in Germany for more than 1000 years (56). Initially the disease was associated with sausage and other foods of animal origin, but this has changed and current problems usually are associated with foods of plant origin or with fish. The organism is widely distributed in nature, particularly in soil and in the aquatic environment. Thus it is easy for plants and fish to become contaminated with spores and/or vegetative cells of *C. botulinum*.

Although botulism in humans is a serious disease that often is fatal, fortunately its incidence is low. As shown in Table 2.4, there have been 766 reported outbreaks of botulism in the United States in 78 years. According to Gunn (55), there were 1961 cases (sick persons) involved in the 766 outbreaks. Of the outbreaks since 1950, about 28% were caused by type A toxin, 12% by type B, 10% by type E, less than 1% by type F, and 49% in which the toxin was not characterized (55). Data for 1970 to 1977, when the number of outbreaks resulting from toxin that was not characterized decreased markedly, indicate that

TABLE 2.4 Number of Outbreaks of Foodborne Botulism in the United States, 1899 to 1977

Years	Source of Food		
	Home Processed	Commercially Processed	Unknown
1899	1	0	0
1900-1909	1	1	0
1910-1919	48	14	8
1920-1929	77	26	13
1930-1939	135	6	13
1940-1949	120	1	13
1950-1959	50	2	51
1960-1969	42	10	26
1970-1977	74	6	19
TOTAL	548	66	152

<u>Source</u>. Adopted from Gunn (55).

about 51% of the outbreaks were caused by type A toxin, 21% by type B, 12% by type E, none by type F, and 17% in which the toxin was not characterized (55). It is evident from these data that type A toxin is most often involved with outbreaks of botulism. The combined total for types B and E (associated with fish) is likely to be less than that of type A.

Earlier in the discussion, wound botulism and infant botulism were mentioned. These conditions are described briefly in the following paragraphs, whereas the more conventional foodborne intoxication is considered in the remainder of the discussion of botulism.

Wound botulism results when a wound becomes infected with *C. botulinum* that produced toxin during growth in the wound. Symptoms of wound botulism are essentially the same as those of foodborne botulism (see later discussion), except that gastrointestinal symptoms are likely to be missing (56). Through 1977 only 18 cases of wound botulism have been reported; type A toxin was responsible for the illness in three cases and type B in one case, and information for the others is not available (55).

Infant botulism results from intraintestinal production of botulinum toxin and seems to occur in infants less than 1 year of age and most often in infants less than 6 months of age. The disease was first recognized in 1976. It is characterized by constipation, followed by neuromuscular paralysis that begins with the cranial nerves and progresses to peripheral and respiratory musculature (55). Through 1977, cases were reported to the Center for Disease Control from 15 states, with most cases noted in California. Of the 58 reported cases, 33 were caused by type A toxin and 25 by type B toxin (55). The particular source or sources of spores ingested by the infants is unclear. Dust in the home has been implicated as the source in one outbreak, whereas honey has been implicated in several other cases (55). However, no source of spores could be found in still other cases. Recently Sugiyama et al. (57) tested 55 retail and 186 producer samples of honey and found spores of *C. botulinum* in two retail samples (type A in one and type B in one) and in 18 producer samples (type A in 11 and type B in 8). Because the spores occur widely in the natural environment, the real importance of contaminated honey in infant botulism remains to be determined. It is likely that ingested spores of *C. botulinum* cause disease in infants and not in adults because infants do not have an adequately developed protective intestinal microflora.

Symptoms and Treatment of the Disease

Symptoms

In most cases the early symptoms of botulism are typical of a gastrointestinal ailment and include nausea, vomiting, and diarrhea (56). The gastrointestinal symptoms are particularly common in type E botulism but are absent in wound botulism. These symptoms commonly appear within 12 to 36 hours after consumption of the toxic food. In rare instances, symptoms have occurred as soon as 2 hours or as long as 14 days after food containing the toxin was ingested. Onset of symptoms of type E botulism seems to be more rapid than onset of other types of botulism (56).

Gastrointestinal disturbances are followed by neurologic symptoms; weakness, lassitude, dizziness, and vertigo often develop early during the course of the disease. Ophthalmologic symptoms appear frequently and include blurred vision, diplopia, dilated and fixed pupils, and impaired reflection to light. Other symptoms that have been noted include ptosis (drooping) of eyelids, weakness of facial muscles, difficulty in speech and swallowing because of pharyngolaryngeal paralysis; constipation; dryness of mouth, tongue, and throat; muscle weakness; and decreased blood pressure. Respiratory muscles and the diaphragm become paralyzed as the disease progresses. Death most often results from respiratory failure and airway obstruction (56). Without treatment, the mortality rate can be as high as 67%. Improvements in treatment of botulism resulted in reduction of the mortality rate to about 23% between 1970 and 1973.

Treatment

Treatment consists of administering a suitable antitoxin to the patient. When this is done, care must be exercised to ensure that the potential recipient will not suffer from serum sensitivity since the antitoxins are of equine origin. Other approaches to treatment include (*a*) use of a tracheostomy or mechanical means to aid respiration if the patient's respiration is impaired and (*b*) removal of the toxin from the intestinal tract through use of enemas, a cathartic, gastric lavage, or administration of sodium bicarbonate. (Complete diagnostic and clinical information and details on the Center for Disease Control's antitoxin supplies are contained in *Botulism*, available from the Center for Disease Control, Atlanta, Georgia.)

The Organism

Clostridium botulinum is a strictly anaerobic, rod-shaped bacterium about 0.8 to 1.2 μm by 4 to 6 μm. The organism is gram-positive and produces heat-resistant endospores. Various strains of *C. botulinum* can be grouped on the basis of the toxin–antitoxin neutralization test.

Spores of the various types of *C. botulinum* are widespread in nature and can be regularly recovered from soil in different parts of the world and less frequently from water (58). Spores of type E *C. botulinum* have been found in marine sediments and the intestinal contents of fish from waters throughout the northern hemisphere (58). There is some evidence that a certain type may predominate in one geographic region, whereas another type may be predominant in a different region (e.g., type A is believed to predominate in the western United States and type B in the midwestern and eastern regions of the United States) (58).

Factors That Affect the Organism

Temperature

Types A and B of *C. botulinum* grow over a temperature range of approximately 10 to 50°C. Type E can grow and produce toxin at 3.3°C (31 to 45 days are needed); the upper limit for its growth is about 45°C (59). Some isolates have been found that can grow at temperatures several degrees below the minimums just indicated, but they seem to be exceptional strains.

Heat can be used to effectively inactivate spores of *C. botulinum.* Spores of types A and B have a *D* value (the time in minutes of heating at a given temperature required for a 90% reduction in the count of viable spores) of 0.1 to 0.21 at 121.1°C. The *D* value of type E spores ranges from 0.33 to 3.3 at 80°C, and in whitefish chubs it is 1.6 to 4.3 at 80°C. Spores of *C. botulinum* are highly resistant to destruction by freezing, but vegetative cells are nearly as susceptible to injury, as are cells of gram-negative bacteria (59).

Irradiation

Spores of *C. botulinum* type B (a particularly resistant strain) can be inactivated by radiation with 3.5 (at 20°C) to 5.28 (at −196°C) Mrad. Actually, these treatments reduced the population of spores by 10^{12} (59). Inactivation by radiation can be affected by temperature, condition of spores, presence of salts in the treating menstruum, and water activity of the substrate. Spores tend to be more resistant to irradiation when they are in food rather than in a laboratory medium (59).

Water Activity

Outgrowth of spores and synthesis of toxin by *C. botulinum* type A can occur at an a_w (water activity) value of 0.95, by type B at 0.94, and by type E at 0.97 when other conditions are optimal (59). Given the right environment (broth, pH 7, and 20 to 30°C), spores of types A, B, and E can germinate at an a_w value of 0.89 if obtained by adding glycerol or at 0.93 if obtained by adding NaCl to the broth (59). Growth of vegetative cells also responds to the a_w of the medium. For example, maximum growth of type E occurs at an a_w value of 0.995 (15 to 30°C), but such growth is reduced by 35% when the a_w is reduced to 0.98 (59).

Curing Salts

There is abundant experimental evidence to indicate that *C. botulinum* can grow and produce toxin in meat products that do not contain nitrite, provided the meat products are held sufficiently long at temperatures (above refrigeration) that allow growth of the organism. Because *C. botulinum* can produce toxin in the absence of nitrite or in the presence of small amounts (when compared to traditional usage) of nitrite, there has been and continues to be worldwide concern about the potential hazard of eliminating or reducing the nitrite content of cured meat products. Efforts to reduce the amount of nitrite used in these foods have been prompted by the occasional appearance of nitrosamine (a carcinogen) in cooked bacon and also by the suspicion that nitrite itself might be a carcinogen (see Chapter 6).

Acids

Clostridium botulinum types A and B fail to grow in foods with a pH of 4.6 or lower (59). Tolerance of *C. botulinum* in acid is reduced when NaCl or other inhibitory agents are in the acidified medium. *Clostridium botulinum* type E is more sensitive to acids than are the other types. The minimal pH permitting outgrowth of type E spores is 5.0 to 5.4 at 30°C, 5.4 at 15.6°C, 5.7 to 5.9 at 8°C, and 6.2 at 5°C.

Chlorine

It has been shown (59) that 4.5 ppm of free available chlorine (pH 6.5 and 25°C) can inactivate spores of *C. botulinum* type A in 6 to 8 minutes, type B in 3 to 8 minutes, and type E in 4 to 6 minutes (59).

Oxidation–Reduction Potential

Outgrowth of spores and growth of vegetative cells of *C. botulinum* type E will occur when the E_h of the substrate is less than 0 to 100 mV (56). Presence of some aerobic microorganisms may accelerate growth of *C. botulinum* by reducing the oxidation–reduction potential of the substrate (56).

Botulinum Toxins

The toxins are simple proteins composed only of amino acids. Molecules of the toxins consist of a toxic component (6 to 7*S*) and a nontoxic component of equal or larger (12 to 16*S*) size (56). The molecular weight of type E toxin is about 350,000, that of type F is about 238,000, and that of type A is greater than the molecular weight of either type E or F (56). Some of the toxins can be activated by trypsin or a trypsin-like enzyme. This was first observed with type E toxin. The toxin is completely inactivated in food by heating it at 80°C for 30 minutes and in correspondingly less time at 100°C (56).

Foods Involved in Outbreaks of Botulism

Foods responsible for outbreaks of botulism most commonly are processed in the home rather than in a commercial establishment (Table 2.4). This is true because low-acid foods often receive an inadequate heat treatment when they are canned in the home. Vegetables likely to have been inadequately processed account for about 50% of the outbreaks of botulism that occurred in the United States from 1899 to 1977 (Table 2.5). Fish and fish products were involved in about 15% of the outbreaks, fruits in about 10% and condiments in about 8% (Table 2.5). Fewer outbreaks were associated with beef, milk and milk products, pork, or poultry.

PERFRINGENS POISONING

The association of *Clostridium perfringens* with mild diarrheal illness was first suggested in 1895 and again in 1899 (60). In 1943 Knox and MacDonald (61) reported that this organism could cause food poisoning and described outbreaks in which children became ill after consuming food at school. The problem resulted from gravy, made the previous day, that was heavily contaminated with anaerobic spore-forming bacteria, including *C. perfringens.* This report was followed 2 years later by that of McClung (62), who described several outbreaks of food poisoning that resulted after chickens steamed the previous day were consumed. *Clostridium perfringens* was isolated from the chickens. Since these reports of nearly 40 years ago, numerous other outbreaks of perfringens food poisoning have been recorded in the literature. Currently *C. perfringens* annually accounts for 6 to 16 reported outbreaks of foodborne illness in the United States (Table 2.1). Since this illness commonly is associated with the food service industry (Table 2.2), the number of cases in any given outbreak could be quite large.

Symptoms of the Disease

Perfringens food poisoning is characterized by severe diarrhea and cramps in the lower abdomen, usually without vomiting. Pyrexia, shivering, headache, and other symptoms of infection are usually absent (60). The incubation period is normally 8 to 24 hours, but it may be shorter or longer. Recovery is usually rapid, with the illness commonly lasting no more than 12 to 24 hours, but fatalities of elderly debilitated persons have occurred (60).

The Organism

Description

Actively growing vegetative cells of *C. perfringens* usually appear as straight, plump rods with blunt ends (63). The cells commonly are 2 to

TABLE 2.5 Outbreaks of Botulism Associated with Various Foods, 1899 to 1977[a]

Type of Food	Type of Toxin						
	A	B	E	F	A & B	Unknown[b]	Total
Vegetables	115	31	1	0	2	2	151
Fish and fish products	11	4	25	0	0	1	41
Fruits	22	7	0	0	0	0	29
Condiments[c]	17	5	0	0	0	1	23
Beef[d]	6	1	0	1	0	0	8
Milk and milk products	3	2	0	0	0	0	5
Pork	2	1	0	0	0	0	3
Poultry	2	2	0	0	0	0	4
Other	8	3	3	0	0	0	14
Unknown[b]	9	3	1	0	0	6	19
TOTAL	195	59	30	1	2	10	297

Source. Adopted from Gunn (55).

[a] For 1899-1973 includes only outbreaks in which toxic type was determined; for 1974-1977 includes all outbreaks.

[b] Category applicable only for 1974-1977.

[c] Includes outbreaks traced to tomato relish, chili peppers, chili sauce and salad dressing.

[d] Includes one outbreak of type F in venison and one outbreak of type A in mutton.

6 by 0.8 to 1.5 μm. Size is dependent on strain of the bacterium, age of the culture, and nature of the substrate (63). Cells may be very short and appear cubical when the bacterium is in the exponential phase of growth. Stained cells from food usually appear shorter and fatter than do cells that grew in a laboratory medium.

The bacterium is anaerobic and gram-positive and produces spores, although they are usually not seen in stained cells grown in food. *Clostridium perfringens* will produce subterminal spores when cultured in a suitable medium. Protoplasts appear in sporulating cultures of some strains of *C. perfringens*. The bacterium is nonmotile, and some strains produce a capsule composed of polysaccharides complexed with protein (63).

Clostridium perfringens produces both smooth and rough colonies. Smooth colonies are 1 to 3 mm in diameter, low convex, opaque or grayish white, glossy, and with entire margins (63). Transparent sectors or protruberances appear on colonies during the second day of incubation; mutants can be isolated from such areas. Rough colonies are 3 to 5 mm in diameter and have a slightly raised, translucent surface with lobate margins (63). Mucoid strains produce raised colonies that are opaque and glistening and have regular margins. Cells from such colonies are heavily capsulated.

Metabolism

Clostridium perfringens is exacting in its nutrition requirements. A synthetic minimal medium that supports growth of the bacterium contains 13 to 14 amino acids, adenine, biotin, calcium pantothenate, pyridoxine, ammonium chloride, magnesium chloride, sodium–potassium buffer, and glucose (63). Foods that are nutritionally adequate for *C. perfringens* include meats and meat dishes, gravies, fish, milk and legumes.

Clostridium perfringens is saccharolytic and thus requires carbohydrates to produce energy. The bacterium is heterofermentative and metabolizes glucose to produce lactate, acetate, butyrate, ethyl alcohol, carbon dioxide, and hydrogen (63). The Embden–Meyerhof pathway and the lactic acid dehydrogenase and pyruvate–clastic system are operative in this bacterium, but the conventional hexose–monophosphate pathway apparently is not involved as a major means of glycolysis (63).

Clostridium perfringens can ferment glucose, fructose, galactose, mannose, maltose, lactose, sucrose, ribose, xylose, trehalose, starch, dextrin, and glycogen. Fermentation of salicin and glycerol is variable; mannitol is not fermented.

Hydrogen sulfide is and indole is not produced by *C. perfringens*. Most strains reduce nitrate to nitrite and hydrolyze gelatin, but not hemoglobin, casein, coagulated albumin, or serum (63). Some strains produce a collagenase that can hydrolyze native collagen. Phospholipase C (lecithinase) is produced and is employed diagnostically.

Factors That Affect the Organism

Many of the factors that contribute to growth or survival of other bacteria also affect *C. perfringens*. The most important of the factors are discussed in the following paragraphs.

Temperature

Clostridium perfringens grows at temperatures in the range of 15 to 50°C. The optimum temperature for most rapid growth (generation time of about 10 minutes) is approximately 45°C. At this temperature cells of many strains will lyse after the logarithmic phase of growth has been completed (63).

Freezing or extended refrigeration will cause a decrease in the viable count of *C. perfringens* (63). This decrease may range from 99 to 99.9%. Vegetative cells are more sensitive to low temperatures (susceptible to cold shock at all stages of growth) than are spores.

Spores from different strains may vary considerably in resistance to heat. For example, spores from a group of heat-sensitive strains had *D* values at 90°C of 3 to 5 minutes, whereas spores from heat-resistant strains had *D* values at 90°C of 15 to 145 minutes.

Acids

Clostridium perfringens grows rapidly at pH values between 6.0 and 7.5, with more rapid growth at pH 6.5 than 7.0 (63). The extremes for growth are pH 5 to 9, but this may differ in foods. Growth of *C. perfringens* in a glucose-containing medium results in acid production and an attendant decrease in pH to below 5. Cells of some strains do not survive more than 4 to 5 days at such low pH values.

Water Activity

The minimum a_w allowing growth of *C. perfringens* is 0.95 to 0.96 (at pH values of 5.5 to 7.0) when glucose is used to adjust the a_w value. Substitution of potassium chloride or sodium chloride for glucose results in 0.97 as the minimum a_w value that allows growth of the organism.

Irradiation

Spores from different strains of *C. perfringens* may vary in their resistance to γ radiation. The *D* value of resistant strains irradiated in an aqueous suspension ranged from 0.26 to 0.34 Mrad, whereas the *D* value of sensitive strains ranged from 0.12 to 0.32 Mrad (64).

Oxidation–Reduction Potential

Clostridium perfringens is not a strict anaerobe, but its ability to initiate growth is dependent on the oxidation–reduction condition of the medium. The upper E_h value for growth is in the range of +31 mV at pH 7.7 to +230 mV at pH 6.0. The optimum E_h value for growth has been reported as −166 mV and −200 mV (60). Undoubtedly, this is affected by strain of *C. perfringens*, pH, inoculum size, and metabolic activity of cells.

Perfringens Enterotoxin

The enterotoxin of *C. perfringens* is believed to be a protein with a molecular weight of 36,000 ±4000 and an isoelectric point of 4.3 (65). The toxin contains 19 amino acids with aspartic acid, serine, leucine, and glutamic acid being predominant. Inactivation of the toxin is achieved with pronase and the protease of *Bacillus subtilis,* but not with trypsin, chymotrypsin, papain, bromelain, or carboxypeptidase (60). The toxin is heat labile, with a *D* value of 4 at 60°C.

The enterotoxin acts by causing a reversal in transport of fluid, electrolytes, and glucose in the intestine. This is accompanied by intestinal cellular damage (66).

Enterotoxin is formed when cells of *C. perfringens* sporulate. Hence illness occurs after consumption of large numbers of vegetative cells, which sporulate in the intestine (67). When cells of *C. perfringens* have formed spores and when mature spores are released from sporangia, enterotoxin is released with the spore (67). The enterotoxin also can be produced in foods (rather than in the intestinal tract), provided that the product in question allows for sporulation of *C. perfringens* (68). Foods with detectable enterotoxin are likely to contain 10^4 to 10^6 spores of *C. perfringens* per gram (68).

Foods Involved in Perfringens Intoxications

Clostridium perfringens is widely distributed in nature and hence can contaminate a variety of foods, although foods of animal origin are most commonly involved in outbreaks of perfringens poisoning. Beef and beef-containing items are the most common vehicles of perfringens poisoning. Poultry products, including turkey and chicken, are the second most common vehicles. Other foods that have been involved less often include pork, lamb, fish, shrimp, crab, beans, potato salad, macaroni and cheese, and olives (63).

Holding foods too long at temperatures near the maximum for growth of *C. perfringens* (as might happen in steam or cooling tables) or cooling foods too slowly (as might happen when a large mass of food, including gravies and sauces, is placed in an undersized refrigerator) are the most common mistakes in food handling that can lead to perfringens poisoning. This is true because these conditions allow for development of large numbers of vegetative cells that, unless inactivated, are ingested when the improperly handled food is served. It is evident from this discussion that conditions needed for outbreaks of perfringens poisoning sometimes exist in restaurants, cafeterias, or other mass feeding operations. Hence it is not surprising that most outbreaks of this illness are associated with the food service industry (Table 2.2).

SALMONELLOSIS

Typhoid fever was studied in 1856 by William Budd, who concluded the disease was infectious, that the causative agent was excreted in the feces of patients, and that contaminated milk and water were important in its dissemination (69). It was not until 1880 that Eberth observed the typhoid bacillus in tissues of dead patients. Four years later, in 1884, Gaffky isolated and cultivated this organism. Another year later, Salmon and Smith isolated an organism from cases of swine fever that they considered to be the causative agent and that they named *Bacillus cholerae suis* (69). It is now known that swine fever is caused by a virus, and the bacillus that Salmon and Smith isolated was probably present as a secondary invader. Nevertheless, the bacteria in the typhoic–paratyphoid–enteritis group were given the generic name *Salmonella,* on the recommendation of Lignieres, in honor of the American bacteriologist, D. E. Salmon, first chief of the U.S. Bureau of Animal Industry (69).

In 1888, Gaertner isolated *Salmonella enteritidis* from a patient who died after consuming contaminated meat, and soon afterward Durham and de Noeble described *Salmonella typhimurium* that was also recovered from patients ill with gastroenteritis following ingestion of infected meat (69). Loeffler, in 1892, identified the causative agent of mouse typhoid as a member of this group, and soon many additional related organisms were described by other investigators.

Salmonellosis continues to be a leading form of foodborne illness in the United States. In fact, in 1976 and 1977 there were more outbreaks of salmonellosis than of any other single form of foodborne illness (Table 2.1). Problems in food service establishments and homes account for most outbreaks of salmonellosis (Table 2.2). Because of the importance of salmonellosis, this form of foodborne illness is considered in some detail.

Symptoms and Treatment of the Disease

There are three main types of salmonellosis: enteric (typhoid) fever, gastroenteritis, and a localizing type with foci in one or more organs accompanied by septicemia. Every *Salmonella* strain is potentially able to produce any of these three clinical types of infection. Each of these types is described briefly.

Enteric Fevers

Typhoid fever is the classic example among the enteric fevers. The incubation period is 7 to 14 days, and onset of the disease is insidious, usually beginning with malaise, anorexia, and a headache. This is usually followed by a fever that, in a stepwise manner, rises to an average of 40°C. The pulse rate tends to be slow in relation to the degree of fever, and nosebleeds may occur at this stage of the disease.

During the first week the patient usually is prostrate and may have diarrhea, although constipation is even more common. Either condition is accompanied by abdominal tenderness and distention. A cough and bronchitis also may be present at this time. Rose spots frequently appear during the first or second week. Splenomegaly is common, and the temperature remains elevated. In severe cases the patient may become delirious and show the so-called typhoid state for which the disease was named (typhoid fever is derived from the Greek and originally designated a state of irrationality and coma). After the third week, the temperature curve shows morning remissions and returns to a normal level by gradual lysis.

Blood cultures taken during the first and second weeks often yield the typhoid bacilli, but less frequently when taken during the third week. The organisms may appear in stool cultures from the beginning and continue to do so until convalescence is completed. Organisms may also be recovered from urine (during second and third weeks), bone marrow, and rose spots.

Relapses occur in about 10% of the cases and probably represent a reinvasion of the blood stream by organisms multiplying in areas such as lymphoid tissue, bone marrow, the spleen, and the biliary system. The mortality rate in untreated patients is about 10%, and death generally results from intestinal hemorrhage or perforation. In some instances salmonellae may establish themselves in the tissue of the host to produce a carrier state after recovery. This is most likely to occur after typhoid fever, where about 3% of the cases are found to excrete *S. typhi* in their stools for over a year after recovery from the disease. The carrier state is observed less frequently with *Salmonella paratyphi* and *Salmonella schottmüelleri* than with *S. typhi*, and its duration is much shorter.

Other enteric fevers usually have a shorter incubation period (1 to 10 days) than typhoid fever and are not as severe. Fever and malaise are the dominant symptoms and usually last from 1 to 3 weeks.

Gastroenteritis

This form of salmonellosis has an incubation period of from 3 to 72 hours, with most outbreaks occurring within 12 to 24 hours after the organisms have been ingested. The principal symptoms of a *Salmonella* gastrointestinal infection are nausea, vomiting, abdominal pain, and diarrhea that usually appears suddenly. Their occurrence may be preceded by a headache and chills. Additional symptoms often associated with the disease include watery, greenish, foul-smelling stools; prostration; muscular weakness; faintness; a moderate fever; restlessness; and twitching and drowsiness.

Severity and duration of the disease vary with the amount of food (and hence salmonellae) consumed, the kind of *Salmonella,* and the resistance of the individual. Intensity varies from slight discomfort and diarrhea to death in 2 to 6 days. Usually, symptoms persist for 2 to 3 days, followed by an uncomplicated recovery. In some instances, however, symptoms may linger for weeks or months. Some patients (0.2 to 5%) become carriers of the *Salmonella* organism that caused their infection. The mortality rate is generally less than 1%.

Septicemias

Septicemias caused by salmonellae are characterized by a high remittent fever and blood cultures that yield the causative organism. Intestinal involvement is usually absent in adults, but in children the septicemia may occur as a complication of gastroenteritis. Organisms may localize in any tissue of the body and may produce local abscesses in the perineal and pelvic regions, cholecystitis, pyleonephritis, endocarditis, pericarditis, meningitis, arthritis, or pneumonia. *Salmonella choleraesuis* is one of the most common organisms found in this type of infection. The mortality in *Salmonella* septicemia ranges from 5 to 20%.

Treatment of Salmonellosis

Formerly, treatment for typhoid fever and other *Salmonella* infections was largely supportive and consisted of maintaining the fluid balance and nutritional state of the patient. More recently, sulfonamide drugs have been found beneficial in the treatment of certain *Salmonella* infections, but their use to treat typhoid fever has been disappointing. A combination of sulfonamides with larger than ordinary doses of penicillin has been found to be of limited therapeutic value.

Streptomycin, although active against salmonellae *in vitro,* has not produced beneficial results when used to treat typhoid fever. Oral administration is accompanied by a marked reduction in number of typhoid bacilli in stools, but the bacteria reappear when streptomycin is discontinued.

Chloramphenicol is effective in treatment of typhoid fever, but patients do not become afebrile until about the fourth day after administration of the antibiotic is started. Patients may become carriers in spite of adequate therapy with this antibiotic. Treatment of other *Salmonella* infections with chloramphenicol has been even less satisfactory. It is thought that the intracellular location of the typhoid organisms accounts for the slow response of this infection to antibiotic therapy.

The Organisms

Bacteria in the genus *Salmonella* are gram-negative, asporogenous, facultative short rods that are usually motile by means of peritrichous flagella, although nonmotile forms may occur (69). They are easily cultivated on ordinary media and are able to produce acid (and usually gas) from glucose, mannitol, maltose, and sorbitol. Lactose, sucrose, salicin, and adonitol are seldom attacked. Fermentation of carbohydrates other than those listed is variable. Salmonellae generally fail to produce acetoin or hydrolyze urea. They do, however, produce nitrite from nitrate.

Salmonellae average about 2 to 3 μm in length and about 0.6 μm in width but may vary in size under different environmental conditions. Young cultures on agar may form a predominance of coccobacillary cells, whereas filamentous forms are occasionally seen in cultures grown in liquid media. Capsules are seldom formed by salmonellae grown at 37°C, but most species give rise to mucoid colonies consisting of encapsulated cells, especially when grown at 20°C.

On ordinary agar media, salmonellae produce colonies, averaging 2 to 3 mm in diameter, difficult to distinguish from those of coliform bacteria. Freshly isolated strains almost invariably produce circular, smooth, glossy colonies that are more translucent and have a more delicate texture than those of *E. coli*. Colonies formed by strains that have frequently been subcultured on artificial media tend to be rough, with a granular surface and an irregular edge. This variation, which also occurs in some other bacteria, is associated with a loss of virulence and of the somatic O antigen.

Cells of salmonellae possess antigens that fall into three main categories. The K (from the German word *Kapsel)* or envelope antigens are thought to surround the cell and are chemically similar to O antigens. Generally, the K antigens are heat labile and tend to mask the somatic antigens of the cells, thus making live cells inagglutinable by O antisera. The Vi antigen of *Salmonella typhi* is an example of the K antigen. Cells of *S. typhi* may undergo a loss of the Vi antigen, at which time they can be agglutinated by both O and Vi antisera.

The O (from the German *ohne Hauch*) or somatic antigens are located in the body of the cell (presumably near the surface), are phospholipid protein polysaccharide complexes, and are heat stable. Chemically, the polysaccharide moiety of the antigen is extremely complex and consists of a variety of sugars, including hexosamines, heptoses, pentoses, hexoses, and desoxyhexoses. Numerous serologically distinct O antigens have been recognized. Most species possess more than one O antigen; thus many species share the same O antigens. Species can be divided into a limited number of groups on the basis of their O antigen composition, with each group characterized by possession of an O antigen not found in the other groups. In this way, most salmonellae can be divided into nine O antigen groups, with most of those commonly encountered falling into the first four of these, designated as A, B, C, and D. Cells can lose the normal O antigens and then are not agglutinable by O antiserum.

The H (from the German *Hauch*) or flagellar antigens are located in the flagella, are proteinaceous in nature, and are heat labile. Many *Salmonella* species may have one or the other of two sets of flagellar antigens and are designated as diphasic in regard to this characteristic. These sets of antigens are known by the terms phase 1 and phase 2. When a diphasic species is grown, one of the two phases predominates, and more often than not it is phase 1. Phase 1 antigens are more and phase 2 antigens less specific than are O antigens. Loss of flagella by a cell is accompanied by loss of the H antigens.

The presence of antigens in salmonellae has led to development of the Kauffmann–White schema for identification of the different serotypes. In this schema *Salmonella* serotypes are arranged in subgroups on the basis of their O antigens, whereas the H antigens represent the type. Use of these serological procedures has led to the identification of hundreds of serotypes, all considered pathogenic to humans.

Factors That Affect the Organisms

Salmonellae are not very hardy bacteria since they do not produce spores. Hence adverse conditions generally have a more profound effect on this group of bacteria than on some of the other microorganisms discussed in this chapter. Some of the major factors that govern growth and survival of salmonellae are discussed in the following paragraphs.

Temperature

Although under laboratory conditions most salmonellae grow best at 37°C, some of these bacteria can grow at 5.5 to 45°C. Matches and Liston (70) determined the minimum growth temperatures of 10 serotypes by means of a temperature-gradient block. *Salmonella heidelberg* and *S. typhimurium* were able to grow in a liquid medium at 6.6°C, but not at 6°C. The pH of the substrate influenced the minimum temperature at which growth was observed. Lowest growth temperatures were associated with pH values of 7.0 to 8.0. Length of incubation is another factor to be considered when the organisms are held at low temperatures. These same investigators found that the minimum temperature for growth of *S. heidelberg* was between 6.0 and 7.0°C when observations were made after 7 days and that it dropped to between 4.0 and 5.7°C after 15 days. Comparable results obtained for *S. typhimurium* and *S. derby* at 7 days were 8.5 to 9.0°C and 8.2 to 9.0°C and at 15 days, 7.8 to 8.2°C and 6.6 to 7.0°C, respectively. Initiation of growth by salmonellae at low temperatures can be important when foods are stored for long periods at temperatures of 4°C or above.

Growth of *S. enteritidis* in a variety of foods at 22°C and 37°C was studied by Segalove and Dack (71). They inoculated cans of asparagus, spinach, string beans, tomato juice, peaches, shrimp, salmon, corn, and peas. Duplicate cans of each food were incubated at 22°C and 37°C. Growth was observed at 22°C in all foods except peaches and at 37°C in all but peaches and asparagus. In nearly all instances, growth was greater at 22°C than at 37°C. Subramanian and Marth (72) compared growth of *S. typhimurium* in skim milk at 22°C and 37°C. They observed numbers approaching 1×10^9 per milliliter after 12 hours at 37°C, with little additional growth during the next 4 hours at the same temperature. In contrast to this, at 22°C numbers slightly in excess of 1.0×10^8 per milliliter were attained after 16 hours of incubation.

Growth of *S. typhi* at 45°C has been reported by several investigators. Spencer and Melroy (73) noted that this organism grew well for 36 transfers at 45°C, and for at least 148 transfers with alternating temperatures of 45°C and 37°C. According to Ware (74), growth of *S. typhi* at high temperatures was enhanced by adding certain amino acids to the culture medium. A simple glucose–salts medium supported growth of *S. typhi* at 37°C, but not at 40°C. Growth occurred at 40°C when the medium was fortified with L-arginine, L-glutamic acid, thymine, or L-lysine. Addition of L-glutamic acid plus L-arginine and thiamin or L-lysine permitted cell production at 43°C in 24 hours. Elimination of L-arginine was accompanied by cell multiplication in 48 but not 24 hours at 43°C. Read et al. (75) isolated *S. anatum*, *S. meleagridis*, *S. new-brunswick*, and *S. tennessee* from dry milk. They then studied these organisms and *S. senftenberg* 775W for their heat resistance to determine if they would survive pasteurization of milk. Thermal inactivation tests were made on washed cells of test organisms that were resuspended in sterile whole milk. Excluding *S. senftenberg*, *D* values ranged from 3.6 to 5.7 seconds at 62.8°C from 1.1 to 1.8 seconds at 65.6°C and from 0.28 to 0.52 seconds at 68.3°C. Similar values for *S. senftenberg* were 34.0, 10.0, 1.2, and 0.6 seconds, respectively, for exposures of 65.6, 68.3, 71.7, and 74.0°C, respectively. Results of these tests suggest that recommended pasteurization processes are adequate to inactivate all seven strains of salmonellae studied, provided the initial concentration does not exceed a calculated 3×10^{12} salmonellae per milliliter of milk.

The *z* values (degrees Fahrenheit required for passage of a decimal reduction time curve through one log cycle) for rough and smooth variants of *S. senftenberg* 775W in various media were determined by Thomas et al. (76). They found the rough variant had *z* values of 11.494, 10.989, 10.638, and 10.204 in 0.5% NaCl, skim milk, beef bouillon, and green pea soup, respectively. The smooth variant yielded values of 10.753, 10.417, and 10.753 in 0.5% NaC1, skim milk, and green pea soup, respectively.

Elliot and Heiniger (77) determined the *D* values of *S. senftenberg* 775W and *S. typhimurium* at relatively low temperatures by means of a temperature-gradient incubator. They recorded a *D* value of 24 for *S. senftenberg* at 55°C and of 8.5 for *S. typhimurium* at the same temperature.

Numerous studies have been conducted on the thermal resistance of salmonellae in various poultry products. Bayne et al. (78) determined the heat resistance of *S. typhimurium* and *S. senftenberg* 775W in ground chicken muscle heated to four different temperatures in the range of 55 to 75°C. Multiple one gram samples of meat containing 3×10^8 cells of *S. typhimurium,* after exposure for 5 minutes at 60°C, contained no viable cells. The more heat resistant *S. senftenberg* 775W required an exposure of 10 to 15 minutes at 65°C to kill an equal number of cells.

Destruction of salmonellae on egg shells during washing was investigated by Bierer and Barnett (79). They found that *S. pullorum, S. gallinarum,* and *S. typhimurium* on egg shells were killed by washing the eggs at 65.6°C for 3 minutes. This procedure, however, resulted in slight albumin coagulation on the inner surface of the shell. Coagulation did not occur when eggs were washed at 65.6°C for 1 minute, although salmonellae were recovered from one egg out of 600 that were washed. The salmonellae used in this study were destroyed in wash water at 53.3°C. The data just cited may be applicable to destruction of salmonellae on equipment surfaces by use of wash water.

The times required to free hard-boiled eggs from salmonellae were calculated by Licciardello et al. (80). When raw eggs were placed directly into boiling water, they indicated that 5.6, 8.4, 8.7, and 9.4 minutes of exposure are required to free small-, medium-, large-, and jumbo-sized eggs, respectively, from *S. senftenberg.* Values for *S. typhimurium* were found to be 4.5, 7.2, 7.3, and 7.8 minutes, respectively, for egg sizes as just listed. Exposure periods of 14.1, 16.0, and 20.6 minutes are required to destroy *S. senftenberg* in medium-, large-, and jumbo-sized eggs, respectively, when they are placed into water at 20°C, which is then brought to a boil and simmered. Values of *S. typhimurium* under the same conditions are 12.6 minutes, 14.8 minutes, and 18.8 minutes.

Destruction of salmonellae in dry products requires a higher treatment temperature, a longer exposure time, or both. Rasmussen et al. (81) used modified thermal death time tubes to determine the heat resistance of salmonellae in naturally contaminated meat meals containing 8 to 10% moisture. A temperature of 82.2°C for 7 minutes was sufficient to consistently destroy all salmonellae in these meals. A third meal containing 13% fat required an exposure of 7 minutes at 90.6°C to consistently free it from viable salmonellae.

The resistance of salmonellae to thermal destruction is 600 to 700 times higher in dried egg white than in liquid egg white, according to data cited by Prost and Riemann (82). They also reported that 7 days of heating at 49 to 50°C was required to eliminate salmonellae from naturally contaminated dry eggs.

Acids

Most investigations on the relationship between salmonellae and acids have dealt with destruction of the organisms. Only relatively few experiments have considered the effect of different acid concentrations on the growth of these organisms in a variety of substrates.

Stearn and Stearn (83) conducted tests to determine the effect of pH on colonial characteristics, morphology, motility, and staining behavior of *S. enteritidis*. When grown on nutrient agar adjusted to pH values in the range of 5.16 to 6.0, colonies appeared opaque, discrete, granular, iridescent, and had raised centers that flattened and appeared lysed and in which were formed secondary colonies with toothed margins. At pH values of 6.1 to 6.3, colonies appeared circular, glistening, white, moist, smooth, and entire. Normal translucent colonies developed at pH values in the range of 6.3 to 7.6, whereas at pH values between 7.8 and 8.4 colonies appeared very thin, spreading, and translucent. Changes in the pH of a broth medium were accompanied by morphological changes in the cells of *S. enteritidis*. During the first 24-hour incubation period at pH 5.15 to 5.6, chains and clumps of cells were evident and individual cells varied greatly in size and shape, ranging from coccoid forms to long, curved rods. At pH values of 5.6 to 7.0, cells were generally short and plump and exhibited marked bipolar staining. An increase in pH to 8.4 was accompanied by a greater tendency toward evenly stained, slender rods, which appeared gram-variable. This organism lost its motility when cultivated at pH values below 6.0 but regained it after subculturing in a neutral medium. At acid pH values, cells stained a deep pink with safranine. The color was lighter at a neutral pH value and cells became gram-variable when the pH entered the alkaline range.

The effects of lactic, acetic, and hydrochloric acids on *Salmonella aertrycke* were studied by Levine and Fellers (84). When hydrochloric acid was added to nutrient broth, growth was not inhibited until a pH value of 4.0 was attained. Inhibition with lactic acid also occurred at pH 4.0, whereas with acetic acid it was observed at pH 4.9. The organism was destroyed during a 48-hour incubation period when the medium was adjusted to pH 3.1 with hydrochloric acid, pH 4.0 with lactic acid, and pH 4.5 with acetic acid. Subramanian and Marth (72) determined the effect of citric, lactic, and hydrochloric acids, when added to milk in increments over a 16-hour period, on growth of *S. typhimurium.* During incubation at 37°C, a slight reduction in growth became evident after 8 hours, regardless of acid added, when a pH value in the range of 5.05 to 5.35 was attained in the milk through gradual addition of acid. Additional incubation was accompanied by further inhibition, with citric acid most inhibitory, followed in order by lactic and hydrochloric acids. Inhibitory effects of all acids were greater at 22°C than at 37°C with citric the most active, followed in order by lactic and hydrochloric acids.

Since acetic acid is commonly used to prepare mayonnaise and salad dressings, and since these products can become contaminated with salmonellae from eggs, numerous experiments have been conducted on the survival of these bacteria in acetic acid at various concentrations. Wethington and Fabian (85) added 1 ml of 24-hour cultures of various salmonella to salad dressing and mayonnaise and held the inoculated products at room temperature and 37°C. At both temperatures, survival in mayonnaise (0.48% acid, pH 3.80) was 12, 12, 6, 6, 1, and 6 hours, respectively, for *S. schottmüelleri, S. typhimurium, S. paratyphi, S. enteritidis, S. choleraesuis,* and *S. pullorum.* In salad dressing (1.10% acid, pH 3.20) survival was 6, 1, 6, 1, 1, and 1 hour, respectively, for the organisms in the sequence just listed.

When mayonnaise was made to contain 0.15% acetic acid (pH 5.0), survival was 144, 144, 156, 156, 156, and 132 hours, respectively, for the organisms listed previously. Tests using salad dressings with 0.4% acetic acid (pH 4.4) yielded survival periods of 144, 120, 24, 96, 96, and 96 hours, respectively, for the salmonellae in the preceding sequence.

Since carbon dioxide forms carbonic acid when in solution, its effect on salmonellae is considered here. Schillinglaw and Levine (86) exposed *S. typhi* cells to carbon dioxide (42 psi) and observed that 90% of the bacteria were destroyed in 6 hours at 30°C. For comparative purposes, 28 hours were required to achieve a similar reduction in number of *E. coli* cells.

The effect of high pH values on survival of *S. typhi* and *S. montivideo* was studied by Riehl et al. (87). They reported that at a pH of 11.0 to 11.5 and a temperature of 15°C, most cells were destroyed in 4 hours. Additional tests indicated prolonged exposure to alkaline water, regardless of other components, killed many of the bacteria.

Moisture, Salts, and Sugar

The moisture requirements for growth of salmonellae can best be expressed in terms of water activity. Christian and Schott (88) studied the a_w requirements of 16 *Salmonella* serotypes at 30°C. Growth of 15 motile strains occurred in liquid media at a_w values between 0.945 and 0.999. In foods, the lower limit for growth was slightly less than in culture media. Anaerobic growth was only slightly less than aerobic growth at each a_w value. The single nonmotile strain grew more slowly over a smaller range of a_w values.

The a_w requirement for growth of *Salmonella oranienburg* in liquid media was investigated by Christian (89). He adjusted the a_w of 0.25 strength brain heart infusion broth, nutrient broth, and a casamino acids–yeast extract–casitone broth with a salts mixture (NaCl, KCl, and Na_2SO_4) or sucrose. In all instances growth was observed at an a_w value of 0.95, but not at 0.94. Substitution of a glucose–inorganic salts broth for the media listed was accompanied by multiplication at an a_w value of 0.97, but not at 0.96, regardless of the substance used to control water activity. Christian (89) further noted that the minimum a_w for growth in the glucose–inorganic salts medium could be reduced to 0.96 by addition of five amino acids, including methionine, histidine, proline, serine, and glutamic acid. Addition of eight vitamins (thiamin, riboflavin, biotin, folic acid, pyridoxine, calcium pantothenate, nicotinic acid, and paraaminobenzoic acid) plus the five amino acids was accompanied by a further reduction in the minimum a_w, permitting growth at 0.95. Addition of the vitamins without the amino acids had no effect in reducing the a_w required for growth.

In other experiments Christian (90) used sucrose, glucose, glycerol, NaCl, and KCl to control the a_w in the glucose–inorganic salts medium and then studied behavior of *S. oranienburg* in these media. Use of glycerol to adjust the a_w of the medium permitted growth of the organism at 0.96 in comparison to the 0.97 required when other compounds were added. Respiration of cells was inhibited less by glycerol than by the salts or sugars used to control the a_w. Control of a_w by addition of glucose or sucrose was accompanied by accumulation of potassium (but not sodium) in *S. oranienburg* cells. Accumulation of potassium was greatest at an a_w value of 0.975. When NaCl served to adjust the a_w, accumulation of potassium was small, and none was concentrated when glycerol replaced NaCl.

The influence of amino acids on a_w requirements of *S. oranienburg* was examined further by Christian and Waldo (91). They observed that a reduction in the a_w value by adding NaCl or sucrose induced a lag and then decreased the rate of glucose oxidation by the organism. At a relatively low value (0.970 a_w), addition of amino acids such as proline, aspartic acid, aspargine, glutamic acid, glutamine, and cysteine caused an appreciable synergistic increase in respiration rate. Proline was the most stimulatory of the amino acids tested, and at an a_w value of 0.960 only this amino acid and its analogue azetidine-2-carboxylic acid gave appreciable stimulation. Proline was also stimulatory when glucose was replaced by pyruvate or succinate. These authors believe that proline is stimulatory by increasing the amino acid pool of the organism and hence decreasing internal water activity.

The behavior of freshly isolated salmonellae in the presence of sodium chloride was described by Bergmann and Seidel (92). They cultured the salmonella on agar fortified with 5 to 10, 15, 20 and 30% sodium chloride and in broth containing 5 to 10% and 15% of the chemical at 37°C for 1 day and at room temperature for 3 days. Visual inspection showed that 7 to 10% sodium chloride in agar and 8 to 10% in broth caused impaired viability or killed salmonellae. The authors also observed that a few strains of salmonellae recovered after they were inhibited by 15 to 30% sodium chloride in broth. Somewhat comparable results were reported by Severens and Tanner (93), who isolated strains of *S. pullorum*, *S. typhi*, and *S. schottmüelleri* able to grow in the presence of 8, 6, and 8% sodium chloride, respectively. Addition of lactic acid failed to increase the sensitivity of this bacterium to sodium chloride. Experiments by Geopfert et al. (94) demonstrated that *S. typhimurium* was able to initiate growth in skim milk adjusted to pH 4.9 with lactic acid and fortified with 3% sodium chloride. Raising the salt content to 4.5% in the same medium was associated with slow death of *S. typhimurium* when held at 7.5 or 13°C for 5 weeks. A further addition of 0.1% acetic acid to the medium enhanced death of the organism.

Tarr (95) demonstrated that addition of 0.02% sodium nitrite to fish digest broth at an acid pH caused inhibition of *S. typhi*. According to data by Severens and Tanner (93), selected strains of *S. pullorum*, *S. typhi*, and *S. schottmüelleri* were able to grow in the presence of 1:800, 1:800, and 1:600 solutions of copper sulfate, respectively, and 1:25,000, 1:50,000, and 1:25,000 solutions of mercuric chloride, respectively. Normal strains of these organisms were inhibited by a 1:4000 solution of copper sulfate and a 1:300,000 solution of mercuric chloride.

Other Chemicals

According to experimental evidence presented by Friberg and Hammarström (96) at 6°C and pH 7.2, a 1000-fold reduction in number of *S. typhi* cells required 0.025 to 0.05 mg free available chlorine per liter. *Salmonella typhimurium* was somewhat more resistant, and hence 0.10 to 0.15 mg of available chlorine per liter was needed for its destruction. Bacteriophages active against *S. typhimurium* were more sensitive to the action of chlorine than were the host cells. Kabler et al. (97) found that water containing 0.22 to 0.23 ppm of chlorine required 0.5 to 1.5 hours at 25 to 26°C and 2 to 15 hours at 1.5 to 3°C to kill cells of *S. typhi*. These investigators also observed that freshly isolated strains of *S. typhi* required approximately 2 hours of exposure to 0.22 to 0.23 ppm of chlorine for destruction, but after a year on laboratory media, the same cultures required only 0.9 hours of exposure for destruction.

The effect of pH and temperature on destruction of *S. typhi* by chlorine was determined by Butterfield et al. (98). According to their data, at pH 7.0 and 2.5°C, 100% destruction was accomplished after 10 minutes by 0.02 ppm, after 3 minutes by 0.03 to 0.06 ppm, and after 1 minute by 0.08 ppm. An increase in pH to 9.8 at the same temperature was accompanied by 100% destruction after 60 minutes by 0.15 ppm, after 20 minutes by 0.40 ppm, after 10 minutes by 0.74 ppm, and after 5 minutes by 1.0 ppm of chlorine. The concentration of chlorine required for 100% destruction at 20 to 25°C was somewhat less at each time period than noted previously.

The phenol coefficients of several quaternary ammonium compounds when acting against *S. typhi* were determined by Lane (99) and Croxall and Melamed (100). They reported phenol coefficients of 335, 75, 310, and 345 for *N*-ammonium chloride, *N*-bis (trimethylphenylpentenyl)-*N* ammonium chloride, *N*-dimethylammonium chloride, and (4-hydroxy-2-butynl)dimethyl(pentadecylbenzyl)ammonium chloride, respectively.

Goetchius and Grinsfelder (101) exposed *S. typhi* for 10 minutes to various concentrations of alkyl tolyl methyl trimethyl ammonium chloride. Their results indicated 50, 80, 90, 95, 98, and 99.5% destruction by concentrations of 6.25×10^{-5}, 7.6×10^{-5}, 8.4×10^{-5}, 9.2×10^{-5}, 10.2×10^{-5}, 10.8×10^{-5}, and 11.5×10^{-5}%, respectively. Stedman et al. (102, 103), working with *S. schottmüelleri*, found that a 1:2,000 concentration of diisobutyl phenoxy ethoxy ethyl dimethyl benzyl ammonium chloride destroyed 99.99% of the organisms in 10 minutes and that a 1:1,000 concentration was necessary to do the same job when organic matter (serum) was present. It was further observed that concentrations of 1:500, 1:100, and less than 1:25 were required to destroy the organism on the surfaces of stainless steel, asphalt tile, and linoleum, respectively. Similar tests with a mixture of alkyl dimethyl 3,4-dichloro benzyl ammonium chloride and alkenyl dimethyl ethyl ammonium bromide revealed that slightly higher concentrations were needed to accomplish the same task.

According to Ross et al. (104), quarternary ammonium compounds with 12 to 14 carbon atoms in the alkyl group showed maximum antibacterial activity against *S. typhi*. Incorporation of more polar substituents in the benzyl group increased the germicidal action, provided the alkyl group contained less than 12 carbon atoms. If the alkyl group contained 14 or more carbon atoms, antibacterial action decreased.

Satta (105) added *S. typhi* (30,000 cells per milliliter) to milk and treated it with hydrogen peroxide in an attempt to destroy the bacteria. Use of 0.25 to 0.30% peroxide destroyed the salmonellae in 4 to 5 hours at 17 to 32°C, whereas a 0.2% concentration of the chemical required 8 to 9 hours to kill the bacteria. Raw milks were inoculated with *S. typhi*, treated with 0.2% hydrogen peroxide, and incubated for 14 to

24 hours at 20 to 22°C and 28 to 30°C in experiments by Monaci (106). When compared to untreated controls, peroxide caused a reduction of 74 to 96% in the number of bacteria present in incubated milks. *Salmonella typhi* was recovered from one sample.

In contrast to the results just cited, Roushdy (107) was able to destroy *S. typhimurium* in milk by exposing it to 0.03% and 0.075% hydrogen peroxide for 4 hours and 1 hour, respectively. Sadilek and Stepanek (108) inoculated milk with *S. enteritidis*, warmed it to 57°C, added 150 to 200 mg of hydrogen peroxide per 100 ml, and treated the milk with catalase. Neither level of peroxide destroyed the organisms under the conditions described.

Salmonellae also can be affected by sorbic acid, lysozyme ultraviolet radiation, ozone, antioxidants, ethylene oxide, β-propiolactone, lecithin, carotene, riboflavin, spices, essential oils, and sulfhydryl compounds. The effects of these materials on the bacteria are discussed briefly by Marth (69).

Cause of Illness

Bryan et al. (109) have summarized current thinking on how salmonellae act to cause gastroenteritis. Apparently salmonellae reach the small bowel and invade the lumen, where they multiply. They then penetrate the ileum and, to a lesser extent, the colon, where an inflamatory reaction occurs. During this process lymphoid follicles can become enlarged and may ulcerate, and mesenteric nodes often become swollen. Salmonellae sometimes overwhelm mucosal and lymphatic barriers, enter the blood stream and cause a septicemia (see earlier discussion).

It has been suggested that endotoxins and lipopolysaccharide constituents of gram-negative cell walls are responsible for the fever often associated with salmonellosis. Some experimental evidence suggests that certain salmonellae can produce an enterotoxin.

The number of salmonellae that must be ingested before illness results was once thought to be large (e.g., 10^6 cells). Silliker (110) summarized information obtained from outbreaks, and the number needed for initiation of illness appears to be quite small, as few as 100 cells under the right circumstances.

Foods Involved in Salmonellosis

Foods of animal origin are nearly always associated with outbreaks of salmonellosis. Poultry meats were associated with 17% of the 500 reported outbreaks that occurred in the United States from 1966 to 1975 (110). In contrast to this, red meats were associated with 13% of the outbreaks, eggs with 6%, and dairy products with 4%. Person-to-person contact was involved in 10% and pets in 3% of the outbreaks.

The remainder of the outbreaks were associated with an assortment of vehicles (19%) or the vehicle was unknown (28%).

Most outbreaks of salmonellosis (1973 to 1976) resulted from mistakes in food handling that were made in food service establishments or the home (110). Problems accounting for most of the outbreaks were inadequate cooling of food, inadequate reheating of food, inadequate hot storage of foods, cross-contamination, use of contaminated raw ingredients, and inadequate cleaning of equipment (110).

OTHER FORMS OF FOODBORNE ILLNESS

Although staphylococcal poisoning, perfringens poisoning, and salmonellosis together have accounted for 80, 65, and 71% of the outbreaks of foodborne illness during 1975, 1976, and 1977, respectively (Table 2.1), small numbers of outbreaks are caused by several bacteria other than *C. botulinum* and by viruses. This section considers illness caused by some of those bacteria and by viruses.

Vibrio parahaemolyticus

Vibrio parahaemolyticus is a gram-negative, rod-shaped bacterium that is facultatively anaerobic and halophilic (111). The bacterium can grow in media with 1 to 8% NaCl but does best when media contain 2 to 4% NaCl. *Vibrio parahaemolyticus* grows best at pH 7.6 to 8.6 (range for growth 5.6 to 9.6) and at 30 to 35°C (range for growth, somewhat above 5 to 42°C). *Vibrio parahaemolyticus* is more sensitive to low temperatures than is *E. coli*. It is killed in 15 minutes or less at 60°C and is also killed by drying, by vinegar in 1 hour and by distilled water in 1 minute (111). The bacterium inhabits the marine environment, especially coastal and estuarine waters and, therefore, is associated with sea fish (111).

The association between *V. parahaemolyticus* and illness was first made in Japan, where raw and semiprocessed fish products are consumed regularly. Since then, outbreaks of foodborne illness caused by *V. parahaemolyticus* have been observed in Southeast Asia, Africa, Australia, Europe, and the United States. It is evident that this problem can be encountered throughout the world.

Illness caused by *V. parahaemolyticus* results from an infection that commonly prompts development of gastroenteritis (111). Symptoms generally occur about 12 hours after infected food is consumed, although the incubation period can range from 2 to 48 hours. The disease is characterized by severe abdominal pain, diarrhea, nausea, and vomiting. A mild fever and headache may be present. Recovery is usually complete in a few days, and the mortality rate is low, with most deaths occurring among older, debilitated persons.

Strains of *V. parahaemolyticus* that cause gastroenteritis generally also produce a heat-stable, extracellular substance that is hemolytic on a medium containing human red blood cells. This is designated as the Kanagawa phenomenon (111). It is thought that the Kanagawa hemolysin may be important in causing illness, although other factors are also likely to be involved.

Enteropathogenic E. Coli

Enteropathogenic *E. coli* can be defined as any strain of *E. coli* that has the potential to cause diarrheal disease. Enteropathogenic *E. coli* strains (EEC) have been divided into two groups, based on the type of disease produced. Those causing a disease with cholera-like symptoms (watery diarrhea leading to dehydration and shock) also produce enterotoxins and thus are called *toxigenic EEC*. These strains have been implicated as the cause of "infantile diarrhea" and "traveler's diarrhea" (112, 113). Those strains causing a *Shigella*-like illness (diarrhea with stools containing blood and mucous) are called *invasive EEC* because of their ability to penetrate the epithelial cells of the colonic mucosa. These strains do not produce an enterotoxin. Invasive EEC are associated with dysentery-like disease in people of all ages.

During November and December 1971 at least 227 persons in 96 separate outbreaks in several states in the United States became ill with acute gastroenteritis about 24 hours after consuming imported French Camembert or Brie cheese (114, 115). *Escherichia coli* of serogroup 0.124:B17 was isolated from stools of several patients and from samples of cheese believed to have caused the illness.

Incidence of E. coli and Coliforms in Cheese

Presence of coliforms in cheese has been the subject of research for over 70 years. Early investigations were concerned with prevention of gassy defects caused by coliform bacteria in curd and cheese (116–118). Since coliforms must reach numbers close to 10^7 per gram to cause gassiness in Cheddar cheese, cheese of normal appearance can still have substantial numbers of *E. coli* present (119). Even with pasteurization, postpasteurization contamination of milk with coliforms can be great enough to cause cheese to become gassy (120). Yale (121) found that Cheddar cheese of high quality could contain up to 57,000 coliforms per gram of curd.

A general survey of coliform bacteria in Canadian pasteurized dairy products by Jones et al. (122) showed that 18.7% of the coliforms isolated from these products were of intestinal origin. Three serotypes or 2% of the *E. coli* isolates were enteropathogenic serotypes. Lightbody (123) found that 97% of Queensland Cheddar cheese contained coliforms after 2 to 3 weeks of aging. *Escherichia coli* biotype I was found in 70% of the samples. Cheese samples with more than 10^6

coliforms per gram were of poor quality. Some high-grade cheese had more than 1000 coliforms per gram. In further studies of Queensland Cheddar cheese, Dommet (124) reported that improper pasteurization of milk, unsanitary equipment, and contaminated starter cultures were all responsible for coliform contamination of the cheese. In recent surveys of Canadian cheese varieties, Elliott and Millard (125) noted that 15% of retail cheese samples contained over 1500 coliforms per gram, and Collins-Thompson et al. (126) found that 18.1% of soft cheeses and 13.6% of semisoft cheese exceeded 1600 total coliforms per gram.

Recently Frank and Marth (127) examined 106 samples of commercial cheese for the presence of fecal coliforms and EEC. Included in their survey were samples of Camembert, Brie, brick, Muenster, and Colby cheeses. Of the samples tested, 58% contained less than 100 fecal coliforms per gram, but 17% contained more than 10,000 per gram. No EEC serotypes were found in any of the cheeses. A similar survey by Glatz and Brudvig (128) also demonstrated the absence of EEC from commercial cheese.

Inhibition of E. coli by Lactic Acid Bacteria

Frank and Marth (129, 130) examined the effects of lactic acid bacteria on *E. coli* when both organisms were in skim milk. With no lactic acid bacteria present, the generation times of pathogenic and nonpathogenic strains of *E. coli* ranged from 28 to 35 minutes when incubation was at 32°C and from 66 to 109 minutes at 21°C. Addition of 0.25 or 2.0% of a commercial starter together with *E. coli* served to completely inhibit growth of *E. coli* by 6 to 9 hours of incubation at 32°C. At 21°C, *E. coli* often had difficulty initiating growth in the presence of the lactic acid bacteria. *Streptococcus cremoris* and *Streptococcus lactis* were equally inhibitory to *E. coli* at 32°C. At 21°C, *S. cremoris* was more inhibitory to *E. coli* than was *S. lactis*, but a commercial mixed-strain lactic starter culture was more inhibitory than was either of the pure cultures.

Behavior of EEC in Cheese

The episode of foodborne illness described earlier prompted Park et al. (131) and Frank et al. (132) to study the fate of EEC during manufacture and ripening of Camembert cheese. They observed the following: (*a*) growth of *E. coli* sometimes was minimal until after curd was cut and hooped; (*b*) populations of approximately 10^4 *E. coli* per gram appeared in some cheese 5 to 6 hours after the cheesemaking process began when milk initially contained about 10^2 *E. coli* per milliliter; (*c*) there was a demise of *E. coli* during ripening, with some strains disappearing from cheese during the first 2 weeks and others surviving for 4 to 6 weeks; and (*d*) no growth of *E. coli* was observed in ripe cheese at a pH of 6.7, but rapid growth of *E. coli* occurred on the surface of the cheese.

The fate of EEC during manufacture and ripening of brick cheese also was determined by Frank et al. (133). Results differed from those obtained with Camembert cheese in that (*a*) somewhat larger populations of EEC developed initially during manufacture of brick cheese, (*b*) inactivation of EEC was slower in brick cheese with 10^3 to 10^4 per gram remaining after 7 weeks, and (*c*) growth of EEC on the surface of brick cheese was more limited.

Yersinia enterocolitica

Yersinia enterocolitica is a gram-negative, rod-shaped bacterium; sometimes the cells can be pleomorphic or ovoid in shape (134). The bacterium is generally nonmotile at 37°C; at lower temperatures it is motile with peritrichous flagellae. *Yersinia enterocolitica* grows best at 22 to 29°C, but it also grows well at 0 to 1°C. Hence *Y. enterocolitica* is one of only a few species of bacteria causing foodborne illness that can grow on food while it is being refrigerated.

Yersinia enterocolitica produces a heat-stable enterotoxin that can cause illness that is characterized by severe abdominal pain, fever, headache, diarrhea, malaise, vomiting, nausea, and chills (134). Recovery usually occurs in 3 days. In children, *Y. enterocolitica* can cause acute mesenteric lymphadenitis that has symptoms that resemble those of appendicitis; thus unnecessary appendectomies are sometimes done.

Raw and chocolate-flavored milk have been associated with outbreaks of foodborne illness caused by *Y. enterocolitica.* The incident involving chocolate milk probably resulted from postpasteurization contamination of the milk, perhaps with the chocolate syrup. The organism also has been isolated from an assortment of foods that did not cause illness. Included are pork, beef, chicken, lamb, cakes, meat products, milk, oysters, and water (134).

Bacillus cereus

Bacillus cereus is a large gram-positive, rod-shaped, aerobic sporeforming bacterium that can grow anaerobically (135). Vegetative cells are 1.0 to 1.2 μm by 3.0 to 5.0 μm, and spores are ellipsoidal, central, or paracentral. *Bacillus cereus* grows at 10 to 48°C, but it grows best at 28 to 35°C.

There are two types of illness caused by *B. cereus*, one characterized by diarrhea and the other by vomiting. Symptoms of the diarrheal form include abdominal pain, profuse watery diarrhea, rectal tenesmus, and moderate nausea that seldom gives raise to vomiting. These symptoms seldom last more than 12 hours.

An outbreak of the vomiting form usually occurs 1 to 5 hours after contaminated food is consumed, but sometimes in as little as 15 to 30 minutes and other times in 6 to 12 hours. Symptoms include an acute attack of nausea and vomiting and usually abdominal pain. Mild diarrhea can occur in about 30% of the affected persons. Recovery is rapid, usually in 6 to 24 hours after onset of the illness (135).

Bacillus cereus produces a variety of extracellular metabolites primarily during the logarithmic phase of growth. Included are proteases, β-lactamases, peptide antibiotics, phospholipases, hemolysins, a toxin lethal to mice, and an enterotoxin (135). The enterotoxin, a protein, has a molecular weight of 55,000 to 60,000 and is heat labile, inactivated in 5 minutes at 56°C (inactivation in 20 minutes at 60°C has also been reported). A second enterotoxin that is stable to heat, extremes of pH, and proteolytic enzymes has been described. This toxin has a molecular weight of less than 5000 and has emetic properties. Production of two enterotoxins by *B. cereus* likely accounts for the two forms of foodborne illness caused by the bacterium.

Foods involved in outbreaks of food poisoning caused by *B. cereus* include pudding, sauce, soup, mashed potatoes, vegetables, minced meat, liver sausage, Indonesian rice dishes, cream pastries, pork casserole, fish and tomato sauce, cooked meats, milk, ice cream, fried and boiled rice, cooked chicken, green bean salad, and pasta dishes. This form of foodborne illness occurred with some frequency in Great Britian during the early to mid-1970s and resulted because boiled or fried rice was mishandled in some Chinese-type restaurants and "carry-out" shops (135).

Viruses

Viruses, usually originating in the human intestine, sometimes can contaminate food, which, if not treated to inactivate the viruses, becomes a vehicle for transmission of the viruses to persons who consume it. Raw foods from some animals (e.g., goats, cattle, and sheep) can be infected with viruses, but they generally are of little concern to human health. In contrast, shellfish can become contaminated with viruses of concern to human health if the shellfish grow in water polluted with human feces. Similarly, it is likely that vegetables in the field can become contaminated with viruses if polluted water is used to irrigate the crop or if human wastes are used as fertilizer (136). As with other agents of foodborne illness, viruses can contaminate food during processing, storage, distribution, and final preparation, if the opportunity to do so is provided. Examples of foodborne viruses of concern are those that cause infectious hepatitis (136) or poliomyelitis and other enteroviruses (136). Viruses in foods can be inactivated by a low pH (e.g., 3) or by relatively mild heat treatments (e.g., 65°C for less than 1 minute). Radiations and disinfectants such as chlorine and iodine also inactivate viruses (136).

Additional Causes of Foodborne Illness

From time to time an outbreak of foodborne illness is caused by a microorganism other than those discussed thus far in this chapter. Examples of such organisms include *Vibrio cholerae*, *Shigella*, *Streptococcus pyogenes*, *Corynebacterium diphtheriae*, *Brucella*, *Coxiella burnetii*, *Bacillus anthracis*, *Francisella tularensis*, *Arizona*, *Citrobacter*, *Proteus*, *Enterobacter*, *Pseudomonas aeruginosa*, and *Campylobacter*. The reader who desires more information on these and some other causes of foodborne illness should consult the discussions by Bryan (134) and Sakazaki (111).

CONTROL OF FOODBORNE ILLNESS

Thus far in this chapter little has been said regarding control of foodborne illness. This was done to minimize the repetition that would have resulted if control had been mentioned with each form of illness that was discussed.

To further reduce the incidence of foodborne illness in the United States, primary attention must be directed to the food service industry and the home—the two places where most mistakes in food handling are made. Although the food processing industry is responsible for relatively few outbreaks of foodborne illness, this segment of the food industry must also receive attention. This is particularly true because the food processing industry is likely to be innovative, and the innovations can sometimes lead to microbiological problems, including foodborne illness, if the changes in processing are not fully understood.

The first step in effective control of foodborne illness is education. Training in proper handling of food should be provided to and required of every student in our nation's secondary schools. This would help to reduce the incidence of foodborne illness from mistakes made in the home and also in the food service industry, since persons hired by this industry would have the basic information on food handling, whereas that is not true today.

In the meantime, education must be provided to people today. One program to do so is that of the National Institute for the Food Service Industry. This program provides training in food sanitation to managers of food service operations, with the expectation that a manager will train workers in his or her establishment. Some food service operations experience more than 100% turnover in employees each year, which makes the need for training an ongoing matter that may not always be handled adequately by managers, even if they have received the needed training. Nevertheless, continuous education programs in the food service industry is our best hope for reducing the incidence of foodborne illness from mistakes made by workers in this industry.

Such education is and must continue to be encouraged by appropriate regulatory agencies.

Regulatory agencies also have a role in controlling foodborne illness through inspection and surveillance of the various segments of the food industry. For example, in addition to the inspection activity for meat and poultry products conducted by the U.S. Department of Agriculture (USDA) the U.S. Food and Drug Administration (FDA) has established several regulations and conducts a variety of programs aimed at controlling foodborne disease associated with processed foods. The FDA regulations include those specifying "good manufacturing practice in manufacturing, processing, packing, or holding human food" and processing conditions for "thermally processed low-acid foods packaged in hermetically sealed containers." These regulations appear in Title 21, Parts 110 and 113, respectively, of The Code of Federal Regulations (137). Federal–state cooperative programs are also conducted by the FDA in the food service, milk safety and shellfish safety areas. The importance of foodborne hazards is shown by the fact that over half of the FDA's food budget is devoted to the food sanitation and cooperative food safety programs. Such regulatory activity, when done appropriately, can undergird the education effort that was mentioned earlier.

Aside from its role in handling food, the general public has another role in facilitating control of foodborne illness, namely, that of promptly reporting an illness thought to be foodborne in origin. Such reporting will enable an investigation of the outbreak in the hope of learning the cause(s) so that future problems can be prevented. Methods for making such an investigation are described in the publication, *Procedures to Investigate Foodborne Illness* (138).

The basic principles in control of foodborne illness are to (*a*) recognize the raw products that are likely to be contaminated with bacteria able to cause foodborne illness, (*b*) handle such products so they do not contaminate processed foods, thus minimizing growth of bacteria, (*c*) process the raw food promptly and prevent postprocessing contamination, (*d*) hold foods being served hot at a temperature sufficiently high (e.g., 62°C) to prevent multiplication of bacteria, (*e*) cool hot, prepared, left-over foods rapidly so that bacterial growth does not occur during the cooling period, and (*f*) use appropriate means to sanitize equipment and ensure that food handlers use good habits of personal hygiene. The specific details for applying these principles varies with the type of food processing or food service operation.

REFERENCES

1. F. R. Healy and D. Juranek, "Parasitic Infections," in H. Riemann and F. L. Bryan, Eds., *Food-Borne Infections and Intoxications,* 2nd ed., Academic, New York, 1979, pp. 343–385.

2. Center for Disease Control, *Foodborne and Waterborne Disease Surveillance.* Annual Summary—1977, Center for Disease Control, Atlanta, 1979.
3. C. C. Dauer and D. J. Davids, *Public Health Rep.*, 75, 915 (1960).
4. C. C. Dauer, *Public Health Rep.*, 76, 915 (1961).
5. L. Buchbinder, *Public Health Rep.*, 76, 515 (1961).
6. Committee on *Salmonella, An Evaluation of the Salmonella Problem*, National Academy of Sciences, Washington, D.C., 1969.
7. A. H. W. Hauschild and F. L. Bryan, *J. Food Prot.*, 43, in press (1980).
8. F. L. Bryan, *J. Food Prot.*, 43, 140 (1980).
9. V. C. Vaughn, *Public Health*, 10, 241 (1884).
10. M. A. Barber, *Philipp. J. Sci.*, 9, 515 (1914).
11. G. M. Dack, W. E. Cary, O. Wolpert, and H. J. Wiggins, *J. Prev. Med.*, 4, 167 (1930).
12. E. O. Jordan, *J. Am. Med. Assoc.*, 94, 1648 (1930).
13. E. O. Jordan, *J. Am. Med. Assoc.*, 97, 1704 (1931).
14. T. E. Minor and E. H. Marth, *J. Milk Food Technol.*, 34, 557 (1971).
15. T. E. Minor and E. H. Marth, *J. Milk Food Technol.*, 35, 21 (1972).
16. T. E. Minor and E. H. Marth, *J. Milk Food Technol.*, 35, 77 (1972).
17. T. E. Minor and E. H. Marth, *J. Milk Food Technol.*, 35, 228 (1972).
18. T. E. Minor and E. H. Marth, *Staphylococci and Their Significance in Foods*, Elsevier, Amsterdam, 1976.
19. F. L. Bryan, "*Staphylococcus aureus*," in M. P. DeFigueiredo and D. F. Splittstoesser, Eds., *Food Microbiology: Public Health and Spoilage Aspects*, AVI, Westport, Conn., 1976, pp. 12–128.
20. M. S. Bergdoll, "Staphylococcal Intoxications," in H. Riemann and F. L. Bryan, Eds., *Food-Borne Infections and Intoxications*, 2nd ed., Academic, New York, 1979, pp. 443–494.
21. R. E. Buchanan and N. E. Gibbons, *Bergey's Manual of Determinative Bacteriology*, 8th ed., Williams and Wilkins, Baltimore, 1974.
22. A. C. Peterson, J. J. Black, and M. F. Gunderson, *Appl. Microbiol.*, 10, 23 (1962).
23. T. R. Oberhofer and W. C. Frazier, *J. Milk Food Technol.*, 31, 335 (1961).
24. C. T. Kao and W. C. Frazier, *Appl. Microbiol.*, 14, 251 (1966).
25. R. S. Dahiya and M. L. Speck, *J. Dairy Sci.*, 51, 1568 (1968).
26. J. J. Iandolo, C. W. Clark, L. Bluhm, and Z. J. Ordal, *Appl. Microbiol.*, 13, 646 (1965).
27. E. H. Marth and R. V. Hussong, *J. Dairy Sci.*, 46, 1033 (1963).
28. W. C. Haines and L. G. Harmon, *Appl. Microbiol.*, 25, 436 (1973).
29. J. A. Troller and W. C. Frazier, *Appl. Microbiol.*, 11, 163 (1963).
30. Z. J. Ordal, *J. Milk Food Technol.*, 34, 548 (1971).
31. T. E. Minor and E. H. Marth, *J. Milk Food Technol.*, 35, 548 (1972).
32. D. L. Scheusner, L. L. Hood, and L. G. Harmon, *J. Milk Food Technol.*, 36, 249 (1973).

33. T. D. Nunheimer and F. W. Fabian, *Am. J. Pub. Health,* 30, 1040 (1940).
34. T. E. Minor and E. H. Marth, *J. Milk Food Technol.,* 35, 191 (1972).
35. W. J. Scott, *Aust. J. Biol. Sci.,* 6, 549 (1953).
36. C. Genigeorgis and W. W. Sadler, *J. Bacteriol.,* 92, 1383 (1966).
37. S. A. Gojvat and J. Jackson, *Can. Inst. Food Technol. J.,* 2, 56 (1969).
38. J. J. Iandolo, Z. J. Ordal, and L. D. Witter, *Can. J. Microbiol.,* 10, 808 (1964).
39. G. J. Hucker and W. C. Haynes, *Am. J. Pub. Health,* 27, 590 (1937).
40. E. P. Casman, *J. Bacteriol.,* 79, 849 (1960).
41. M. S. Bergdoll, H. Sugiyama, and G. M. Dack, *Arch. Biochem. Biophys.,* 85, 62 (1959).
42. M. S. Bergdoll, C. R. Borja, and R. M. Avena, *J. Bacteriol.,* 90, 1481 (1965).
43. E. P. Casman, R. W. Bennett, A. E. Dorsey, and J. A. Issa, *J. Bacteriol.,* 94, 1875 (1967).
44. M. S. Bergdoll, C. R. Borja, R. Robbins, and K. F. Weiss, *Infect. Immunol.,* 4, 593 (1971).
45. C. R. Borja and M. S. Bergdoll, *Biochemistry,* 6, 1467 (1967).
46. R. M. Avena and M. S. Bergdoll, *Biochemistry,* 6, 1474 (1967).
47. E. J. Schantz, W. G. Roessler, J. Wagman, L. Spero, D. A. Dunnery, and M. S. Bergdoll, *Biochemistry,* 4, 1011 (1965).
48. F. S. Chu, K. Thadhani, E. J. Schantz, and M. S. Bergdoll, *Biochemistry,* 5, 3281 (1966).
49. I.-Y. Huang and M. S. Bergdoll, *J. Biol. Chem.,* 245, 3493 (1970).
50. I.-Y. Huang and M. S. Bergdoll, *J. Biol. Chem.,* 245, 3511 (1970).
51. I.-Y. Huang and M. S. Bergdoll, *J. Biol. Chem.,* 245, 3518 (1970).
52. E. Davison and G. M. Dack, *J. Infect. Dis.,* 64, 302 (1939).
53. M. S. Bergdoll, "The Enterotoxins," in J. O. Cohen, Ed., *The Staphylococci,* Wiley, New York, 1972.
54. M. S. Bergdoll, "Enterotoxins," in T. C. Montie, S. Kadis, and S. J. Ajl, Eds., *Microbial Toxins,* Vol. 3, Academic, New York, 1970.
55. R. A. Gunn, *Botulism in the United States, 1899–1977,* Center for Disease Control, Atlanta, 1979.
56. G. Sakaguchi, "Botulism," in H. Riemann and F. L. Bryan, Eds., *Food-Borne Infections and Intoxications,* 2nd ed., Academic, New York, 1979, pp. 389–442.
57. H. Sugiyama, D. C. Mills, and L-J. C. Kuo, *J. Food Prot.,* 41, 848 (1978).
58. R. B. Tompkin and L. M. Christiansen, "*Clostridium botulinum,*" in M. P. DeFigueiredo and D. F. Splittstoesser, Eds., *Food Microbiology: Public Health and Spoilage Aspects,* AVI, Westport, Conn., 1976, pp. 156–169.
59. C. Genigeorgis and H. Riemann, "Food Processing and Hygiene," in H. Riemann and F. L. Bryan, Eds., *Food-Borne Infections and Intoxications,* 2nd ed., Academic, New York, 1979, pp. 613–713.

60. B. C. Hobbs, "*Clostridium perfringens* Gastroenteritis," in H. Riemann and F. L. Bryan, Eds., *Food-Borne Infections and Intoxications*, 2nd ed., Academic, New York, 1979.

61. R. Knox and E. K. MacDonald, *Med. Off.*, 69, 21 (1943).

62. L. S. McClung, *J. Bacteriol.*, 50, 229 (1945).

63. C. L. Duncan, "*Clostridum perfringens*," in M. P. DeFigueiredo and D. F. Splittstoesser, Eds., *Food Microbiology: Public Health and Spoilage Aspects*, AVI, Westport, Conn., 1976.

64. T. A. Roberts, *J. Appl. Bacteriol.*, 31, 133 (1968).

65. R. L. Stark and C. L. Duncan, *Infect. Immunol.*, 5, 147 (1972.)

66. J. L. McDonel, *Food Technol.*, 34 (4), 91 (1980).

67. R. G. Labbe, *Food Technol.*, 34 (4), 88 (1980).

68. S. E. Craven, *Food Technol.*, 34 (4), 80 (1980).

69. E. H. Marth, *J. Dairy Sci.*, 52, 283 (1969).

70. J. Matches and J. Liston, *Proc. 27th Annu. Meeting, Inst. Food Technol.* (1967), p. 82.

71. M. Segalove and G. M. Dack, *Food Res.*, 9, 1 (1944).

72. C. S. Subramanian and E. H. Marth, *J. Milk Food Technol.*, 31, 323 (1968).

73. R. R. Spencer and M. B. Melroy, *J. Nat. Cancer Inst.*, 3, 1 (1942).

74. G. C. Ware, *J. Gen. Microbiol.*, 11, 398 (1954).

75. R. B. Read, Jr., J. G. Bradshaw, R. W. Dickerson, Jr., and J. T. Peeler, *Bacteriol. Proc.*, 1968, 9 (1968).

76. C. Thomas, J. C. White, and K. Longree, *Appl. Microbiol.*, 14, 815 (1966).

77. R. P. Elliot and P. K. Heiniger, *Appl. Microbiol.*, 13, 73 (1965).

78. H. G. Bayne, J. A. Garibaldi, and H. Lineweaver, *Poultry Sci.*, 44, 1281 (1965).

79. B. W. Bierer and B. D. Barnett, *J. Am. Vet. Med. Assoc.*, 146, 735 (1965).

80. J. J. Licciardello, J. T. R. Nickerson, and S. A. Goldblith, *Am. J. Public Health*, 55, 1622 (1965).

81. O. G. Rasmussen, R. Hansen, N. J. Jacobs, and O. H. M. Wilder, *Poultry Sci.*, 43, 1151 (1964).

82. E. Prost and H. Riemann, *Annu. Rev. Microbiol.*, 21, 495 (1967).

83. E. W. Stearn and A. E. Stearn, *J. Bacteriol.*, 26, 9 (1933).

84. A. S. Levine and C. R. Fellers, *J. Bacteriol.*, 39, 499 (1940).

85. M. C. Wethington and F. W. Fabian, *Food Res.*, 15, 125 (1950).

86. C. A. Schillinglaw and M. Levine, *Food Res.*, 8, 464 (1943).

87. M. L. Riehl, H. H. Weiser, and B. T. Rheins, *J. Am. Water Works Assoc.*, 44, 466 (1952).

88. J. H. B. Christian and W. J. Schott, *Aust. J. Biol. Sci.*, 6, 565 (1953).

89. J. H. B. Christian, *Aust. J. Biol. Sci.*, 8, 75 (1955).

90. J. H. B. Christian, *Aust. J. Biol. Sci.*, 8, 490 (1955).

91. J. H. B. Christian and J. A. Waldo, *J. Gen. Microbiol.*, 43, 345 (1966).
92. G. Bergmann and G. Seidel, *Lebensmittelhygiene*, 8, 30 (1957).
93. J. M. Severens and F. W. Tanner, *J. Bacteriol.*, 49, 383 (1945).
94. J. M. Geopfert, N. F. Olson, and E. H. Marth, *Appl. Microbiol.*, 16, 862 (1968).
95. H. L. A. Tarr, *J. Fish. Res. Board Can.*, 6, 74 (1942).
96. L. Friberg and G. Hammerström, *Acta Pathol. Microbiol. Scand.*, 38, 127 (1956).
97. P. Kabler, G. O. Pierce, and G. S. Michaelsen, *J. Bacteriol.*, 37, 1 (1939).
98. C. T. Butterfield, E. Wattie, S. Megregian, and C. W. Chambers, *U.S. Public Health Rep.*, 58, 1837 (1943).
99. E. W. Lane, U.S. Patent No. 2,807,614 (1957).
100. W. J. Croxall and S. Melamed, U.S. Patent No. 2,582,748 (1952).
101. G. R. Goetchius and H. Grinsfelder, *Appl. Microbiol.*, 1, 271 (1953).
102. R. L. Stedman, E. Kravitz, and H. Bell, *Appl. Microbiol.*, 2, 119 (1954).
103. R. L. Stedman, E. Kravitz, and H. Bell, *Appl. Microbiol.*, 2, 322 (1954).
104. S. Ross, C. E. Kwartler, and J. H. Bailey, *J. Colloid Sci.*, 8, 385 (1953).
105. E. Satta, *Med. Biol.*, 3, 333 (1943).
106. V. Monaci, *Boll. Inst. Sieroter.*, 28, 357 (1949).
107. A. Roushdy, *Neth. Milk Dairy J.*, 13, 151 (1959).
108. J. Sadilek and M. Stepanek, *Proceedings of the 16th International Dairy Congress*, C, 1029 (1962).
109. F. L. Bryan, M. J. Fanelli and H. Riemann, "*Salmonella* Infections," in H. Riemann and F. L. Bryan, Eds., *Food-Borne Infections and Intoxications*, 2nd ed., Academic, New York, 1979, pp. 73–130.
110. J. H. Silliker, *J. Food Prot.*, 43, 307 (1980).
111. R. Sakazaki, "*Vibrio* Infections," in H. Riemann and F. L. Bryan, Eds., *Food-Borne Infections and Intoxications*, 2nd ed., Academic, New York, 1979, pp. 173–209.
112. H. I. Dupont, S. B. Formal, R. B. Horneck, M. J. Snyder, J. P. Libonati, D. G. Sheahan, E. H. LaBrie, and J. P. Kalas, *New Engl. J. Med.*, 285, 3 (1971).
113. R. W. Ryder, I. K. Wachsmuth, A. E. Buston, and J. Barrett, *New Engl. J. Med.*, 295, 849 (1976).
114. R. Barnard and W. Callahan, *Morbid. Mortal. Rep.*, 20, 445 (1971).
115. L. W. Schnurrenberger, R. Beck, and J. Pate, *Morbid. Mortal. Rep.*, 20, 427 (1971).
116. F. C. Harison, *Ontario Agr. Coll. Bull.*, No. 141 (1905).
117. C. E. Marshall, *Mich. Agr. Coll. Bull.*, No. 183 (1900).
118. H. L. Russell, *Wisc. Agr. Exp. Sta. Annu. Rep.*, (1895).
119. M. W. Yale and J. C. Marquardt, *N. Y. Agr. Exp. Sta. Tech. Bull.*, No. 270 (1943).
120. C. A. Ernstrom, *Milk Prod. J.*, 45, 21 (1954).

121. M. W. Yale, *J. Dairy Sci.*, 26, 766 (1943).

122. G. A. Jones, D. L. Gibson, and K. J. Cheng, *Can. J. Public Health*, 58, 257 (1967).

123. L. G. Lightbody, *Queensland J. Agr. Sci.*, 19, 305 (1962).

124. T. W. Dommet, *Aust. J. Dairy Technol.*, 25, 54 (1970).

125. J. A. Elliott and G. E. Millard, *Can. Inst. Food Sci. Technol. J.*, 9, 95 (1976).

126. D. L. Collins-Thompson, I. E. Erdman, M. E. Milling, D. M. Bargner, V. T. Purvis, A. Loit, and R. M. Coulter, *J. Food. Prot.*, 40, 406 (1977).

127. J. F. Frank and E. H. Marth, *J. Food Prot.*, 41, 198 (1978).

128. B. A. Glatz and S. A. Brudvig, *J. Food Prot.*, 43, in press (1980).

129. J. F. Frank and E. H. Marth, *J. Food Prot.*, 40, 749 (1977).

130. J. F. Frank and E. H. Marth, *J. Food Prot.*, 40, 754 (1977).

131. H. S. Park, E. H. Marth, and N. F. Olson, *J. Milk Food Technol.*, 36, 543 (1973).

132. J. F. Frank, E. H. Marth, and N. F. Olson, *J. Food Prot.*, 40, 835 (1977).

133. J. F. Frank, E. H. Marth, and N. F. Olson, *J. Food Prot.*, 41, 111 (1978).

134. F. L. Bryan, "Infections and Intoxications Caused by Other Bacteria," in H. Riemann and F. L. Bryan, Eds., *Food-Borne Infections and Intoxications*, 2nd ed., Academic, New York, 1979, pp. 211–297.

135. R. J. Gilbert, "*Bacillus cereus* Gastroenteritis," in H. Riemann and F. L. Bryan, Eds., *Food-Borne Infections and Intoxications*, 2nd ed., Academic, New York, 1979, pp. 495–518.

136. D. O. Cliver, "Viral Infections," in H. Riemann and F. L. Bryan, Eds., *Food-Borne Infections and Intoxications*, 2nd ed., Academic, New York, 1979, pp. 299–342.

137. *Code of Federal Regulations, Title 21: Food and Drugs*, U.S. Government Printing Office, Washington, D.C., 1980.

138. F. L. Bryan, H. W. Anderson, R. R. Anderson, K. J. Baker, H. Matsumura, T. W. McKinley, R. C. Swanson, and E. C. D. Todd, *Procedures to Investigate Foodborne Illness*, 3rd ed., International Association of Milk, Food and Environmental Sanitarians, Ames, Iowa, 1976.

CHAPTER 3

Nutritional Hazards

V. J. STULTS

Nutritional hazards derive their high-risk ranking from their dual role in food safety (1). Adverse impacts can occur from deficiencies as well as from excesses in nutrient intakes. The importance of the nutritional aspects of food safety derives from the fact that deviations from the optimum intake, either deficiencies or excesses, can arise in a wide variety of ways. In addition, there are some circumstances that can lead to impairment of the optimum utilization or retention of nutrients. These nutritional hazards may occur because of a specific component in, or characteristic of, a given food item, but the majority of hazards are associated with improper use of individual foods or improper balance among all the foods in the diet.

Many of the potential hazards posed by significant deficiencies or excesses in nutrient intakes are fairly well recognized. For example, the role of nutrients in such deficiency diseases as scurvy, pellagra, and rickets is widely known, as are some of the acute toxic effects resulting from excess intakes of both vitamins and minerals. Chronic effects of excess intakes are less well known but have been recognized by public health officials as meriting increased attention. The area of nutritional hazards arising from improper use of foods and improper dietary balance is not well understood but is frequently treated as if it were. This is unfortunate since it is undoubtedly one of the most important aspects of nutrition's role in food safety.

The science of nutrition is concerned with the study of diet and health. Such study encompasses nutrient functions, sources, interactions, metabolism, requirements, deficiencies and excesses, balance, and the effects of these factors on health. There is also the component of application—how foods are used—which is especially important in an affluent society. Assimilating all these factors, nutritionists, with the aid of food scientists, could define a near-perfect diet in terms of appeal and nutrient content, but of course humans do not eat this way.

Therefore, nutrition hazards must be considered not only from the point of view of individual nutrients, but also with regard to the uses of foods in the total diet.

This chapter briefly reviews the functions, the ranges of recommended intake levels, and the deficiency diseases or symptoms for the major nutrients confirmed to be essential. Those nutrients for which the potential for toxicity exists are identified and described. Also discussed are proteins; carbohydrates and fat; and the total diet, including dietary trends, the current American diet, fad diets, and the trend toward developing dietary guidelines. In addition, current practices and future considerations relating to nutrient fortification and enrichment of foods are outlined.

ESSENTIAL NUTRIENTS AND THEIR FUNCTIONS

Essential nutrients are organic and inorganic substances required to sustain life. Different organisms require different nutrients for their existence. To sustain life and maintain health, dietary intakes of nutrients must meet certain minimum levels. However, nutrient intakes sufficiently in excess of requirements can lead to a variety of toxic effects, including death. This potential for adverse effects from both deficiencies and excesses makes nutrients very different from other chemicals in foods and underlines the importance of nutritional hazards in food safety.

The essential nutrients for humans are those required by our bodies in specific amounts for the normal growth, maintenance, and repair of body tissues and for reproduction. Aside from the critical requirements of humans and most living organisms for oxygen and water, there are five major categories of essential nutrients: (*a*) proteins, (*b*) carbohydrates, (*c*) fats, (*d*) vitamins, and (*e*) minerals. Human requirements for the essential nutrients are customarily discussed in terms of the recommended dietary allowances (RDAs) that are developed by the Food and Nutrition Board of the National Research Council, National Academy of Sciences. The RDAs represent:

> . . . the levels of intake of essential nutrients considered, in the judgment of the Committee on Dietary Allowances of the Food and Nutrition Board on the basis of available scientific knowledge, to be adequate to meet the known nutritional needs of practically all healthy persons (2).

The first set of RDAs was developed in 1943, and periodic revisions based on new information have been made, the most recent occurring in 1980 (2).

It is important to note that present knowledge of nutritional needs is incomplete. Therefore, the mere attainment of the RDAs is not necessarily a guarantee of nutritional well-being for all persons, nor will nutrient intakes less than the RDAs necessarily result in malnutrition in given individuals. The RDAs should be considered as "targets" for nutrient intake by groups of individuals, but it should be remembered that diets are more than a simple vehicle for ingesting a prescribed combination of nutrients. As the National Academy of Sciences suggests, the RDAs should be provided from as varied a selection of acceptable and palatable foods as possible.

There are several other points of caution that should be recognized in interpreting the RDAs. The first point deals with the variability of nutrient requirements. Recommended daily allowances do not reflect the special and individual needs of those persons subject to metabolic disorders, chronic diseases, or other abnormal health conditions. Again, the reference base for the RDAs is recommended intakes for groups of healthy individuals. Nutrient requirements and hence recommended intake levels also vary with respect to age, sex, weight, and other physiological as well as genetic differences. It is only common sense that nutrient requirements are greater at ages involving rapid growth and for pregnant or lactating women than for situations involving normal maintenance. Thus different RDAs have been established based on age, weight and height, sex, and pregnancy or lactation. One final point of explanation is that the recommended dietary allowances (RDAs) developed by the Food and Nutrition Board should not be confused with the U.S. RDAs. The U.S. RDAs are derived from the RDAs and are used as the regulatory basis for nutritional labeling of foods. The FDA has established four sets of U.S. RDAs: for infants, for children up to 4 years of age, for children over 4 years and adults, and for pregnant and lactating women. Generally speaking, the U.S. RDAs represent the largest of the RDAs in each of the categories (3).

With the preceding explanation, the reader is referred to Table 3.1, where, for convenience and ease of reference, a summary of the major functions, deficiency diseases, or symptoms in humans and the range of recommended daily intakes is presented for the known essential nutrients. For additional information on the functions and deficiency aspects of the nutrients, the reader is referred to standard texts such as Goodhart and Shils (4) or Krause and Mahan (5).

It should be noted that for certain nutrients, there is less information on which to base allowances. For these nutrients, the Food and Nutrition Board has determined ranges of "estimated safe and adequate daily dietary intakes" rather than RDAs. Safe and adequate intakes have been listed for vitamin K. biotin, pantothenic acid, copper, manganese, fluoride, chromium, selenium, molybdenum, sodium, potassium, and chloride (2). The intake ranges in Table 3.1 are based on either the RDA or the safe and adequate range, depending on the nutrient listed.

TABLE 3.1 Recommendations, Functions, and Deficiency Aspects of Essential Nutrients

Nutrient	Recommended Intake Range[a]	Functions	Deficiency Disease or Symptoms
Protein	2.0 g/kg–56 g	Provide essential amino acids required for metabolism, growth and repair of all body tissue	Impaired growth and repair of tissue; lowered resistance to disease
Carbohydrate	None (see text)	Store and provide energy, detoxification and elimination of unwanted substances, regulation of fat and protein metabolism	None known (energy deficits and ketosis with minimal intakes)
Fat	Essential fatty acids 1–2% of total calories (see text)	Store and provide energy, provide essential fatty acid (linoleic acid), cell membrane structure, hormone synthesis, fat soluble vitamins, insulation, transport of lipid precursors	Dry, scaly skin; impaired growth; impaired absorption of fat soluble vitamins

TABLE 3.1 Recommendations, Functions, and Deficiency Aspects of Essential Nutrients (Continued)

Nutrient	Recommended Intake Range[a]	Functions	Deficiency Disease or Symptoms
Vitamins			
Ascorbic Acid (Vitamin C)	35-60 mg	Coenzyme, synthesis of collagen and connective tissue, hormone synthesis	Scurvy
Thiamin (Vitamin B_1)	0.3-1.5 mg	Coenzyme (TPP) in energy yielding and energy transfer reactions	Beriberi; irritability; fatigue; emotional instability; depression; loss of appetite; growth retardation
Riboflavin (Vitamin B_2)	0.4-1.7 mg	Coenzyme (FAD) in energy yielding and energy transfer reactions	Cheilosis; angular stomatitis; glossitis; seborrheic dermatitis; itching and burning of eyes and sensitivity to light

TABLE 3.1 Recommendations, Functions, and Deficiency Aspects of Essential Nutrients (Continued)

Nutrient	Recommended Intake Range[a]	Functions	Deficiency Disease or Symptoms
Niacin	6-19 mg	Coenzyme (NAD) in energy yielding and energy transfer reactions	Pellagra; 4 D's: dermatitis, dementia, diarrhea, death
Pyridoxine (Vitamin B_6)	0.3-2.2 mg	Coenzyme in amino acid metabolism and linoleic acid metabolism	Cheilosis; nervous system disorders; depression; neuritis; seborrheic dermatitis; vomiting; mucous membrane lesions
Folacin	30-400 μg	Coenzyme in the transfer of one carbon units (methyl donor), nucleic acid synthesis formation and maturation of red blood cells	Megaloblastic anemia; glossitis; gastrointestinal disorders

TABLE 3.1 Recommendations, Functions, and Deficiency Aspects of Essential Nutrients (Continued)

Nutrient	Recommended Intake Range[a]	Functions	Deficiency Disease or Symptoms
Cyanocobalamin (Vitamin B_{12})	0.5-3 μg	Synthesis of red blood cell, nucleic acids and nerve myelin sheath	Pernicious anemia; nerve demyelination
Pantothenic Acid	2-7 mg	Coenzyme-A; acyl-transfer in fat, carbohydrate and protein metabolism	Experimentally induced symptoms include: anorexia depression, fainting, nausea, decreased antibody production
Biotin	35-200 μg	Coenzyme in carboxylation reactions and deamination reactions	Experimentally induced symptoms include: dermatitis, depression, muscle pains, anorexia, nausea, anemia

TABLE 3.1 Recommendations, Functions, and Deficiency Aspects of Essential Nutrients (Continued)

Nutrient	Recommended Intake Range[a]	Functions	Deficiency Disease or Symptoms
Retinol (Vitamin A; β-carotene)	400-1000 μg (retinol equivalents)	Regulates protein synthesis, cellular differentiation, vision, growth	Night blindness; xerophthalmia; keratinized epithelium; depressed growth; lowered resistance to infection
Calciferol (Vitamin D)	10-7.5 μg (as cholecalciferol)	Regulates protein synthesis in GI tract; calcium and phosphorus levels in the blood; mineralization of bones and teeth	Rickets in children; osteomalacia in adults
Tocopherols (Vitamin E)	3-10 mg (α-tocopherol equivalents)	Acts as an antioxidant and protects vitamin A and fatty acids from peroxidation. Function in humans unclear	Hemolytic anemia: creatinuria; ceroid deposits in smooth muscles

TABLE 3.1 Recommendations, Functions, and Deficiency Aspects of Essential Nutrients (Continued)

Nutrient	Recommended Intake Range[a]	Functions	Deficiency Disease or Symptoms
Phylloquinone/ Menaquinone (Vitamin K)	12-140 μg	Required for blood clotting	Hemmorhage-impaired blood clotting mechanism
Minerals			
Calcium	360-1200 mg	Crystalline structure of bones and teeth, cell membrane transport, cofactor in enzyme systems, blood clotting, protein synthesis	Abnormal bone development; nerve irritability; tetany
Phosphorus	240-1200 mg	Parallels calcium in structure and crystalization of bones and teeth; essential to cellular energy transfer	Drug-induced symptoms of malaise; anorexia; bone pain and demineralization; negative calcium balance

TABLE 3.1 Recommendations, Functions, and Deficiency Aspects of Essential Nutrients (Continued)

Nutrient	Recommended Intake Range[a]	Functions	Deficiency Disease or Symptoms
Magnesium	50–400 mg	Cofactor for enzymes in energy metabolism, protein synthesis, and an intracellular cation	Clinical induced symptoms include: anorexia, growth failure, impaired nerve and muscle functions
Iron	10–18 mg	Cellular respiration cofactor in enzymes and other large protein complexes	Hypochromic, microcytic anemia
Zinc	3–15 mg	Cofactor for enzymes in energy, protein and nucleic acid metabolism	Depressed growth; anorexia; impaired maturation; loss of sense of taste and smell
Iodine	40–150 μg	Constituent of thyroid hormone thyroxine	Goiter; impaired energy metabolism; listlessness
Fluorine	0.1–4.0 mg	Improved crystallization of bones and teeth	Teeth are more susceptible to decay

TABLE 3.1 Recommendations, Functions, and Deficiency Aspects of Essential Nutrients (Continued)

Nutrient	Recommended Intake Range[a]	Functions	Deficiency Disease or Symptoms
Copper	0.5-3 mg	Cofactor for enzymes in protein and nucleic acid metabolism and red cell formation	Clinically induced symptoms include: hypochromic anemia, bone demineralization, low white cell count
Chromium	0.01-0.2 mg	Cofactor in glucose transport with insulin	Hyperglycemia; impaired growth; glycosuria and increased serum cholesterol
Cobalt	Component of B_{12}	Active constituent of B_{12} molecule	None known or same as B_{12} deficiency
Manganese	0.5-5 mg	Cofactor for enzymes in protein synthesis and energy metabolism	None known
Molybdenum	0.03-0.5 mg	Cofactor for enzyme xanthine oxidase	None known

TABLE 3.1 Recommendations, Functions, and Deficiency Aspects of Essential Nutrients (Continued)

Nutrient	Recommended Intake Range[a]	Functions	Deficiency Disease or Symptoms
Sodium	115-3300 mg	Major extracellular cation, water balance, cell membrane transport and osmotic equilibrium	Hyponatremia; weakness; decreased extracellular fluid volume
Potassium	350-5625 mg	Major intracellular cation, cofactor for protein synthesis and energy metabolism	Hypokalemia; loss of muscle function; weakness and lethargy
Chlorine	275-5100 mg	Formation of HCl, extracellular anion	Deviations from normal induced by vomiting or excess renal excretion; failure to thrive in infancy

[a] Range of RDA's or safe and adequate levels, first entry is lowest value for infants, last entry is highest value for adults, except pregnant and lactating women. See text.

Source. Food and Nutrition Board, Recommended Dietary Allowances, Ninth ed., National Academy of Sciences, Washington, D.C., 1980.

TOXICITY OF VITAMINS

As noted in the preceding, vitamins in relatively small amounts are required by our bodies to regulate important metabolic reactions. Insufficient vitamin intake can lead to the classical deficiency diseases, and continuing insufficiency can be fatal. The critical need for vitamins has, at least in part, led to the current belief in megadose vitamin therapy. Every day millions of Americans ingest up to 100 times the recommended dietary intake levels of vitamins to prevent or reduce the effects of colds, prevent aging, eliminate stress, and increase sexual potency. Recent claims have even advocated megadose vitamin therapy as a cure for cancer. Therefore, it is of fundamental importance to the question of food safety to consider the toxic effects that can occur with excess intakes of vitamins.

Fortunately, toxicity symptoms associated with excess vitamin ingestion from normal food consumption are not seen in the United States because of the wide variety of foods consumed. Toxicity can readily be induced, however, from large doses of vitamin supplements or the misuse of specific foods.

In discussing toxicity it is important to differentiate between the fat-soluble vitamins (A, D, E, and K) and the water-soluble vitamins (C and the B vitamins). The fat-soluble vitamins are stored in body fat, and excess intake, from either excessive consumption of specific foods or from supplements, may result in accumulation and attendant toxic symptoms. By and large, excess intakes of the water-soluble vitamins result only from the use of supplements and merely lead to the body excreting what it does not need in urine and sweat. However, research is beginning to associate some side effects with excess intakes of the water-soluble vitamins.

The following material discusses the toxicity of the fat-soluble vitamins and the complexities produced by their interactions as well as the toxicity of the major water-soluble vitamins.

Vitamin A

Vitamin A is popularly known for its classical deficiency disease, night blindness, which was recognized thousands of years ago by the Egyptians. The toxicity of vitamin A has become known only in modern times but now is described in an extensive body of literature. Currently, the term "vitamin A" includes several biologically active compounds existing in different isomeric forms (retinol, retinal, and retinoic acid). In addition, there are the carotenoids that are provitamin A compounds or precursors of vitamin A, as they are readily converted to vitamin A in the body. The most important precursor is β-carotene, which is widely distributed in plants and is present in some animal tissues.

Although there have been no documented deaths specifically attributed to excessive intake of vitamin A compounds, extremely high levels of intake have resulted in permanent liver damage (6) and stunted growth (7, 8). Generally speaking, other toxic symptoms undergo remission when excessive intake is halted.

Vitamin A toxicity is rarely caused by the ingestion of foods; the exceptions are the intake of such uncommon items as polar bear liver or shark, halibut, and cod liver, in which 13,000 to 100,000 IU per gram can occur (9, 10). (One International Unit equates to 0.3 μg of retinol or 0.6 μg of β-carotene.) Acute toxicity in adults cannot easily be produced from the ingestion of vitamin supplements since a dose of 2 to 5 million IU is required (10, 11). However, in infants, much lower doses (in the range of 75,000 to 300,000 IU) can produce acute toxicity (12). Such doses can readily result from accidental or overzealous use of supplements.

Symptoms of acute toxicity in adults and infants can appear in a matter of hours. Chronic toxicity is seen most commonly in infants and children, occurring within a few months. In adults, chronic symptoms appear within several months. Vitamin A toxicity characteristics are summarized in Table 3.2.

The β-carotene precursor of vitamin A accumulates in the liver and adipose tissue and can result in a yellowing of the skin but does not produce the toxic symptoms that the biologically active vitamin A compounds do. Absorption and conversion of β-carotene to the active retinol by humans is relatively low, amounting to only 20 to 30% of that ingested. The active form of vitamin A, retinol, also accumulates in the liver and adipose tissue, mainly as esters of acetate and palmitate. It has been suggested that in hypervitaminosis A the retinoids interfere with cellular membrane integrity by splitting open membranes and allowing the release of hydrolytic enzymes (18, 19).

A comprehensive review of vitamin A safety was recently reported by the Select Committee on GRAS Substances (SCOGS) of the Federation of American Societies for Experimental Biology (20). The SCOGS summarized the manifestations of hypervitaminosis A as skin dryness, anorexia, headache, weakness, hair loss, joint pain, vomiting, irritability, enlarged liver and spleen, and increased intracranial pressure. According to the SCOGS, the lowest adverse effect level for humans is in the range of 700 to 3000 IU/kg/day. Also noted was the indication that doses on the order of 700 to 800 IU/kg/day during most of pregnancy may lead to abnormalities in human offspring (20).

It can be concluded from the available evidence that intakes of vitamin A should be limited to the neighborhood of the recommended dietary allowances unless specifically prescribed for treatment of a particular vitamin A deficiency disease. The American Academy of Pediatrics has noted that daily ingestion of retinol supplements exceeding 10,000 IU by infants and children is recommended only under the di-

TABLE 3.2 Vitamin A Toxicity Symptoms

Symptoms	Intake Levels	Reference
Acute		
Infants (12 hours after ingestion) - anorexia, bulging fontanelles, hyperirritability, vomiting. (Several days after ingestion) - cutaneous desquamation.	75,000 to 300,000 IU	10,12,13
Adults (6-8 hours after ingestion) - headache, dizziness, drowsiness, nausea, vomiting. (12-20 hours after ingestion) - redness and erythematous swelling of skin, peeling of skin.	2-5 million IU	9,10,11
Chronic		
Anorexia, headache, blurred vision, sore muscles, loss of hair, bleeding lips, cracking and pealing skin, nose bleed, liver and spleen enlargement, anemia.	Infant - less than 1 year: 18,000-60,000 IU/day, 1-5 years: 80,000-500,000 IU/day	14,15
	Adult - 100,000 IU/day, 3,000 IU/kg/day; several months to years	6,16,17

rection of a physician (21). According to the National Nutrition Consortium, intakes of vitamin A exceeding 5 to 10 times the RDA pose substantial risk (22), and the Food and Nutrition Board considers regular ingestion of more than 25,000 IU imprudent (2).

Vitamin D

In several respects, vitamin D (calciferol) is relatively unique among the vitamins. One respect is its toxicity. Excess intake from supplemental sources has resulted in a wide variety of toxic effects including death (23, 24). Vitamin D is popularly known as the "sunshine vitamin." Its common D_3 form (cholecalciferol), unlike other vitamins, is produced in our bodies by the sunlight activation of 7-dehydrocholesterol in the skin. The other common form, vitamin D_2 (ergocalciferol), can also be produced by ultraviolet light, in this case from ergosterol. In the body the major circulating form of vitamin D is the active metabolite 25-hydroxycholecalciferol (25), which is further hydroxylated to related metabolites, the most potent of which for intestinal calcium transport is 1,25-dihydroxycholecalciferol.

Tanning of the skin, by creating a filter for ultraviolet light, prevents the sunlight production of vitamin D from proceeding to toxic levels. Thus the toxic hazards of vitamin D have been produced only by the ingestion of exogenous natural sources such as fish liver oils or vitamin supplements. Negroid populations, with their increased skin pigmentation, appear to have greater requirements for, and less susceptibility to the toxic effects of, vitamin D (26).

There is some animal evidence that the common forms of vitamin D differ in their toxicity, but similar evidence in humans is lacking. Toxicity symptoms of vitamin D are fairly nonspecific. Excess ingestion can lead to hypercalcemia, membrane damage and subsequent hypertension, cardiac insufficiency, renal failure, and hypochromic anemia. In many cases the effects of hypervitaminosis D can be reversed by withdrawal of vitamin D or treatment with glucocorticoids (27); but, as noted, death can result from excess ingestion. A summary of observed toxic symptoms and the associated levels of ingestion considered dangerous is given in Table 3.3.

Concern about the toxicity of vitamin D has led the Food and Nutrition Board to advise against consumption in excess of the largest RDA of 400 IU (10 μg of cholecalciferol = 400 IU of vitamin D), which can be obtained from fortified milk. To discourage excessive use of vitamin D, the Committee on Nutrition Misinformation of the Food and Nutrition Board (35) issued a statement of caution:

> Excessive amounts of vitamin D are hazardous and only individuals with disease affecting vitamin D absorption or metabolism require more than 400 IU per day. Such additional needs should be established by clinical

TABLE 3.3 Vitamin D Toxicity Symptoms

Symptoms	Intake Levels	Reference
Acute (2-8 days after ingestion) - anorexia, nausea, vomiting, diarrhea, headache, polyuria, polydipsia	1,000-3,000 IU/kg/day	28,29
Chronic weight loss, pallor, constipation, fever, hypercalcemia,	10,000 IU/day (1-4 mo.) to 200,000 IU/day (2 wks.)	30-32
calcium deposits in soft tissues	100,000-500,000 IU/day (2-4 yrs)	33,34

evaluation, and treatment should be specifically recommended and supervised by physicians.

In its 1980 revision of the RDAs (2), the Food and Nutrition Board recommends 10 μg/day for infants and children through 18 years of age, 7.5 μg/day for 19 to 22 years of age, 5 μg/day above 23 years of age, and +5 μg/day for pregnant and lactating women, and cautions against exceeding these levels.

Vitamin E

In addition to α-tocopherol, the most active form and the form most often termed "vitamin E," there are seven other tocopherols. The tocopherols occur naturally in the lipid fraction of plants, and the most important sources are the various vegetable oils. Cottonseed, safflower, and sunflower oils, for example, have relatively high contents of α-tocopherol.

The potential toxicity of vitamin E is not completely known. In general; it appears that fairly large amounts of the vitamin E compounds can be ingested with little or no evident harm, work by Farrell and Bieri (36) has suggested that vitamin E supplements in the range of 100 to 1000 IU/day produce no adverse effects. Other results are somewhat conflicting, and those ingesting excessive amounts may either have suffered no adverse effects or perhaps not recognized any symptoms as being associated with vitamin E. Suppression of normal hematologic response to parenteral iron has been reported in anemic children given large doses of α-tocopherol (37). The Food and Nutrition Board (2) and the Select Committee on GRAS Substances (38) have both characterized vitamin E as relatively nontoxic. Some of the symptoms that have been associated with excess intake of vitamin E are shown in Table 3.4. It has also been suggested that vitamin E is an

TABLE 3.4 Vitamin E Toxicity Symptoms

Symptoms	Intake Levels
Headache, nausea, fatigue, giddiness, blurred vision	Approximately 300 mg/day
Suspected disruption of gonadal function	4-12g/day prolonged use
Slight creatinuria, chapping, cheilosis, angular stomatitis, gastrointenstinal disturbances, muscle weakness	2-4g/day for 3 months

Source. K. C. Hayes and D. M. Hegstead, "Toxicity of the Vitamins," in National Academy of Sciences, Toxicants Occurring Naturally in Foods, 2nd ed., Washington, D.C., 1973.

antagonist of vitamin K and may lead to impairment of blood coagulation (39).

It is important to note that, for normal, healthy individuals, there are no proven health or other benefits attributable to levels of vitamin E in excess of the RDAs that range from 3 mg (infants) to 10 mg (men) of α-tocopherol equivalents. [The activity of naturally occurring α-tocopherol is 1.49 IU/mg (2).] Tappel (40) has reviewed the wide range of conditions popularly believed to be alleviated by supplemental doses of vitamin E and concluded that such supplemental usage is not only wasteful, but likely has kept thousands from proper medical treatment. The consumption of vitamin E for therapeutic reasons should not exceed 10 times the RDA (41).

Vitamin K

The two general forms of vitamin K are K_1, or phylloquinone, and K_2, the menaquinone family of compounds. Phylloquinone is synthesized by a wide variety of plants, especially green leafy vegetables, and alfalfa and tobacco are particularly rich sources. Menaquinone-7, the specific form referred to as K_2, is one of a large family of homologues. A related form, menaquinone-4, is synthesized in animals from menadione, which is sometimes referred to as "vitamin K_3" (42).

Vitamin K preparations are often administered to newborn infants to prevent hemorrhage resulting from lack of vitamin K-dependent coagulation factors (43). Several forms of vitamin K are available for therapeutic uses. The fat-soluble, K_1 form, derived from plant sources, is preferred over the K_3 analogues (which can produce hemolytic anemia and liver damage) for those conditions requiring pharmacological amounts such as liver disease and severe malabsorption syndromes (44).

What little is known about the toxicity of vitamin K has come out of experience with its therapeutic use. Vitamin K_1 has not been shown to

TABLE 3.5 Toxicity Symptoms of Water-Soluble Vitamin K

Symptoms	Dose Levels
Pregnant Women	
None observed	10-25 mg/day
Infants	
Hemolytic anemia, hyperbilirubinemia, kernicterus	5-10 mg/day, parenterally
Premature Infants	
Kernicterus, death	10 mg, 3 times/day for 3 days

Source. Committee on Nutrition, Pediatrics, 28, 501 (1961).

produce toxic symptoms at the usual therapeutic dose levels of 10 to 25 mg. However, the water-soluble forms in some cases produce vomiting with oral administration and hemolytic anemia, kernicterus, and even death at a 5 to 10-mg dose level, especially in premature infants (45). A summary of the toxic effects observed with the water-soluble vitamin K compounds is shown in Table 3.5.

Vitamin K requirements in humans are provided in approximately equal portions from plant sources and from microbiological synthesis in the gut. Normal diets in the United States provide on the order of 300 to 500 μg/day, well above the safe and adequate adult level of 70 to 140 μg. Thus there appears to be no need for additional supplementation in normal, healthy individuals. Excess intake should occur only for therapeutic reasons under the care of a physician (46).

Interactions of Vitamins A, D, E, and K

The toxicity of the fat-soluble vitamins is difficult to completely elucidate in certain cases because of their interactions (39, 47, 48). High levels of one of these vitamins may well lessen the toxicity symptoms of another. Thus an otherwise toxic level of a fat-soluble vitamin may result in reduced adverse effects if high, or at least adequate, levels of another fat-soluble vitamin are present. For example, the varying time of onset for the toxic effects associated with excess consumption of cod liver oil may be due to the relative proportions of vitamins A and D present rather than solely to the vitamin A. Some of the possible protective interactions among the fat-soluble vitamins are summarized in Table 3.6.

The converse of the protective interactions should also be noted. Deficiency of one or more of these vitamins may extend the toxic effects of others. For example, the toxic effects of vitamins A and D can be exacerbated in the presence of vitamin E and/or K deficiencies, and vitamins A and E are mutually antagonistic.

TABLE 3.6 Protective Interactions of the Fat-Soluble Vitamins

Adequate or High Levels of Vitamin	Protects Against High Levels of Vitamin	Reference
A or D	D or A	42,47
E	A	40,49
E	A or D	50
E	D_3 and D_2	51
K	A	48,52

Whatever the exact nature of the interactions among the fat-soluble vitamins may be, prudence still dictates avoidance of excess levels of consumption except for specific therapeutic uses under the direction of a physician (2).

Water-Soluble Vitamins

The water-soluble vitamins in general have fewer toxic effects than do the fat-soluble vitamins primarily because they are not retained in the body to the same extent as are the fat-soluble vitamins. Excesses of these vitamins above threshold levels are eliminated quite rapidly (normally on the order of 24 hours), with urine the principal route of excretion. These elimination characteristics can permit accidental or deliberate ingestion of excesses, as in megavitamin supplementation, without obvious discernible harm to the individual. Nonetheless, several adverse effects associated with excessive intake of some water-soluble vitamins have been reported, and consumption of levels of these vitamins in excess of RDAs, except under the direction of a physician, is not recommended by nutritionists. Ingestion of large doses of any vitamin for other than treatment of specific deficiency conditions may pose a serious risk either by masking a disease condition or by substituting for far more appropriate medical care of a disease condition.

A summary of the available information on adverse effects associated with the water-soluble vitamins is given in the following.

Vitamin C

For the past several years debate has continued over the efficacy of vitamin C (ascorbic acid) in the treatment of other than the deficiency disease, scurvy. This debate has focused particularly on the advocacy of large doses ($\leq$10 to 15 g/day) of vitamin C to prevent or diminish effects of the common cold, as proposed by Dr. Linus Pauling (53). Megadose ingestion of vitamin C has been rapidly embraced by many food faddists and criticized by many nutritional scientists. Trials to investigate the efficacy of vitamin C in preventing colds or alleviating cold symptoms have included a University of Toronto study (54) from which it was later concluded (55) that:

> There is no evidence that prophylactic doses (of ascorbate) in excess of 250 mg per day offer any advantage, and the optimum daily dose may well be lower than this, since tissue saturation is probably the limiting factor for a prophylactic effect.

The Toronto and other studies of vitamin C efficacy have provided some incidental knowledge of adverse effects from high doses, the most common of which appears to be gastrointestinal disturbances. The wide variety of other effects observed in human and animal studies are summarized in Table 3.7.

It is interesting that vitamin C is generally considered beneficial relative to the absorption of inorganic elements. Iron absorption, for example, has been shown to be enhanced at levels of 1 g (61). However, the opposite is true of copper, whose absorption is depressed by high intake levels of vitamin C (62).

There has been a great deal of research (68) as well as debate on vitamin C, and both are continuing. In a 1979 article (69) a new theory on the biological function of vitamin C, formulated by its discoverer, Dr. Szent-Gyorgyi, was presented. This theory, which attributes a fundamental role to ascorbate at the cellular level, will likely promote both debate and research on vitamin C. In the meantime it appears doubtful that large supplemental doses of vitamin C are of significant benefit and that caution is warranted at levels of 2 to 8 g/day. Levels greater than that may be distinctly hazardous (70).

Niacin

The adverse effects of niacin (nicotinic acid) are primarily the side effects associated with large therapeutic doses prescribed for nutritional deficiency. Nicotinic acid has marked vasodilative effects on administration of large doses (100 to 300 mg orally or 20 mg intravenously). Flushing reactions, headache, cramps, and nausea have been recorded (71). A transient impact on serum cholesterol has also been associated with nicotinic acid. Reduction of the levels of cholesterol, β-lipoproteins, and triglycerides in the blood have occurred at doses of 3 to 6 g/day (72). Daily ingestion of 3 g or more resulted in decreased serum lipids, increased utilization of muscle glycogen stores, and decreased mobilization of fatty acids from adipose tissue during exercise (73). Parsons and Flinn (74) reported a 10 to 15% drop in serum cholesterol following daily doses of 2 to 6 g/day. Other side effects included pruritis, desquamation, and pigmented dermatosis. All such side effects usually disappear when ingestion is halted, and apparently no remaining harm is induced.

Because of the side effects of nicotinic acid, the physiologically active amide form (nicotinamide) is often used in deficiency therapy. Nicotinamide has no side effects at therapeutic dose levels, nor does it produce lowered serum cholesterol. Rats, however, showed growth in-

TABLE 3.7 Adverse Effects From Excessive Intake of Vitamin C

Effect	Intake Levels	Reference
Nausea, diarrhea, gastrointestinal disturbances, flatus	1g/day for 17 weeks; 10-80 g/day for 6-10 days	56,57,58
Increased oxalates, potential stone formation	4-9g/day	59,60
Decreased copper absorption	Above 1g/day	61,62
Destruction of B_{12}	500mg	63
Interference with anticoagulant therapy - heparin and coumadin drugs	2mg (.01mg heparin)	64,65
Menstrual bleeding (in attempted therapeutic abortions)	6g/day for 3 days	45
Increased dependency		
- in adults	continued excess	65,66
- in newborn infants	400mg/day for pregnant mother	67

hibition and fatty livers when fed at 1% of a diet high in fat and low in choline (75).

Niacin requirements of normal individuals are easily met with usual U.S. diets. Therefore, except for therapeutic treatment of pellagra, supplemental ingestion of niacin is seldom needed, and prolonged high-dose ingestion of supplements should be avoided.

Folacin

Dose levels of folacin (folic acid or pteroylglutamic acid) many times in excess of the RDA have rarely been associated with adverse effects. Folic acid has been reported to interfere with some antiepileptic drugs, resulting in increased seizure frequency in epileptic patients (76). Evidence of other effects is generally lacking. Adult doses of up to 400 mg/day for 5 months or 10 mg/day for 5 years without effects suggest the relative safety of the usual therapeutic dose level of 5 mg/day (71).

There is one important aspect of folic acid therapy that should be emphasized. In doses above 1 mg per day, folic acid can alleviate the pernicious anemia resulting from vitamin B_{12} deficiency but will not impact other B_{12} deficiency disease symptoms. Doses of folic acid sufficient to correct the hematological aspects of pernicious anemia (25 to 250 μg) can delay recognition of the neurological aspects of pernicious anemia arising from vitamin B_{12} deficiency. Failure to diagnose and treat the B_{12} deficiency can, in turn, lead to severe neurological damage (76, 77). For this reason most nutritional scientists believe that excessive intake of folic acid, except under a physician's care, should be avoided.

Other Water-Soluble Vitamins

Among the remaining water-soluble vitamins, pantothenic acid, pyridoxine (vitamin B_6), and thiamin (vitamin B_1), there have been few indications of harmful effects and none of dietary origin. A summary of available information on these vitamins is given in Table 3.8.

ANTIVITAMINS OR VITAMIN ANTAGONISTS

In addition to the food safety problems posed by deficiencies of vitamins and other essential nutrients, we have considered the toxicity potential of excessive ingestion of vitamins and do likewise in subsequent pages for the trace elements. At this point it is useful to consider the adverse effects of those compounds that inhibit the activity of a vitamin in a metabolic reaction. The collection of compounds involved in these phenomena are quite diverse and do not lend themselves to simple characterization. In the 1950s such compounds were observed to have similar structure, to produce symptoms similar to those produced by deficiencies of specific vitamins, and to compete with specific vitamins. They were, therefore, termed "antivitamins" (81, 82). Subsequently, it was noted that the antivitamins were of two basic types: (*a*) structurally similar antimetabolites and (*b*) structure-modifying compounds that inhibit the effect of a vitamin by forming inactive complexes with, or otherwise modifying, the molecule (83).

Although the precise mechanism of action is not known in many cases, the antivitamins or vitamin antagonists are now recognized as a varied class of compounds that cancel or reduce the effect of a vitamin. The most obvious vitamin antagonists are certain vitamins themselves. As noted previously, vitamins A and E are mutually antagonistic, and vitamin A is also an antagonist of vitamin K. Some other examples of the antivitamins and their characteristics, compiled from comprehensive reviews by Somogyi (83, 84), are shown in Table 3.9.

Most of the vitamin antagonists are of little significance in human nutrition in the United States because of the varied diets consumed. However, the current trend toward self-prescribed megavitamin therapy could lead to serious consequences because of the complex and imperfectly understood interactions involved, especially in the case of the fat-soluble vitamins.

One final consideration related to the general area of vitamin antagonism is the problem of drug-nutrient interactions. Such interactions are far too extensive and complex to describe here but can be indicated by

TABLE 3.8 Effects of Pantothenic Acid, Pyridoxine, and Thiamin

Vitamin	Effects/Dose	Reference
Pantothenic Acid	Occasional diarrhea, 10-20 g (calcium salt); no effects at 100 mg/kg	71
Pyridoxine (vitamin B_6)	No toxic effects, I.V. dose of 200 mg; no side effects, oral dose of 100-300 mg/day	71,78
Thiamin (vitamin B_1)	No toxic effects from oral doses (maximum absorption is 5 mg/dose); anaphylactic shock, vasodilation, nausea, tachycardia and dyspnea from repeated parenteral administration of 10-100 mg	79,80

noting the nutrients involved. In a handbook on drug and nutrient interactions produced by the American Dietetic Association (85), the following vitamins are noted:

Vitamin A	Folacin
Vitamin C	Niacin
Vitamin D	Pantothenic acid
Vitamin E	Pyridoxine
Vitamin K	Riboflavin
Vitamin B_{12}	Thiamin

(In addition, calcium, iron, magnesium, phosphorus, potassium, and zinc are listed as being involved in nutrient-drug interactions.)

VITAMIN AND MINERAL SUPPLEMENTS

A vast array and quantity of vitamin and mineral preparations are purchased annually in the United States. Most of these products are legally designated by the FDA as dietary supplements, but if therapeutic claims are made in the labeling, such products are legally designated as nonprescription or "over-the-counter" (OTC) drugs. As part of a safety and effectiveness evaluation of all OTC drugs, an expert panel convened by FDA reviewed OTC vitamin and mineral products and reported their results in 1979 (86). This comprehensive review of vit-

TABLE 3.9 Antivitamin Examples

Antagonist (Vitamin)	Type of Inhibition	Source
Thiaminase (Thiamin)	Modifies structure, deactivated by heat	Varied, including plants, fish and shellfish
Hypoglycin A (Riboflavin)	Mechanism unknown, counteracted by riboflavin	Ackee plum
Niacytin and leucine (Niacin)	Mechanism unknown, reversed by alkaline medium	Wheat bran, millet
Avidin (Biotin)	Forms complex action prevented by heat	Egg white
Linatine (Pyridoxine)	Alters structure of all three pyridoxine forms	Flaxseed
Large amounts of β-carotene (Vitamin D)	Mechanism unknown	Varied
Polyenic acids (Vitamin E)	Mechanism unknown, counteracted by vitamin E	Varied
Dicumarol and related synthetic products (Vitamin K)	Mechanism unknown, may be competition in prothrombin synthesis	Anticoagulants

amins and minerals provides an excellent summary of the toxicity aspects discussed in the previous section of this chapter. Among the conclusions and recommendations of the expert panel were the following:

- Vitamin C, vitamin B_{12}, folic acid, niacin, vitamin B_6, riboflavin, thiamin, vitamin A, and vitamin D could be sold as single-ingredient OTC drugs with specified labeling.
- Biotin, choline, vitamin E, and pantothenic acid should not be sold in single-ingredient form, because deficiencies are virtually nonexistent.
- Vitamin K, particularly dangerous for those using anticoagulant drugs, should be available only by prescription.
- Calcium, iron, and zinc could be sold as single-ingredient preparations with specified labeling.
- Copper, fluoride, iodine, magnesium, manganese, phosphorus, and potassium should not be sold as OTC products because deficiencies are rare.
- Combinations containing only the fat-soluble vitamins should not be permitted.

The expert panel in each case recommended dosages for prevention and treatment of deficiency diseases. Cautioning about the hazards of excess intakes of vitamins and minerals, the panel said that labels should state that the products are for use in deficiency therapy "when the need for such therapy has been determined by a physician." The panel also recommended in many cases specific labeling that would provide contraindications and warnings about excess intake. Similarly, recommendations prohibiting medical claims, other than those relative to deficiencies, were specifically made in several cases.

PROTEINS AND AMINO ACIDS

Proteins, found in all living cells, consist of various complex combinations of amino acids. More than 20 different amino acids have been identified as components of protein. The almost infinite possibilities for combinations and arrangements of the amino acids explain the wide variety and different characteristics of plant and animal proteins (87).

Requirements

As noted at the beginning of this chapter, protein provides the essential amino acids needed by the body for growth and repair of tissue. Protein is also a major component of enzymes required in metabolic reactions and serves as an energy source supplemental to fats and carbohydrates. There are eight amino acids that are generally agreed to be essential for humans:

Isoleucine	Threonine
Leucine	Trypotophan
Lysine	Valine
Methionine	Phenylalanine

In addition, histidine is an essential amino acid for infants and may also be essential for adults (2, 88). Arginine, although not essential for adults, cannot be synthesized rapidly enough in infants to meet the requirements, and thus a dietary source is essential for the very young (89). Two other amino acids, cystine and tyrosine, might be termed "semiessential." Cystine can be synthesized only from methionine, and tyrosine only from phenylalanine. When cystine and tyrosine are present in the diet, requirements for methionine and phenylalanine are reduced. Any of the remaining required amino acids can be manufactured in the body in adequate amounts.

Because different proteins have different types and levels of essential amino acids, it is customary to refer to their "quality" in terms of their amino acid profile relative to human requirements. The most common method of measuring quality is the protein efficiency ratio (PER), which employs casein as the reference standard. Proteins from animal sources (meat, fish, poultry, eggs, and milk) contain all the essential amino acids in profiles similar to human requirements, and thus animal protein sources have higher PERs than do plant protein sources. However, proper combinations of plant protein sources can provide adequate levels of the essential amino acids.

The essential amino acids are not stored by humans, but normal dietary intakes of animal proteins together with plant proteins easily provide required levels. Normal dietary intakes that include a variety of protein sources have not been associated with protein-deficiency symptoms. Deficiency problems can arise when diets are limited in individual amino acids, however. For example, corn, a major food item in many parts of the world, is a poor source of tryptophan and lysine. Pellagra, which has been endemic in predominantly corn-eating areas of the world for many years, is associated with a deficiency of both nicotinic acid and its precursor tryptophan (90).

Toxicity

Food safety problems have not generally been associated with excess intakes of natural sources of protein other than perhaps those problems arising from hypersensitivity or allergic reactions. Similarly, hazards have rarely been attributed to excessive ingestion of individual amino acids. As always, there are qualifications and exceptions to such general conclusions, however, and these must be considered to put the food safety role of these nutrients in proper perspective.

Proteins

The wide availability of both plant and animal sources of protein in the U.S. food supply can easily lead to consumption of protein well in excess of daily requirements. In addition, the "if a little bit is good, a lot is better" philosophy has led to even greater excesses of protein intake in fad diets and by misinformed athletes seeking improved performance. Although significant overt symptoms of toxicity have seldom been associated with such consumption, a number of cautionary factors should be considered. For example, animal studies have shown liver and kidney hypertrophy with excess ingestion of protein and cystine, and human studies have shown increased calcium excretion with high protein ingestion (91). The latter effects, especially, indicate some disruption of normal physiological function and thus point to the advisability of avoiding large excesses in protein intake.

Food safety cautions relative to protein are important not only in the current context, but also with respect to domestic dietary trends and world food supply problems. The U.S. production of both animal and plant protein foods continues to meet domestic needs and provide excesses for export. However, rising costs, shortages in many parts of the world, and recognition of future requirements have caused a great deal of attention to be focused on this area. Increasing domestic emphasis is being placed on plant protein products as extenders and partial or total replacements for traditional animal protein foods. Substantial research is also being devoted to improving protein quality and developing novel protein sources to meet foreign shortages. Along with this focus on protein foods has come increased attention on the food safety aspects. The need for such attention is illustrated, for example, in an article by Martinez (92), who has discussed two classes of antinutritional factors associated with edible proteins: (*a*) the toxic effects of naturally introduced (e.g., gossypol in cottonseed protein) and environmentally introduced (e.g., selenium and cadmium) contaminants as well as the effects of constituents concentrated in processing (e.g., fluoride and nucleic acids) and (*b*) factors impacting on bioavailability of nutrients such as the effect of phytates on trace mineral availability.

Jaffé (93) has discussed toxic proteins and peptides and has noted that nearly all foods contain protein, and certain proteins, including bacterial toxins and plant toxins such as ricin, are highly toxic. However, humans and most other animals have developed detoxification mechanisms against foreign proteins; the major protective mechanism is digestion in the alimentary tract. For those cases in which whole proteins do enter the circulatory system, our immune defense mechanisms are stimulated. The modes of action of toxic proteins have not been clearly identified but certain reactions are known. Among the toxic proteins are hemagglutinins found mostly in plants, which cause red blood cells to agglutinate. Other proteins contain toxic elements such as selenium or are biologically active enzymes that decompose essential constituents in the diet. Yet another category, about which we know even less, are the toxic peptides. These are found in such foods as certain species of mushrooms and certain molds.

According to Jaffé (93), many individuals show allergic reactions after the ingestion of certain types of food. In fact, most foods have reportedly resulted in some sort of adverse reaction, however mild. The nature of these reactions remains unclear but common allergens include grains, milk, eggs, fish, crustaceans, tomatoes, strawberries, nuts, and chocolate. The specific agent is unknown for most food allergies, but frequently they are proteins that can be denatured by heating (120°C for 30 minutes). The best method of prevention is to eliminate the source from the diet. (The importance of certain proteins and protein constituents in food safety is further discussed in Chapters 4, 5, and 6.)

Some reassurance of protein safety has been provided by a detailed evaluation of protein hydrolyzates performed by SCOGS for FDA (94). The SCOGS specifically examined the safety of acid hydrolyzed protein, autolyzed yeast extracts, and soy sauces as flavor enhancers in foods. Conclusions from the evaluation were that there is no existing evidence of hazard for current levels of use of these products but that the need for additional research is indicated.

Significant hazards have also been associated with misuse of protein products, however. In late 1977 reports began to surface of a variety of side effects and even death associated with the use of certain protein preparations used for weight reduction (95). By the end of 1978, 60 deaths had been associated with the use of protein preparations as the principal or sole source of calories for rapid weight loss. The consumption of partially hydrolyzed protein of low biological quality, the so-called predigested liquid protein products, was associated with most of the deaths. In 17 of the 60 cases, death was ascribed to cardiac causes (acute myocarditis, intractable cardiac arrhythmias, and related cardiac abnormalities), and no evidence of any significant disease state, other than obesity, existed. There were, in addition, over 225 reports of adverse reactions attributed to such protein preparations. Although anecdotal in most cases, widely diverse reactions ranging from nausea, vomiting, and diarrhea to cardiac arrhythmias, myocardial infarction, hemorrhage, and anaphylaxis were reported (95).

The details associated with the 17 deaths appear to indicate that neither the physical form nor the biological quality of the protein involved was a major factor. The relative contributions of the severely restricted caloric intake (200 to 400 calories/day) and the use of protein per se as the sole or primary source of calories are not clear. However, it is significant that no other foods promoted for weight reduction have been associated with similar hazards, and the FDA has moved to require warning labels relative to the use of various protein products in weight reduction diets (95).

Amino Acids

As with proteins, food safety problems have rarely been associated with excess intakes of individual amino acids from naturally occurring food sources. The amount of any given amino acid ingested from such sources would be limited and would likely be present along with a balanced pattern of other amino acids. Adult diets in the United States at times or even frequently may contain on the order of 10 times the required levels of individual amino acids. Although these intakes are apparently tolerated without obvious adverse effects, excesses of essential amino acids are less well tolerated than are excesses of nonessential forms. This relates to their physical and chemical properties and to the functions for which they are required (96).

There has been relatively little systematic study of excess intakes of individual amino acids in humans. What information does exist has by and large resulted from studies of potential therapeutic effects. Generally, these studies show little by way of adverse effects at amino acid levels of 10 or more times the required intakes; the symptoms that have been observed include gastric distress, nausea, headache, flushing, and disorientation. Amino acid imbalance and resulting depression in food intake and growth has been shown in animals on purified diets, but the extent to which it is relevant in humans is not clear (96). However, the use of amino acid supplements in cases where the diet is low in protein or essential amino acids to begin with is cause for concern. As a general rule, regular doses of any therapeutic agent should be planned so that nutrient needs are met in all regards.

It is of interest to note that some individuals suffer from genetic defects in which the initial enzyme for the metabolism of a given amino acid is lacking. This lack leads to accumulation of the affected amino acid even at otherwise normal levels of ingestion (97). For example, a significant number of infants are born with an inability to metabolize the essential amino acid phenylalanine, a disease termed *phenylketonuria* (PKU). Unless phenylalanine intake is sufficiently restricted, the PKU infant will accumulate blood levels of phenylalanine sufficiently high to induce brain damage. Phenylketonuria may persist until 5 or 6 years of age or longer.

Increased intakes of phenylalanine that might result from use of a new sweetener, aspartame (a combination of the two amino acids, phenylalanine and aspartic acid), have been questioned on safety grounds, especially in combination with increased intakes of glutamic acid and its salts. These questions arose on the basis of controversial studies indicating hypothalamic lesions in neonatal rats. However, because of the absence of such lesions in infant monkeys, regulatory authorities in the United States, Canada, and Europe have concluded that aspartame is safe for food uses (see Chapter 6).

Glutamic acid and its salts (monoammonium glutamate, monopotassium glutamate, monosodium glutamate, and glutamic acid hydrochloride) were specifically reviewed for the FDA by the Federation of American Societies for Experimental Biology (SCOGS) in 1978 (98). Their evaluation is part of the FDA's GRAS review program (see Chapter 6). Having considered the food uses of these glutamates, the SCOGS concluded that the available information does not indicate a hazard to individuals beyond infancy at current levels of food uses but that it is not possible to determine, without additional data, whether significantly increased usage would constitute a dietary hazard. The SCOGS also considered the evidence concerning adverse effects of glutamates on infants to be somewhat equivocal but endorsed the practice of omitting the addition of glutamic acid and its salts to commercially prepared infant and junior foods. Additional evaluation of glutamates is being made by the SCOGS and by the FDA.

CARBOHYDRATES

Carbohydrates are certainly among the most maligned of dietary constituents, but they are essential human nutrients (i.e., various tissues of the body require carbohydrate under all physiological conditions) (99). Unless carbohydrate is provided from food ingestion in fairly large amounts, the required carbohydrate must be manufactured by the body from other sources. As with other aspects of food safety, much of the popular concern about carbohydrates is not entirely in consonance with nutritional knowledge. This is not to say that carbohydrates are totally without food safety concern—indeed, there is no such food constituent.

Carbohydrates constitute a large and varied class of organic compounds consisting of carbon, hydrogen, and oxygen in the ratio of n–$2n$–n, respectively. They are usually discussed in terms of three types: (*a*) monosaccharides (which include the simple sugars such as galactose, fructose, and glucose), (*b*) disaccharides (sucrose, lactose, and maltose), and, (c) polysaccharides (including digestible starch and dextrins as well as indigestible celluloses).

Requirements

Digestible carbohydrates are the body's primary source of energy, are involved in important metabolic functions, and play protective roles in vital organs. Carbohydrates are digested in the intestinal tract, absorbed as simple sugars, converted to glycogen in the liver, and used for energy or other metabolic processes or stored. The body's carbohydrate stores exist primarily as glycogen in the muscles; the rest occur as glucose in the blood and glycogen in the liver.

As an energy source, carbohydrates are especially important in muscular activity because of their efficiency in that regard, and the heat generated in energy metabolism helps maintain body temperature. Fat and protein can also supply energy requirements for muscular activity, but they are less efficient than carbohydrates for this purpose. This is the case since energy is required to convert fat and protein into forms usable by the body for such activity. During heavy and prolonged muscular activity, carbohydrate reserves can be depleted, resulting in the breakdown by the body of protein and fat. In the absence of additional carbohydrate ingestion, ketones (by-products of fat metabolism) accumulate in the blood. Carbohydrates are even more important to cardiac muscles, which must function continuously. Heart muscles can make use of their glycogen stores, and thus temporary postprandial hypoglycemia is not a problem under normal circumstances. However, for the damaged or diseased heart, hypoglycemia can result in stenocardia, and high-carbohydrate therapy has been used for treatment of both diabetic and nondiabetic patients.

To perform the vital function of detoxifying harmful substances in the body, the liver requires sufficient glycogen content, supplied and maintained by carbohydrate ingestion. Thus adequate carbohydrate ingestion is a human requisite for efficient health maintenance. Glycuronic acid formed from carbohydrates can react in the liver with various chemicals to render them harmless. This reaction, for example, helps regulate the metabolism of steroid hormones and elimination of toxic compounds, thus preventing any harm resulting from excess levels. Acetylation reactions, dependent on carbohydrates are also key to detoxification reactions in the body (99).

Another important function of carbohydrates is in the regulation of protein and fat metabolism. The utilization of protein in tissue maintenance and repair is dependent on the presence of sufficient carbohydrate. If carbohydrate levels are deficient, amino acids are broken down to provide energy rather than serving their body-building function. Similarly, adequate availability of carbohydrate to the liver is required to control the rate of breakdown of fats.

A continuing supply of glucose is critical to the normal CNS functioning, and carbohydrate is the principal source of this fuel. Unlike other body tissues, nerve tissue does not maintain a sufficient reserve of glycogen and thus requires a glucose source, provided by carbohydrates. The importance of carbohydrates in this regard is illustrated by CNS susceptibility to hypoglycemic shock and to irreversible damage if the hypoglycemia is prolonged.

Controversy about carbohydrates is probably greatest in regard to how much should be consumed. No quantitative daily requirement has been established for carbohydrates. Normal diets providing the total caloric needs of individuals provide ample intakes of carbohydrates. For ease of reference, the energy intakes (total calories from protein, fat, and carbohydrates) recommended by the Food and Nutrition Board (2) are shown in Table 3.10. These caloric intakes are based on observed median heights and weights of children and desirable weights for adults (18 to 34 years of age) at mean observed heights. The calorie ranges for adults are based on a ±400-calorie variation and for children on the tenth and ninetieth percentiles, indicating the wide ranges of energy intake, depending on physical activity and other factors, appropriate for any subpopulation.

Toxicity

There are relatively few examples of adverse effects associated with carbohydrate consumption. Population groups throughout the world whose diets consist almost entirely of carbohydrates are not uncommon. Such groups apparently do not suffer adverse effects if sufficient food is available. Temporary adverse effects such as diarrhea and intestinal cramps can be associated with marked changes in the form and/or the amount of carbohydrates ingested, such as sudden consumption of high-fiber foods. Also, individuals with impaired carbohydrate metabolism can evidence abnormal glucose tolerance curves and associated symptoms on the sudden ingestion of concentrated sugar sources.

Of course, natural foods with relatively high carbohydrate levels may contain naturally occurring toxicants. Examples include the toxic alkaloids in the white potato and the mycotoxins in grains. (These are discussed in Chapter 5.) Another interesting example is honey, which is much extolled by the so-called natural food advocates. It is not surprising that honey can contain materials toxic to humans considering the varieties of plants that may be visited by the honey bees. The oldest recorded incident of toxic honey is the description by Xenophon of the mass poisoning of the expedition of Cyrus in 401 B.C. in Asia Minor, subsequently attributed to a toxic substance called *andromedotoxin* (100). Commercial honey producers take great pains to avoid such contamination, however.

Although there is little by way of significant toxicity associated with the general class of carbohydrates per se, there is one interesting exception to the rule. The exception involves particular intolerances. Classic among these is lactose intolerance, first reported as a congenital defect by Holzel et al. in 1959 (101). Lactose (milk sugar) intolerance is especially prevalent in the Far East, Middle East, and Africa, primarily among adults. Such individuals have difficulty digesting lactose and can experience diarrhea and gastric distress with ingestion of lactose-containing foods such as milk.

TABLE 3.10 Recommended Energy Intake

Category	Age (years)	Height (inches)	Weight (pounds)	Energy Needs (Calories)
Infants	0-.5	24	13	lbs x (43-66)
	.5-1	28	20	lbs x (36-61)
Children	1-3	35	29	1300 (900-1800)
	4-6	44	44	1700 (1300-2300)
	7-10	52	62	2400 (1650-3300)
Males	11-14	62	99	2700 (2000-3700)
	15-18	69	145	2800 (2100-3900)
	19-22	70	154	2900 (2500-3300)
	23-50	70	154	2700 (2300-3100)
	51-75	70	154	2400 (2000-2800)
	76+	70	154	2050 (1650-2450)
Females	11-14	62	101	2200 (1500-3000)
	15-18	64	120	2100 (1200-3000)
	19-22	64	120	2100 (1700-2500)
	23-50	64	120	2000 (1600-2400)
	51-75	64	120	1800 (1400-2200)
	76+	64	120	1600 (1200-2000)
Pregnancy				+300
Lactation				+500

Source. Food and Nutrition Board, Recommended Dietary Allowances, Ninth ed., National Academy of Sciences, Washington, D.C., 1980.

Lactose intolerance results from a deficiency of the enzyme, lactase, which functions in the digestive tract to break lactose down to monosaccharides that can be absorbed. In intolerant individuals, lactose remains in the gut, leading to osmotic catharsis. Its subsequent fermentation in the colon to lactic and related acids contributes to diarrhea and gastric distress. Intolerance for sucrose and other disaccharides has also been recognized, again associated with deficiencies in a particular enzyme or a combination of enzymes. Disaccharide intolerance can occur as a congenital defect in infants and also occurs, as noted, among many adults. Lactose intolerance for some reason, is by far the most common.

Most of the current popular concern about carbohydrates focuses not so much on the complex carbohydrates (the polysaccharides) as on sugar (i.e., sucrose), which is the recipient of the majority of attention. There is an extensive body of scientific literature on the sugars and sugar alcohols (see Chapter 6). In the case of sucrose a comprehensive review and evaluation of the literature up through 1976 has been made by SCOGS of the Federation of American Societies for Experimental Biology (FASEB) (102).

The SCOGS review of sucrose examined absorption and metabolism; acute and chronic studies; special studies including carcinogenicity, teratogenicity, and atherosclerosis; diabetes; and dental caries. Their overall summary conclusion is of interest:

> Other than the contribution made to dental caries, there is no clear evidence in the available information on sucrose that demonstrates a hazard to the public when used at the levels that are now current and in the manner now practiced. However, it is not possible to determine without additional data whether an increase in sugar consumption—that would result if there were a significant increase in the total of sucrose, corn sugar, corn syrup and invert sugar, added to foods—would constitute a dietary hazard.

It should be noted that a separate evaluation has been made by the SCOGS of corn sugar (dextrose), corn syrup, and invert sugar as food ingredients (103). In addition, the FASEB has specifically discounted the alleged connection between sucrose and the incidence of tumors in laboratory animals (104).

In essence, the SCOGS has pointed out, after careful evaluation, that sucrose, like any other natural or added constituent of foods, is not totally without food safety concern, but that popular association of sucrose in the etiology of serious chronic diseases is not warranted by existing scientific information.

The SCOGS review of sucrose to some extent leaves open the question of sucrose and dental caries—a question deserving of at least brief mention here. There is a strong popular notion that sucrose is the sole cause of dental caries and that reduction of sucrose ingestion would largely solve that problem. Of course, none of the questions in food safety are that simple—all are much more complex. Actually, the etiology of dental caries is imperfectly understood but is clearly multifactorial in nature. Dental caries is a chronic infectious disease of the teeth involving demineralization of the tooth surface over time by organic acids produced from bacterial fermentation of carbohydrate deposits (105). The development of dental caries requires the interaction of four principal factors: (*a*) host (susceptible teeth and oral environment conducive to caries formation), (*b*) agent (presence of viable cariogenic microorganisms), (*c*) environment (diet conductive to caries, deficient fluoride exposure and oral hygiene), and (*d*) time (sufficient duration of conducive environment).

The presence of sugar, or actually any fermentable carbohydrate, is thus only one factor and cannot be indicted as the sole cause of dental caries. Contrary to much popular belief, substantial reductions in dietary intake of fermentable carbohydrates (even if such were possible) is not considered by most health professionals as the sole answer to dental caries. For example, the 1979 Surgeon General's report (106) states:

> Fluoridation is one of the most effective—and cost-effective—preventive measures known. By making teeth less susceptible to decay through increasing resistance to the action of bacteria-produced acid (it may have an antibacterial effect as well), optimal level fluoridation of drinking water can prevent 65 percent of decay that would otherwise occur. And it can do so for 10 to 40 cents per person per year, depending upon the size of the community.

In the absence of fluoridated community water supplies, the Surgeon General recommends fluoridation of school drinking water, fluoride mouth washes (shown in clinical studies to reduce decay by 35%), dietary fluoride supplements, fluoride toothpastes, and application of fluoride to the teeth by dental professionals.

Alcohol

Although alcohol is not technically a carbohydrate, it is convenient to briefly discuss this important compound at this point. Ethyl alcohol at low levels is a normal component of many foods, but its nutritional and food safety importance derives from its consumption in beer, wine, and distilled spirits. A concentrated energy source providing 7 calories/g, alcohol is rapidly absorbed in the mouth, esophagus, and stomach.

Excessive alcohol intake is associated with three kinds of adverse effects: (*a*) the replacement of other dietary components (many persons consume 10 to 20% of their calories as alcohol, primarily at the expense of fat and carbohydrate intake), (*b*) direct toxic effects on several body organs and tissues, and (*c*) especially in alcoholism, the possibility of severe nutritional deficiencies accompanying chronic excess ingestion.

In too many individuals, alcohol consumption adds to an otherwise normal or even excessive consumption of calories. For others, there is partial substitution of alcohol for food, and often in the alcoholic, drinking is typically to the exclusion of eating. Therefore, dietary insufficiency is common, with folic acid as the most common deficiency, followed by thiamin, niacin, and vitamin B_6 (107). The characteristic tremors of the alcoholic are directly related to insufficient consumption of the B vitamins. These problems are compounded by the structural and functional changes that excessive alcohol intake induces on the intestinal tract (108). In addition to low nutrient intake and poor absorption of the nutrients that are consumed, the alcoholic is commonly subject to symptoms such as nausea, anorexia, and gastritis, which further deter adequate food intake. Decreased pancreatic secretion in the alcoholic contributes to gastritis, impaired digestion of food, and poor absorption of nutrients. Alcoholics also suffer megaloblastic anemia, which relates to folate deficiency and/or impaired metabolism of folate at the level of red blood cell production. Low platelet counts are also experienced by alcoholics (109–111).

The major organs in which adverse effects of excessive alcohol consumption are commonly seen are the liver, the gastrointestinal tract, the pancreas, and the blood. Excessive alcohol intake produces adverse effects on the liver in two ways: (*a*) direct toxic effect on liver function and (*b*) food intake and nutrient absorption, thereby resulting in malnutrition and thus impairing liver function. Although the direct relationships between the fatty deposits, hepatitis, and cirrhosis found in the livers of alcoholics have not been completely elucidated, it is clear that death results if these conditions advance to the stage at which the liver can no longer perform its vital functions (112).

There is some indication that increased caloric intake from alcohol consumption in the United States is accompanying the rather surprising decrease in average caloric intake from other foods. For example, Van Itallie (113) has noted that on the average about 10% of calories consumed in the United States comes from alcohol. This trend should be taken into account in formulating food safety policy and health promotion–disease prevention programs. In the 1980 Dietary Guidelines for Americans, prepared by USDA and the Department of Health, Education, and Welfare (HEW), the advice is given, "One or two drinks daily appear to cause no harm in adults. If you drink you should do so in moderation" (114).

FATS AND OTHER LIPIDS

The lipids are a varied and diverse class of biologically important compounds that have in common their solubility in organic solvents and usually contain a fatty acid or a fatty acid derivative. Fats in the diet may be of either animal or vegetable origin and consist primarily of triglycerides, the fat molecules normally found in human tissues. Triglycerides are composed of a molecule of glycerol and three fatty acids that may be saturated, unsaturated, or polyunsaturated. Unsaturated fats (e.g., those in many vegetables oils) are normally liquid at room temperature and chemically differ from saturated fats by the presence of double bonds in their structure, which makes them more susceptible to oxidative breakdown both in foods and in the body. The class of lipids also includes sterols and steroids, in particular the important cholesterol found in animal fats but not in vegetable fats, and the phospholipids. The fat-soluble vitamins A, D, E, and K are also classified as lipids but are discussed separately in this chapter.

Requirements

The body utilizes fats along with carbohydrates to meet energy requirements. Fat is the most concentrated energy source, providing approximately 9 calories/g (carbohydrates provide 4 calories/g) and is the only source of energy that can be stored by the body in quantity. When required for energy, stored fat is readily made available by the body. Normally, U.S. diets provide about 40% of the caloric intake from fats. In addition to providing a direct source of energy, fats and the other lipids play a wide variety of important and highly complex nutritional roles (115).

One fundamental role of fats is that of helping make food palatable, and to most individuals, minimal fat content is necessary to assure acceptance. Much of the flavor and the aroma of foods is due to fat-soluble compounds. Furthermore, fats decelerate the digestive process, thus slowing the recurrence of hunger sensations. A balanced intake of fat-containing foods has a protein-sparing effect, provides essential fatty acids, and permits utilization of the fat-soluble vitamins (116). These latter two needs can be met by a diet containing 15 to 25 g of appropriate food fats (2).

Among the essential polyunsaturated fatty acids, those that are biosynthesized inadequately or not at all, the most important to humans is linoleic acid. Other fatty acids of importance include linolenic and arachidonic acids, which are essential but can be synthesized by the body. Deficiencies of essential fatty acids in infants are associated with eczematous dermatitis and reduced weight gain. Essential fatty acid deficiencies result in alterations in cell membranes, may also result in imparied transport and metabolism of cholesterol, and impair the production of prostaglandins.

Adults, presumably because of reserves in adipose tissue, do not normally experience deficiencies of essential fatty acids, but such deficiency has been produced by parenteral feeding of saturated fat (115). Deficiencies have been treated by repeated feedings that mobilize fat stores and release essential fatty acids from adipose tissue. Requirements for the essential fatty acids are not precisely known but, based on linoleic acid deficiency, may be on the order of 1 to 2% of calories. The Food and Nutrition Board considers 3% of total calories as linoleic acid satisfactory for groups whose fat intakes contribute less than 25% of calories (2). Considerable controversy exists relative to the percent of total calories that should be provided by fats and to the requirements and/or limitations on saturated and unsaturated fats and cholesterol (116–118). The Food and Nutrition Board recommends that especially in diets below 2000 calories, fat should not contribute more than 35% of energy for those known or suspected to be in a high-risk category for certain diseases. For such high-risk individuals, 8 to 10% of total calories should be from essential polyunsaturated fatty acids (2).

Toxicity

As is the case with all the nutrients, there are natural foods for which potential toxicity has been associated with their lipid contents. The perhaps classic example in this case is the erucic acid found in a number of plants and studied especially in rapeseed oil and mustardseed oil (119). Again, however, excess consumption of lipid-containing foods have seldom been associated with acute toxicity.

The primary toxicity question in the case of the lipids is an extremely complicated and, therefore, a controversial one—heart disease. Although the death rate for heart disease has decreased since the 1960s (a decrease of 22% between 1968 and 1977), it is still the leading cause of death in males over 40 years and is also the leading cause of death in the total population, accounting for about 37% of the death rate. Heart disease was responsible for 700,000 deaths in 1977. Stroke and arteriosclerosis added approximately 183,000 and 29,000 deaths, respectively, in 1977. Heart disease is not only the leading cause of death, but also a primary cause of permanent disability and reduced activity and leads to more days of hospitalization than does any other disorder (106). The quest for a simple answer to heart disease has resulted in a variety of proposals ranging from the ingestion of vitamin E and lecithin supplements (especially by those marketing such supplements) to the more common reduction of dietary intakes of cholesterol and saturated fats.

Unfortunately, research on the etiology of heart disease has led in many cases to apparently conflicting results, generating more controversy than conclusions. This is especially true relative to the role of fats and cholesterol in heart disease.

The controversy over the role of dietary cholesterol and fats (118, 120) in heart disease is understandable when the complexity of the issue is considered. As the Food and Nutrition Board has noted, "It should be emphasized that most chronic or degenerative diseases have a number of contributing factors, only one of which may be diet" (2). Risk factors that may be causally associated with heart disease include genetic predisposition, obesity, sex, inactivity, smoking, hypertension, diabetes, stress, diet, and probably others as well (106, 114, 118, 121, 122).

It is clear that those individuals with high blood cholesterol levels have a greater chance of having a heart attack. However, a high blood cholesterol level does not necessarily lead to a heart attack, nor does an already low or a reduced blood cholesterol level necessarily prevent one. Further, there is wide variability among individuals with respect to the association between (*a*) dietary intake of saturated fat and cholesterol and (*b*) blood cholesterol levels. Some persons can ingest diets high in fat and cholesterol yet maintain normal blood cholesterol; others have a high blood cholesterol level even with low dietary intakes. Such characteristics make the cholesterol question difficult to elucidate. Then, too, there are many other questions such as the importance of the low-density and high-density lipoproteins, transfatty acids, and cholesterol oxidation products.

The association between fat and cholesterol and heart disease has led to a variety of recommendations for dietary modifications by both federal and private organizations. Recommendations have been made by the Senate Select Committee (120), the Surgeon General (106), the Senate Subcommittee on Nutrition (123), the American Council on Science and Health (122), the Food and Nutrition Board (2), and the USDA/HEW (114). In light of our incomplete knowledge in this area, drastic dietary modifications for the general public do not appear warranted. However, moderation relative to dietary intake of fat and cholesterol is warranted, as indeed is moderation for the intake of all dietary components. For those individuals at high risk levels because of age, sex, family history, smoking, overweight, high blood pressure, diabetes, or other factors, individual dietary advice from a physician is a prudent course of action.

TOXICITY OF TRACE ELEMENTS

The Food and Nutrition Board has established RDAs for the trace elements iron, zinc, and iodine and has listed ranges of "estimated safe and adequate daily dietary intakes" for copper, manganese, fluoride, chromium, selenium, and molybdenum and for the electrolytes sodium, potassium, and chloride (2). Other trace elements of interest are tin, vanadium, silicon, and nickel, which are probably also essential human nutrients, but this has not been definitely established. These trace elements have seldom been considered to present significant hazards, especially in comparison with arsenic, lead, cadmium, and mercury, which are specifically noted for their toxic manifestations. (Arsenic and lead may also be essential nutrients for some mammals, including humans.) However, all the trace elements are toxic, and the degree of toxicity depends on a variety of factors. Their toxicity is illustrated, for example, by the specific caution of the Food and Nutrition Board against habitually exceeding the upper levels of intake listed for copper, manganese, fluoride, chromium, selenium, and molybdenum (2).

In discussing the hazards of the trace elements it is important to note that the absolute level of intake is not the only factor involved with either acute or chronic toxicity. For some of the elements, toxic intake levels can vary considerably with individual circumstances. Thus there are, in effect, different "safe" or "toxic" levels of each of the trace elements depending on the particular conditions prevailing.

In a discussion of essentiality versus toxicity, Schwarz (124) emphasizes the circumstances under which the differences between the required level and the minimum chronic toxic dose are only several orders of magnitude. For example, an element that is readily stored may have accumulated in tissues over time so that the minimum intake required to produce a toxic effect has been steadily reduced. As the accumulated levels in tissues increase, the minimum chronic toxic dose approaches the requirement level. Characteristics like this make it very difficult to define chronic toxic levels for certain trace elements in terms of an overall population average. Similarly, an acute toxic dose may, in general, be well above the required level or chronic toxic dose yet may vary considerably with the health and the nutrient status of the individual.

Both acute and chronic toxic doses are also affected by control mechanisms that are triggered as a particular element enters the body. The two most important controls are decreased absorption and/or increased excretion. The importance of each varies from one trace element to another. For those elements that are absorbed readily, excretion is generally rapid and efficient. Likewise, for those that are not readily excreted, absorption is generally limited.

A third control mechanism is the immobilization or the storage of the toxic element in less active tissue such as bone. A fourth mechanism is detoxification, followed by secretion of the detoxified complex.

These control mechanisms and the resultant toxicity are affected by the physical and chemical state of the element. The duration of exposure and the age, sex, nutritional status, immune status, physical activity, and exposure to other toxicants (e.g., alcohol and tobacco) of the individual also affect toxicity.

Probably the most important factor impacting on mineral toxicity is nutrient interaction. In comparison to situations in which diets are replete with essential trace elements, the toxicity from excessive intake of a given element can be greatly enhanced when intake of one or more other elements is marginal or deficient. For example, Hamilton, et al. (125) found greater toxicity from zinc in Japanese quail when dietary copper was marginally deficient than when it was adequate or slightly elevated.

Diets containing nutrients at well above required levels have been used in many animal studies of trace element toxicity. Such diets produce much higher estimates of safe levels for humans than would diets containing only required levels or deficient levels. Animal studies of this type have supported the historical lack of concern about the toxicity of the trace elements but do not reflect the dietary status of many humans today.

Trace element deficiencies (as well as other nutrient deficiencies) are often difficult to recognize in humans since symptoms may be nonspecific and adequate means for detecting trace element status are lacking in many cases. However, it is generally accepted that a significant segment of our population suffers from frank or latent nutrition deficiencies. Trace element toxicity should be considered not only with respect to classic acute and chronic toxicity, but also from the viewpoint of nutrient interactions, and food safety assessment should take into account marginal or deficient trace element (and other nutrient) status. In many cases, perhaps, the current popularity of vitamin and mineral supplements being consumed by individuals with health and/or nutritional status that may enhance the toxicity of the supplements makes the further study of nutrient interactions of fundamental importance.

Little is known about the actual mechanisms of action resulting in toxic effects from ingestion of various levels of the trace elements, and it is likely that each may have more than one pathway by which the metabolism of the host is affected. It is important to emphasize that the mechanisms of toxicity are not necessarily the same as those of essentiality. Mickelson et al. (126) have characterized the toxic effects of the minerals as being related to the inhibition of enzyme activity, substitution resulting in altered enzyme activity, catalyzing the degradation of essential metabolites, and replacement of structurally or electrochemically important elements.

Unlike the situation with vitamin toxicity where intake is limited to oral ingestion or parenteral administration, trace elements may enter the body as a result of air and water pollution and dermatological exposure, as well as through the diet. These latter exposures frequently vary on a geographical basis. Trace element levels in a given region may be naturally high in the soil, thus affecting crops grown there, or in the local water supply. In addition, as noted in Chapter 4, industrial sources may significantly impact on exposure.

In the following, the toxicity aspects of several of the major trace elements are briefly reviewed. (Lead, cadmium, arsenic, and mercury are discussed in Chapter 4.)

Iodine

Adverse effects from the intake of iodine have traditionally been centered around the deficiency disease, iodine deficiency goiter. In fact, seaweed containing high levels of iodine was used by the ancient Chinese, Egyptians, and Incas for treating goiter. Adverse effects can also occur in certain individuals from excessive iodine intake.

Moderate increases in iodine intake (≤1 mg/day) produce little or no effect in individuals with normal thyroid function. However, in individuals with hyperthyroidism, such intakes lead to a prompt inhibition of thyroxine release. Recognition of this effect has lead to the use of iodide in the treatment of hyperthyroidism. Larger intakes of iodide result in a prompt but transient inhibition of thyroxine synthesis in humans without hyperthyroidism, the so-called Wolff-Chaikoff effect. If excessive intake (200 mg/day) is continued for weeks or months, individuals with normal thyroid function usually adapt, but those with thryoid abnormalities do not and become functionally hypothyroid, often with some thyroid enlargment (127). The outcome in any particular case cannot be predetermined.

The existence of endemic goiter has been traced to very high intakes of iodine (≤ 200 mg/day) from seaweed consumed intermittently by residents of certain coastal regions in Japan. Affecting nearly 10% of these residents, the goiters observed were associated with the Wolff–Chaikoff impairment of thyroxine synthesis. Such effects were reversed when seaweed ingestion was halted and could be reinitiated with administration of iodide (128).

Other effects of excess iodine have also been observed. Hyperthyroidism was reported in Tasmania following the addition of iodate to bread to reduce the incidence of endemic goiter in iodine-deficient areas. Such effects occurred primarily among older women with apparently normal thyroid function but with a history of preexisting goiter (129).

The action of excess iodine in producing thyrotoxicosis is not clearly understood. Wolff (130) has described four levels of iodide excess:

1. Low levels, with a temporary positive iodine balance and formation of organically bound iodine.
2. Larger levels, with inhibition of iodine release from the thyroid gland.

3. Slightly larger levels, with inhibition of organic iodine formation and Wolff-Chaikoff effect.

4. Very high levels, with saturation of iodine transport mechanism and acute pharmacological effects.

The 1979 adult RDA for iodine is set at 150 μg/day (2), and a range of intake from 50 μg minimum to 1000 μg maximum has been described as safe by the Food and Nutrition Board (131). Wolff (130) has suggested that 2000 μg/day should be considered a potentially harmful intake level. Usual dietary intakes do not result in harmful levels of iodine ingestion, and although therapeutic use of potassium iodide or iodine-containing medications can induce myxedema and/or iodide goiter, there is remission of these effects when ingestion is terminated.

There are many possible sources for dietary intake of iodine. Several foods contain significant amounts of naturally occurring iodine, and seafoods are especially noted for their iodine content (132). In addition, iodized salt, iodine-containing food additives (e.g., calcium and potassium iodates), sanitizing agents (e.g., iodophors), and the use of the food color FD&C Red No. 3 (erythrosine), further increase dietary intakes. Another source is meat, eggs, and dairy products from animals receiving iodine-supplemented rations or iodine-containing veterinary drugs. Finally, atmospheric iodine may also be contributing to the intake.

The FASEB (132) reviewed iodine intake and toxicity in a report for the FDA and concluded in part that the iodine consumed by individuals in North America had increased in recent years, exceeding the amounts necessary for nutritional needs. (Actually, intakes of some individuals may be manifold in excess of the RDA.) The FASEB also noted that there is no documented evidence of either endemic iodine deficiency or iodine toxicity or hypersensitivity in the United States but recommended continued scrutiny of iodine intake trends.

Iron

Dietary intakes of iron can vary considerably with the consumption of individual foods. However, iron intake from naturally occurring foods and from iron-enriched breads and cereal products are generally sufficient to meet the requirements of normal healthy males and postmenopausal women (10 to 18 mg/day). Iron intake may be marginal or deficient in many menstruating women and adolescent girls, particularly those on weight-loss diets. The Food and Nutrition Board recommends 30 to 60 mg of supplemental iron during pregnancy and continued supplementation for 2 to 3 months after parturition to replenish iron stores depleted during pregnancy (2).

As with most other nutrients, adverse effects of iron are most often related to the deficiency disease, in this case iron-deficiency anemia. In the absence of hemorrhage or other major blood loss, iron-deficiency anemia is progressive, occurring over a period of months or years. Initially overt anemia is absent, followed by a mild normocytic, normochromic phase and ultimately by severe microcytic, hypochromic anemia.

Adverse effects from excess levels of iron arise primarily from more than normal absorption from the intestinal tract. Such excess absorption does not occur with normal dietary intakes by healthy individuals. However, indiscriminate use of iron and combination supplements, especially in children, can produce acute iron toxicity. According to Knott and Miller (133). "Iron preparations and vitamins are the fourth most frequently ingested toxic products in children under five years of age. Some 2,000 cases are reported annually in the United States." Mortality from iron poisoning has ranged up to 50% in the past but has been significantly reduced with improved treatment methods.

In the disease hemochromatosis, the normal mechanisms for limiting iron absorption are impaired; iron is absorbed and widely dispersed in the body, particularly in the liver and pancreas, with associated liver enlargement and functional insufficiency, portal cirrhosis, bronze pigmentation of the skin, diabetes mellitus, and cardiac failure (134). A rare, idiopathic disease, hemochromatosis is generally believed to be produced by an inborn error of metabolism that causes increased absorption of iron. There is uncertainty about the relationship between dietary intakes and iron deposition as well as disagreement as to whether the effects seen are due to the iron per se or are the combined results of the iron with some other nutritional, toxic, metabolic, or infectious abnormality (135).

Excessive iron stores, hemosiderosis, in the body may also result from repeated transfusions, excessive or prolonged iron therapy, or excessive ingestion of iron. The latter has been reported among the Bantu (136), largely from iron cooking pots and iron drums in which their Kaffir beer is made, resulting in iron ingestion 10 to 20 times greater than that from normal diets. Portal cirrhosis occurs in a majority of Bantu patients, and some develop clinical diabetes. The extent to which prevalent malnutrition or coexistent intakes of other toxic substances are factors in these effects is unknown.

Hemochromatosis provided a major issue in a long-standing controversy over the increased levels of iron in enriched flour and bread proposed by the FDA in 1970 (137). That proposal, based on an industry petition, received major impetus from the concern of some public health nutritionists about iron-deficiency anemia. However, objections to the proposal expressed equal or greater concern about exposing patients with hemochromatosis to increased levels of iron in the diet. A

FASEB review (138) of the proposal concluded that the proposed increase of iron would have little or no effect on the accumulation of iron by normal males and that the extra iron per se would not precipitate hemochromatosis or other hereditary iron storage disorders, but could accelerate the course of such diseases. Consultants involved in the FASEB review were not unanimous in their views as to the significance of the proposed increase on the accumulation of iron in the latent stages of these disorders. The question of bioavailability of various iron sources also arose (139), and the FDA finally decided in late 1977 to withdraw the proposal, at least until further research was done on the need, the efficacy, and the safety of increased iron enrichment in flour and bread. In addition to its impact on iron itself, this long-drawn-out controversy will likely impact on other nutrient fortification measures that might be proposed in the future.

Although excessive iron ingestion does not occur from normal dietary intakes by healthy individuals, caution is still warranted. Public familiarity with iron-deficiency anemia has led to the popularity and the wide availability of iron supplements. Excess ingestion of such supplements or excessive parenteral administration can lead to acute symptoms of nausea, vomiting, diarrhea, acidosis, shock, and even cardiovascular collapse and death (140). Further excess ingestion of iron can interfere with the absorption and utilization of copper, zinc, and manganese, and deficient dietary phosphate enhances the toxicity of excess iron (141).

Zinc

Zinc has been recognized as an essential nutrient for nearly a century. Its occurrence in a wide variety of enzymes indicates its important role in protein and carbohydrate metabolism. Body deposits of zinc are apparently not readily mobilized; therefore, regular dietary intake, especially in periods of rapid growth is required. Fortunately, zinc is widely distributed in many different foods, and there is minimal difficulty in meeting requirements from normal U.S. diets (142).

Unlike other nutrients that are usually identified with a predominant deficiency disease, there are numerous zinc deficiency symptoms, as would be expected from its diverse roles in metabolism, its role in protein and RNA synthesis, and its presence in perhaps as many as 100 enzymes. Characteristic manifestations of zinc deficiency in young animals include growth depression, anorexia, skin lesions, and testicular atrophy. Similar characteristics, including retarded growth, marked hypogonadism, and rough and dry skin, were first associated with zinc deficiency in humans by Prasad and co-workers (143). The nutritional importance of zinc is also illustrated by increased fetal mortality and major malformations in the offspring of rats with even short periods of zinc deficiency during pregnancy. Experimental work in other animal models has indicated a broad spectrum of metabolic alterations and clinical symptoms associated with zinc deficiency (143). A comprehensive review of zinc metabolism, biochemistry, physiology, and deficiency is given in the text by Prasad (143).

Historically, food safety problems from zinc were considered much more likely to derive from deficiencies than from excesses, but there have been cases of zinc poisoning after prolonged consumption of water from galvanized pipes and containers (144, 145). In one case, water containing in excess of 40 ppm zinc led to irritability, muscle pain and stiffness, and nausea. Toxicity studies in the rat indicate anemia, depressed growth, and increased mortality at intakes of 5000 to 10,000 ppm, but no effects at 2500 ppm (142). However, relatively little is known about prolonged excessive intake of zinc in humans. We do know that the nutrient content of the diet affects the toxicity of zinc to a considerable degree; intakes of copper, iron, and other dietary constituents such as phytate are important (146). For example, with marginally deficient levels of copper, growth depression, decreased liver concentrations of iron, copper and manganese, and other effects were observed in Japanese quail at supplemental zinc levels as low as 125 ppm in the diet (125). Zinc antagonism of copper has been shown to be enhanced in the presence of supplemental ascorbic acid in the diet (147, 148). Since human dietary intake of copper rarely exceeds the requirement or may be marginally deficient (125), O'Dell's conclusion that zinc supplements should be used judiciously and with professional advice seems plausible (146).

Copper

Like zinc, copper is present in a wide variety of enzymes. Copper is involved in such major roles as hemoglobin synthesis, bone and elastic tissue development, and CNS functioning. Its deficiency can adversely impact on growth and metabolism. There is no reported clinical evidence of primary copper deficiency in human adults, as would be expected, since copper is widely distributed in foods. However, dietary intakes of trace elements including copper may be marginally deficient in many individuals, a factor of importance in zinc toxicity (125). Although uncommon, a deficiency condition similar to copper and iron deficiency anemia in swine on milk diets occurs in some infants and can be corrected by administration of iron and copper.

Acute symptoms of copper poisoning (copper sulfate) include nausea, vomiting, diarrhea, headache, dizziness, and weakness. In more severe cases tachycardia, hypertension, and coma may occur and can be followed by jaundice, hemolytic anemia, hemoglobinuria, uremia, and death. Copper sulfate doses in reported poisoning cases range from about 1 to 100 g (149). Nicholas (150) reported diarrhea and vomiting from 25 ppm copper in tea, and rashes were reported from 7.6 ppm in drinking water (151). In 1970 the WHO/FAO Joint Expert Committee on Food Additives defined 200 mg/kg body weight as a fatal oral dose of copper salts for humans and maintained 0.5 mg/kg body weight as the maximum acceptable daily intake of copper.

A special form of copper toxicosis occurs in Wilson's disease, a rare genetic disease associated with excessive accumulation of copper in the liver, kidney, brain, and cornea. Toxic manifestations of this disease can derive from normal dietary intake levels of copper (2 to 5 mg daily) but are accelerated by larger intake levels. Copper accumulation leads to cirrhosis of the liver, kidney damage, and brain damage, with the characteristic brown or green corneal rings called *Kayser–Fleischer rings.* Wilson's disease is progressive and ultimately fatal unless treated with chelating agents, such as penicillamine, to remove the copper from tissues and promote urinary excretion (152).

An interesting aspect of copper toxicity, evidenced in animal studies, is its dependence on other elements, especially molybdenum and sulfur, whose presence can inhibit copper retention and hence increase tolerance. Copper also mutually interacts with iron and zinc, and combined deficiencies of zinc, copper, and iron can enhance the toxicity of lead and cadmium (149, 153).

Relatively little is known about chronic copper toxicity in humans, but animal evidence would indicate a wide margin between required and chronic toxic levels, perhaps on the order of a twentyfold difference, indicating a human dietary tolerance as high as 200 ppm (on a dry basis) (142).

Fluorine

Although fluorine has not been demonstrated to be an essential nutrient in the conventional sense of the term, it is the subject of considerable nutritional interest, primarily involving the teeth, since deficient intakes have been associated with dental caries and excesses produce discoloration, mottling, and increased brittleness of tooth enamel. Human intake of fluoride can vary considerably, depending on location, but except for exposures to industrial sources and/or naturally or artificially fluoridated drinking water, the primary intake is from food (142). A concentration of 1 to 1.5 ppm of fluorides in drinking water has been recommended as ideal since less than 1 ppm may result in cariogenic effects and more than 2.5 ppm may induce mottling in developing teeth (154). Since 1945 studies in the United States have shown significant impact on dental caries when inorganic fluorides were added to low-fluoride drinking water supplies. The U.S. Surgeon General has concluded that "optimal level fluoridation of drinking water can prevent 65 percent of decay that would otherwise occur" (106).

Except for increased susceptibility to caries, adverse effects from deficiencies in fluorine have generally not been noted, but Leone (155) reported some relationship between low-fluoride content of drinking water and adverse effects on bone structure. In addition, alleviation of osteoporosis has been accomplished by treatment with high levels of fluoride and calcium (156).

The body protects against potentially toxic levels of fluoride by increased urinary excretion and deposition of retained fluoride in the bones. Gradual accumulation with age in humans can lead, with excessive exposure (on the order of 20 to 80 mg daily for 10 to 20 years), to calcification of joints and in advanced cases to emaciation and death (157, 158). This crippling fluorosis is endemic in certain areas of the world with excessively high fluoride in drinking water supplies.

Detectable osteosclerotic lesions have not been reported when water supplies have about 4 to 5 ppm fluoride, and a similar situation holds with respect to occupational fluoride exposure. Thus there is probably about a four- to fivefold safety margin relative to optimal protection against dental caries (1 to 1.5 ppm) and detectable osteosclerotic lesions, and about a fifteen- to twentyfold safety margin relative to optimal levels and those associated with crippling fluorosis.

Manganese

No diseases nor symptoms specifically attributable to manganese deficiency have been identified in humans, and as yet no RDA has been established for this trace element. However, manganese is a dietary essential, and its deficiency in both mammals and birds is associated with a variety of effects, including depressed growth, bone abnormalities, reproductive dysfunction, ataxia, and disturbed lipid metabolism. Its role as a cofactor for enzymes in protein synthesis and energy metabolism has also been established (142).

Although, as with other trace elements, the toxicity of manganese depends on the levels of other nutrients present, it is considered to have a relatively low order of toxicity. There have not been any reported incidents of manganese toxicity in humans from food sources, a fact consistent with the generally low levels of manganese in the diet (tea, which contains on the order of 0.3 to 1.3 mg per cup, is one of the richest sources). The Food and Nutrition Board lists 2.5 to 5 mg/day as a safe and adequate intake for adults (2). This equates to a maximum safe level of slightly over 10 ppm (dry basis) in the diet, which is much lower than the levels showing effects in animals when other nutrients in the diet are adequate. However, it has been hypothesized that manganese interferes with iron absorption (159), and it is known that manganese absorption is increased in iron-deficient animals (160). Reduced growth rate and anemia have been associated with excessive manganese intake, and those with iron-deficiency anemia are likely to be more susceptible to effects from excessive manganese.

Chronic manganese toxicity has been reported among miners working with manganese ores. In these cases manganese oxide dust enters the lungs as well as the gastrointestinal tract, eventually leading to a schizophrenia-like mental disorder and a neurological disorder similar to Parkinson's disease (161).

Manganese provides an excellent example of the importance of nutrient interactions in trace element toxicity. In addition to the enhancement of iron-deficiency anemia by excess manganese, interaction with calcium, zinc, copper, cadmium, and molybdenum have been reported. Manganese may also interfere with the absorption of vitamin B_{12}, and its absorption from the gut is enhanced by the ingestion of alcohol (162).

Cobalt

Cobalt serves its primary known nutritional function in humans as a component of vitamin B_{12} and to be of nutritional value, must be ingested as vitamin B_{12}. Thus the only known deficiency problem is the pernicious anemia, resulting from poor diet or physiological disorders, associated with vitamin B_{12} itself (142).

Cobalt toxicity is classically identified with several instances of severe cardiac failure in heavy beer drinkers. In addition to congestive heart failure, individuals who drink large amounts of beer, containing 1.2 ppm of cobalt salts added as a foam stabilizer, in many cases also developed polycythemia, pericardial effusion, thyroid epithelial hyperplasia, and neurological abnormalities (163). (Subsequently, the FDA specifically prohibited the use of cobaltous salts in fermented malt beverages.) However, the cobalt sulfate (~8 mg) ingested in these cases is well below levels of up to 300 mg, which have been used therapeutically in the treatment of certain anemic patients, without evidence of cardiac effects. Apparently the effects noted in these beer drinkers resulted from a synergism of alcohol and cobalt and could also have been enhanced by protein or thiamin deficiencies (164, 165). In general, cobalt toxicity is virtually impossible from normal diets because of the small amounts found in foods (142).

Selenium

Selenium for many years was primarily noted for its toxicity in grazing animals. Animals feeding on plants that accumulate selenium from soils high in selenium content can evidence toxic effects such as blindness, muscle paralysis, and death from respiratory failure. Chronic effects in such animals also include hair loss, hoof soreness, anemia, cirrhosis of the liver, and cardiac atrophy. Some implication of selenium with toxicity in humans, involving chronic dermatitis, excessive fatigue, and dizziness, has been associated with areas of seleniferous soils (166). However, humans in seleniferous areas have not generally evidenced these toxic symptoms since they do not consume seleniferous plants and human dietary intake is, of course, composed of foods from many different areas other than local. Further, the selenium contents normally found in human foods are reduced in processing and preparation (142).

Other than being an essential component of erythrocyte glutathione peroxidase, relatively little is known about selenium's biochemical role (167). Deficiency symptoms in animals in most cases result from diets deficient in both selenium and vitamin E. There is some evidence that selenium has a sparing effect on vitamin E relative to its antioxidant functions (168), and a great deal of research interest is centered on the complex interactions between selenium, vitamin E, and polyunsaturated fatty acids.

The mechanism entailed in selenium toxicity is not well understood but clearly is associated with the composition of the rest of the diet, protein, sulfate, and arsenic levels as important factors. The Food and Nutrition Board (2) has listed 0.2 mg as the maximum safe daily intake in adults. Even taking into account the likely variation in selenium content, nutritionally balanced human diets would not be expected to provide more than a fraction of toxic levels. However, as in other cases, departure from a balanced diet and/or indiscriminate use of supplements can pose hazards.

Selenium has also been indicted as an animal carcinogen based on rat studies (and thus would meet the current legal definition of a carcinogen), but this has been scientifically challenged (169). Further, there is even some limited epidemiological evidence interpreted to indicate that selenium may inhibit cancer in humans (170).

Chromium

The essentiality of trivalent chromium for humans is based in part on animal evidence that indicates an important role in carbohydrate and lipid metabolism and glucose utilization. Knowledge about the physiological functions of chromium is incomplete, but it is known that chromium activates several enzymes and stimulates fatty acid and cholesterol synthesis (171). Chromium deficiency has primarily been demonstrated in the rat with depressed growth and severely impaired glucose metabolism noted. Metabolic disorders of carbohydrate metabolism responsive to chromium treatment in human adults has also been reported (172).

Interest in chromium has centered more on its physiological role than on its toxic properties. Much of the research is centered around its relationship to adult-onset diabetes and its role in glucose utilization (173). It is known that the absorption of chromium in the intestinal tract varies widely, from less than 1% to more than 20% of a given dose, and that tissue levels tend to decline with age (174). No adverse effects in humans from dietary excesses of chromium have been reported, but animal studies have indicated growth depression and liver and kidney damage from excess chromate. The hexavalent form is much more toxic than the naturally occurring trivalent chromium. Chronic occupational exposure to chromate dust has been associated with an increased incidence of human lung cancer (175).

There is a dearth of information on the long-term effects of excessive chromium ingestion, but excessive intakes can be expected to contribute to imbalances in other trace elements and should be avoided (173).

Molybdenum

Nutritional interest in molybdenum lies mainly with its enzyme activity, particularly its role in the xanthine oxidase activity of tissues. Molybdenum is also a constituent of various flavin-dependent enzymes. No characteristic deficiency symptoms have been identified, but dietary deficiencies do decrease xanthine oxidase activity. Iron and molybdenum (and perhaps other metals) are important constituents of xanthine oxidase and aldehyde oxidase enzymes, but knowledge about their precise function is incomplete (176). Molybdenum is rapidly absorbed from the intestinal tract and rapidly excreted in urine.

Somewhat like selenium, high levels of molybdenum in pasture soils can induce toxicosis in grazing cattle ranging from diarrhea to debilitation and death. This toxicity in cattle is increased when the copper content of the soil is low, and copper sulfate is customarily used to treat the molybdenosis. Interestingly, other species have a much higher tolerance for molybdenum than do cattle. Toxicity signs in experimental animals include weight loss, anorexia, and anemia, which are also characteristic of copper deficiency. Protection against molybdenum toxicity can be achieved by feeding them additional copper (177). By the same token, molybdenum and sulfate have been shown to be effective against copper toxicity in animals (178). The interaction between molybdenum and manganese has also been noted (162).

There is little information relating molybdenum to toxicity in humans. Molybdenum has been suggested as a factor in gout and multiple sclerosis. A high incidence of gout accompanied by elevated blood and urinary molybdenum was noted in an Armenian region having high levels of soil and plant molybdenum (179). Layton and Sutherland (180) have suggested a relationship between high-risk areas for multiple sclerosis and high molybdenum: copper ratios in the soil of those areas.

Dietary intakes of molybdenum by humans are quite low, perhaps on the order of 0.1 mg/day (181), which is consistent with the safe and adequate range of 0.15 to 0.5 mg listed by the Food and Nutrition Board (2). Such intake levels, coupled with the rapid excretion of molybdenum and protective effects of copper, indicate a lack of hazard from normal dietary sources. However, the Food and Nutrition Board has cautioned against habitually exceeding the upper levels of intake listed for the trace minerals.

SAFETY OF THE DIET AS A WHOLE

The preceding review of individual nutrients indicates that consumption of a varied and balanced diet does not present significant food safety problems for normal, healthy individuals. As we have seen, safety problems are invariably associated with particular deficits or excesses of individual nutrients or combinations of nutrients. The remaining safety question to be addressed is that of the dietary practices currently followed by Americans, the focus of a great deal of criticism by some and a subject of interest to almost everyone. To put the nutritional aspects of food safety in perspective, it is of value to examine overall dietary trends and the current U.S. diet and diet modifications, including fad diets as well as the recommended national dietary goals and dietary guidelines.

Dietary Trends

By most commonly accepted criteria, the U.S. food supply can be considered safe and nutritious—even though its nature has been changing for some time, a characteristic considered by some as detrimental to health. It is important to examine those changes, and this can be done in part by utilizing U.S. food disappearance data collected by the USDA. These disappearance data represent deliveries to food processors, restaurants, and retail outlets. Because of waste, spoilage, and other losses, actual consumption may be as much as 25% less, depending on the food commodity involved. Available USDA disappearance data permit an examination of trends in estimated consumption and hence estimated nutrient intakes from the beginning of the century (1909 to 1913) to 1976. Although most major commodities have remained a stable part of the food supply, there have been several significant changes in this period. One such change is in estimated meat consumption. Per capita meat disappearance, after a decline in the 1920s, rose from 141 pounds in 1909 to 1913 to 165 pounds in 1976. In the same period poultry disappearance increased from 18 to 53 pounds per capita and rose further to 62 pounds in 1979. Part of the change in meat consumption can be associated with the use of ground beef in fast-food outlets, but more important has been the apparent consumer desire and ability to pay for meat. Most of the increase in poultry disappearance has come since the 1950s, when prices became competitive with meat and fast-food uses began to increase.

During this century per capita disappearance of flour and cereal products decreased markedly from 291 to 140 pounds annually. Potato (white and sweet) disappearance also decreased markedly from 205 to

80 pounds. Dairy products (other than butter) have become increasingly available, primarily because of the growing popularity of cheese. Per capita disappearance of fruits and vegetables (other than potatoes) was similar in 1976 to that seen earlier in the century, but there have been shifts from fresh to frozen and canned, a nutritionally positive change from the point of view of availability throughout the year in all parts of the country.

More than with any other commodity, the trends in consumption of sugar have received attention. The data in this case can be misleading if not examined carefully. From 1909 to about 1930 sugar disappearance increased significantly, then fell back, and since 1935 has been fairly constant, except for the decline during World War II. During the 1970s corn sweeteners, especially high-fructose corn syrups, began to a significant degree to replace some of the sugar used in processed foods. The major change relative to sugar during this century has been in the source of ingestion. Early in the 1900s food was usually prepared at home with sugar being added at that point. Later, lifestyle changes led to greater use of processed foods with sugar already incorporated.

With respect to the nutrient content of the diet, one of the most noteworthy changes has been the increase in fat intake. Although part of this increase is associated with increased meat and poultry consumption (and hence animal fat), vegetable fat usage has increased substantially with increases in margarine, vegetable shortening, and salad and cooking oils. These increases have more than compensated for decreased lard and butter consumption and more recent decreases in fluid whole-milk consumption. Since fat provides more than twice as many calories per gram than does protein or carbohydrate, these changes indicate an important shift in the sources of caloric intake.

Changes in vitamin and mineral intake in the last 70 years include the increases in intakes of iron and the B vitamins, provided by the enrichment and fortification of breads and cereal grain products. These steps have resulted in the virtual elimination of the B vitamin deficiency diseases, beriberi and pellagra. Increases in vitamin C intake and the introduction of vitamin D fortified milk and iodized salt have similarly essentially eliminated scurvy, rickets, and iodine-deficiency goiter, respectively.

A somewhat surprising dietary trend is the gradual decrease in total caloric intake from food. As Forbes (182) has pointed out, total caloric intakes are now much lower than they are generally thought to be, peaking at about 2800 calories/day for 18-year-old males and at about 1875 for 10-year-old females (median values). Both the HANES I (Health and Nutrition Examination Survey) data (1971 to 1974) and USDA data (1978 to 1979) indicate this decreasing trend in calorie intakes. Yet, nutrient intakes for the overall population have generally remained above RDA levels, indicating a favorable nutrient density for the current U.S. food supply.

The Current U.S. Diet

From the preceding, it is obvious that the U.S. diet is generally, perhaps surprisingly, low in calories and relatively high in nutrient density. Yet there is a great deal of criticism of the diet and a variety of attempts to change it, in some cases in fairly significant ways. Is the U.S. diet safe? Like other food safety questions, that question does not have a simple answer. Much, of course, depends on the variable practices of individuals. However, practically, we can only look at the diet of the overall population.

Criticism of the overall U.S. diet centers primarily on such constituents as sugar, salt, fat, cholesterol, and the poorly understood class of substances referred to as "food additives." Each has been publically associated with one or more of the major diseases such as obesity, hypertension, heart disease, and cancer. However, any proper evaluation of our diet must start from a broader perspective by first looking at the health status of Americans, the trends in disease incidence, and the risk factors associated with major diseases.

According to the 1979 report of the U.S. Surgeon General, "The health of the American people has never been better" (106). In this century the annual death rate has nearly been halved. Infectious and communicable diseases no longer have their formerly characteristic life-threatening potential. Life expectancy at birth has increased from 47 years in 1900 to 73 years in 1979; in the last 10 years, life expectancy has increased by 2.7 years (106).

Progress against infectious diseases and the near eradication of nutritional deficiency diseases have substantially contributed to longer life spans. Increased life spans have resulted in a larger group of older Americans who inevitably are more susceptible to chronic degenerative diseases such as heart disease, stroke, and cancer, which have become the leading causes of death. However, as previously noted and contrary to popular belief, the death rate for heart disease, the leading cause of death, has declined by 22% between 1968 and 1977 (106). Stroke, the third leading cause of death, has also had a decline in death rate, by 32% in that same period.

Cancer, the second leading cause of death, is more complicated to evaluate because it is a collection of diseases, each with its own incidence rates and death rates, which differ between sexes and between whites and nonwhites. For example, for men, lung and prostate cancer incidences have increased since 1947, but cancer of the rectum has decreased slightly and stomach cancer has decreased markedly since 1947. For women, marked increases have been seen since 1947 in lung cancer incidence and marked decreases in cancer of the stomach and the cervix (106). With respect to overall cancer death rates, the U.S. Surgeon General has noted:

> While the number of cancer victims has increased dramatically in the past 40 years, much of the increase is due to population growth. When changes in the size and age composition of the American population are taken into consideration, overall cancer death rates have increased only slightly for men since 1937 and have actually decreased slightly for women.

Actually, if lung cancer is separated out, cancer death rates have decreased for both sexes. Although there is a popular tendency to do so, cancer is far too complex to be simply linked to the current U.S. diet. Some appreciation of this fact can be gained by reviewing examples of the vast number of papers, books, and symposia on the subject (183–187).

The complexity of attempting to determine the risk factors associated with major diseases is amply illustrated by the 1979 Symposium on the Evidence Relating Six Dietary Factors to the Nation's Health sponsored by the American Society for Clinical Nutrition (188). This symposium treated:

- Dietary cholesterol in relation to arteriosclerotic disease.
- Saturated and unsaturated dietary fat in relation to arteriosclerotic disease.
- Carbohydrate and sucrose in relation to arteriosclerotic disease, diabetes, and dental caries.
- Alcohol consumption in relation to liver disease and arteriosclerotic disease.
- Excess calories in relation to obesity, hypertension, diabetes, and arteriosclerotic disease.
- Dietary sodium in relation to hypertension.

A task force of nine scientists evaluated the evidence of association in each case and also assigned a numerical score (from zero for no evidence to 100 for "rock solid" evidence). The highest median score (88) was for alcohol and liver disease and the lowest (11) for carbohydrate and atherosclerosis. Greatest variability in numerical scores occurred for cholesterol and arteriosclerotic disease. This evaluation amply illustrates the drawbacks in attempting to identify simple dietary causes for major diseases.

Proper evaluation of the U.S. diet must take into account the variety of, and interaction among, the diverse risk factors associated with major diseases. It should first be noted that health hazards vary greatly as a function of age from developmental problems of infants, to the accidents of youth, to the chronic diseases of adults. Further, most major health hazards are associated with the complex interaction of a variety of risk factors.

The U.S. Surgeon General's report (106) classifies risk factors into three categories: (*a*) inherited biological, (*b*) environmental, and (*c*) behavioral. In addition to its role in clearly heritable diseases, genetic makeup determines future biological characteristics that can greatly influence disease development. The role of environmental factors, which include the physical, socioeconomic and family environments, is of major importance in disease development. Contamination of air, water, and food, together with occupational exposure to toxic materials and the stresses of modern life, can interact synergistically to induce high risk in many disease states. Finally, one's personal behavior can also influence disease development. As a matter of fact, most of the leading causes of death could be reduced through improved personal habits with respect to smoking, poor diet, excessive alcohol intake, lack of exercise, and improper use of medication (106).

In evaluating the relative contribution of these risk factors to the 10 leading causes of death, the U.S. Surgeon General emphasized the implications of an analysis that attributed half of the mortality to unhealthy behavior or lifestyle, 20% to environmental factors, 20% to human biological factors, and 10% to health care inadequacies. Smoking, occupational hazards, alcohol and drug abuse, and injuries are specifically singled out for major attention (106). It would thus seem that the major risk relative to food safety is the departure from and not the adherence to the U.S. diet.

Diet Modifications

Historically, humans have been concerned with getting enough to eat and surviving as long as possible. In today's affluent society, however, that is not enough. We are taught to worship the body beautiful, and many will believe almost any claim that might help achieve it. We can afford to take for granted, or question, our convenient, highly-processed food supply. Our diet has changed from one prepared at home—from what we at least thought were familiar ingredients—to one consisting largely of prepared foods, with labels indicating unfamiliar sounding ingredients. The pre-prepared foods may actually be better than what grandma prepared, but nostalgia tells us otherwise.

Many of us are thus led as individuals to try the latest fad diet or experiment with a wide variety of dietary supplements. Collectively, we are debating not only the need for, but also the details of national dietary goals or guidelines. Both of these areas are important to our discussion of food safety.

Fad Diets

One of the main sources of confusion about today's diet is the plethora of nutrition misinformation that far and away exceeds sound nutritional advice. It seems that out of emotional or psychological impetus, too many Americans are foregoing nutritional science in favor of the latest fads. The concern of many nutritionists in this regard is illustrated by Harper (118):

> The worst of the misinformation, and the most widely read and quoted, does not come mainly from magazine and television advertising of food products .... It comes from food supplement promoters, from authors who earn a living by selling sensational nutrition information and from a wide variety of pseudonutrition experts.

Unfortunately, too little is known by the public about nutrition, and there is still too little effort being devoted to nutrition education in primary and secondary schools. Thus most of the food information to which we are exposed relates to fads rather than sound nutrition. One of the few attempts to argue against current fads for the lay reader is that of Stare and Whelan (189), who have formulated a list of ways on "How to Spot a Quack":

1. They always have something to sell; a course of lectures, pills, nature foods, tonics, food supplements, diet plans, rollers (to roll off the fat), books, even pots and pans.

2. They guarantee quick cures—on a money-back basis, "if your arthritis isn't cured in eight days."

3. They often claim to be "medical experts" or "nutritionists" with some secret formula or knowledge that will cure, and they usually boast about membership in a "scientific society" with a high-sounding name.

4. They use testimonials and case histories to prove that their product is miraculous, not facts based on carefully controlled studies, published in reputable medical journals and confirmed by independent workers.

5. They distort scientific data to suit their own ends.

6. They say that what you are doing now is deadly—the food you are eating is poisoned, the way you cook it is all wrong.

7. They claim to be persecuted by medical men or governmental agencies, because, they insist, the medical profession and the Food and Drug Administration are corrupt and influenced by big business.

In addition to damage to the pocketbook by many fad diets, the potential hazards associated with radical departures from normal diets (such as the ingestion of large excesses of vitamins and minerals) have been discussed previously in this chapter. So, too, have the serious effects associated with the use of certain liquid protein preparations in weight-reduction diets. Although apparently frequently forgotten by some of the lay public, the best advice is still to try to achieve a balanced dietary intake from a variety of foods, to use caution in undertaking radical changes in diet, and to avoid excessive dietary supplements of vitamins and minerals except when there is evidence of actual need. With respect to weight reduction and weight maintenance, the best advice is the simple statement in the USDA/HEW Dietary Guidelines:

> It is not well understood why some people can eat much more than others and still maintain normal weight. However, one thing is definite: to lose weight, you must take in fewer calories than you burn. This means that you must either select foods containing fewer calories or you must increase your activity—or both (114).

Dietary Goals and Guidelines

There are a number of nutrition-related activities, occurring on the national level, that have been prompted by the associations between certain components of the diet and major disease states. Such activities have helped expand the debate concerning diet-disease links, which in large measure was initiated with the formulation of the "Dietary Goals for the United States" (120).

In 1977 the then Select Committee on Nutrition and Human Needs of the U.S. Senate released their staff report (120) containing six "U.S. Dietary Goals" as follows:

1. Increase carbohydrate consumption to account for 55 to 60% of the energy (caloric) intake.
2. Reduce overall fat consumption from approximately 40 to 30% of energy intake.
3. Reduce saturated fat consumption to account for about 10% of total energy intake, and balance that with polyunsaturated and monounsaturated fats, which should account for about 10% of energy intake each.
4. Reduce cholesterol consumption to about 300 mg/day.
5. Reduce sugar consumption by about 40% to account for 15% of total energy intake.
6. Reduce salt consumption by about 50 to 85% to approximately 3 g/day (more recently increased to 5 g/day added to food).

The framework within which the goals were placed implied that they were a generally-agreed-on prescription for the prevention of major diseases. However, release of the recommendations generated a variety of disagreement. In one of the many critiques written, Harper (118) concluded:

> The dietary goals report is not scientifically sound: it is a political and moralistic document. It will appeal to those who accept pseudoscientific reasoning about the wisdom of returning to the diet of last century and to that of the peasant of poor countries. Back to nature movements have occurred regularly throughout history when the problems to be solved were complex and solutions for them were not readily attainable.

The American Medical Association (AMA) noted that diets cannot be standardized but must reflect individual needs and thus termed the dietary goals inappropriate. The AMA also noted (190):

> The evidence for assuming the benefits to be derived from the adoption of such universal dietary goals as set forth in the Report is not conclusive and there is a potential for harmful effects from a long term dietary change as would occur through adoption of the proposed national goals.

Proposals aimed at improving health status by diet modification have also been made in Sweden, Norway, Canada, and the United Kingdom (191, 192). In general, these proposals are more modest than the U.S. dietary goals, but they still imply a causal connection between dietary and disease trends.

In early 1980, the USDA and HEW jointly issued *Nutrition and Your Health—Dietary Guidelines for Americans* (114). These guidelines avoid the rigid quantification of the Dietary Goals and instead list in priority order:

1. Eat a variety of foods.
2. Maintain ideal weight.
3. Avoid too much fat, saturated fat, and cholesterol.
4. Eat foods with adequate starch and fiber.
5. Avoid too much sugar.
6. Avoid too much sodium.
7. If you drink alcohol, do so in moderation.

Also in early 1980 the Food and Nutrition Board, as part of the new RDAs, addressed "desirable amounts and proportions of dietary fat and carbohydrate" and "safe and adequate intakes of sodium, potassium and chloride" (2). The board recommended that total fat intake, particularly in diets below 2000 calories, should be reduced so fat is not more than 35% of dietary energy, the reduction preferably coming predominantly in animal fats, but with an upper limit of 10% of dietary energy as polyunsaturated fatty acids. Also recommended was a reduction in intake of refined sugar and maintenance or increase in complex carbohydrate intake. Maximum salt intake would be limited to 8 to 9 g/day for adults (2).

Despite considerable disagreement with the U.S. dietary goals and some disagreement with other guidelines, there probably is a consensus that some guidelines for a healthful diet would be useful. Such guidelines could be formulated and supported if they were flexible enough to reflect individual needs, were based on a consensus of the scientific community, and were not dependent on the search for a simple panacea for major chronic diseases (192). Although the RDAs are not designed for such guideline purposes, they do provide an example of what can be done by careful, periodic review of existing scientific evidence in the nutrition area. The classic food group concept whether expressed as the basic four or the newer five food groups of the USDA, is another example of the possibility of developing dietary guidelines that would be a useful tool for dietary planning. Along with any such approach, however, an increased emphasis on nutrition education is a basic requirement.

Fortification of Foods

A final area of importance in evaluating the safety of the diet as a whole is that of fortification—the addition of nutrients to foods.

The earliest example of a nutrient addition to a food in the United States was the iodine added to table salt in the 1920s (193). This simple process virtually eliminated the widespread problem of iodine-deficiency goiter in areas such as the Great Lakes, where the local food and water contained negligible amounts of iodine.

The second example of fortification occurred in the 1930s, when vitamin D was added to milk (194). Children need adequate amounts of vitamin D from birth through development, to build strong and straight bones. Milk was chosen as the carrier for vitamin D because it is widely consumed by infants and children and contains the calcium and phosphorus, which work with vitamin D in skeletal development. Such fortification has largely eradicated the crippling disease, rickets.

The success of these two measures and the development of methods of isolating and manufacturing nutrients has brought about a new approach to nutrition and a change of philosophy from passive identification of deficiency diseases to active prevention.

During the 1930s, when margarine became available to the U.S. public, it was widely used as an inexpensive substitute for butter (195). Carotene was added to margarine to supply comparable vitamin A levels as well as an appealing color. This was a successful nutrition decision and has also been applied to milk, cheese, and other dairy products.

In May of 1941 the U.S. government established World War II food order number I (196). This order required the addition of thiamin, niacin, and iron to the flour used to make white bread and rolls. The addition was termed *enrichment;* it is still used to refer to bread, rice, hot cereals, pasta, and other refined grain products to which three B vitamins (thiamin, niacin, and riboflavin) and iron are added at specified levels.

In determining the nutrients with which to fortify cereal grain products or other foods, most manufacturers follow the principles outlined by the National Academy of Sciences (NAS) in 1974 (197):

1. The intake of the nutrient is below the desirable level in the diets of a significant number of people.
2. The food used to supply the nutrient is likely to be consumed in quantities that will make a significant contribution to the diet of the population in need.
3. The addition of the nutrient is not likely to create an imbalance of essential nutrients.
4. The nutrient added is stable under proper conditions of storage and use.
5. The nutrient is physiologically available from the food.
6. There is reasonable assurance against excessive intake to a level of toxicity.
7. The additional cost should be reasonable for the intended consumer.

These guidelines have prevented the misuse of fortification and what some nutritionists fear most in allowing nutrients to be added to foods, namely, a "horsepower race" that would result in the indiscriminant addition of nutrients to foods for marketing reasons.

After an extensive review period, the FDA in 1980 issued a fortification policy consistent with the NAS guidelines (198). The FDA document is not a regulation, but rather a statement of what the FDA considers to be "a uniform set of principles that will serve as a model for the rational addition of nutrients to foods." There are basically four situations when it is considered appropriate by the FDA to add nutrients to foods: (*a*) to correct a recognized dietary insufficiency, (*b*) to restore nutrients to levels found in foods prior to storage, handling, and processing, (*c*) to balance vitamin, mineral, and protein content of fabricated foods in proportion to calories, and (*d*) to avoid nutrition inferiority in a food that replaces a traditional food.

Several conditions are attached to these policy recommendations. Foods and nutrients covered by separate regulations are exempted by this policy. Nutrients added to foods should be stable under customary conditions of storage, distribution, and use and should be physiologically available. The levels of nutrients added should be such as to avoid excessive intake from all sources in the diet. Labeling provisions for fortified foods are also provided in the policy statement. The FDA states that it is inappropriate to fortify fresh produce; meat, poultry, or fish products; and sugars or snack foods.

In the case of dietary insufficiencies, nutrient fortification would be subject to review by the FDA to affirm the dietary need, the nutrients and the levels to be used, and the suitability of the selected food item. Restoration of lost nutrients would be limited to those originally present in amounts equal to at least 20% of the U.S. RDA. Fortification of fabricated foods that replace significant portions of the total diet (one serving provides at least 40 calories) should be such as to provide 5% of the U.S. RDA per 100 calories for each of a profile of essential nutrients required to sustain a balance in overall nutrient intake. Fortification of substitute foods to avoid nutritional inferiority would be such as to avoid any reduction in the level of an essential nutrient (except fat or calories) present in the original food in a measurable amount.

The FDA policy provides a logical framework for the rational nutrient fortification of the American food supply and as such will likely be adhered to by the food industry.

Darby (199) has recommended international guidelines for the addition of nutrients to foods. This is particularly appropriate in light of the fact that completely new foods are being developed, including formulated or fabricated foods that could have nutritional values equal to or greater than conventional foods. When these foods are used in place of conventional foods that make a significant nutrient contribution, it is important that they also provide a comparable nutrient benefit. A product designed as a meal or meal replacement would be required to provide "25 to 50% of the National Standard Allowance" (199).

The amount of a nutrient to add or the number of nutrients to add depends on the intended use of the product, the nutrient content of a comparable conventional product, the nutrient content of the unprocessed food or its ingredients, and the specific needs of the population consuming the food.

The principles are basically the same in any fortification program, but consistency in practice is made difficult on an international level because of the extreme variations in foods and eating habits. Thus a major advantage of international guidelines would be the easing of food export and import problems.

Nutrient fortification in the U.S. diet may well become an even more important consideration in the near future. If present U.S. trends toward decreasing caloric intake continue, it will become more and more difficult to meet individual nutrient requirements. This is the case since the requirements for most nutrients are independent of caloric intake (e.g., iron, zinc, or vitamin C requirements remain the same regardless of caloric intake). Forbes (182) has noted the tendency of western peoples to become "incredibly sedentary," a fundamental truth, thus indicating continuance of the trend toward reduced caloric intake. Further, as the U.S. population continues to age and more and more fall in the lower-caloric-intake categories, the nutrient density of the U.S. diet and hence fortification will likely become more of an issue.

CONCLUSIONS

The relatively well known classic diseases and related adverse effects associated with deficient intakes of the essential nutrients illustrate the basic importance of nutritional hazards in food safety. That importance is amplified by the less well known, but nonetheless significant, toxicity produced by excessive intakes of individual nutrients. Furthermore, it is becoming increasingly evident that the complex interactions among nutrients can exacerbate the effects of both deficient as well as excessive nutrient intakes. Interactions between various drugs or excessive alcohol and nutrient intakes present still another facet of nutritional hazards. Fortunately adequate diets present no significant hazards to normal, healthy individuals. However, misuse of individual foods, drastic departures from normal diets for weight loss or other purposes, and excessive ingestion of nutrient supplements in the absence of demonstrated need all can pose substantial hazards.

Although the U.S. food supply has in some respects changed significantly since the beginning of the century, it can be considered safe and nutritious by most commonly accepted standards. Estimated overall average nutrient intake for the U.S. population indicates a surprisingly low level of calories and a relatively high nutrient density. Yet many Americans are overweight because of a sedentary lifestyle, and many have marginally deficient nutrient intakes.

During this century the health status of Americans has improved considerably. Infectious and communicable diseases no longer have their former characteristic mortality. Nutritional deficiency diseases have, in large measure, been eradicated. The annual death rate has nearly been halved, which has led to a larger group of older citizens inevitably more susceptible to chronic, degenerative diseases. In spite of this, the death rate from heart disease, stroke, and most forms of cancer has decreased in recent years. Yet concern about the major diseases has led to seeking a simple solution for their prevention, and many have focused on diet modification as the hoped-for answer. This focus has enhanced the appeal of fad diets and generated proposed national dietary goals and guidelines. Actually, most major health hazards result from the complex interactions of multiple factors, with diet as only one.

The nature of nutritional hazards is such that the best available advice is moderation—eat a variety of foods, avoid excess calories from any source, and avoid overconsumption of any food or nutrient. For those at high risk of major diseases, dietary advice from a qualified source is a prudent course of action. Drastic changes in diet including the excess ingestion of nutrient supplements should be undertaken only when the need to do so has been determined and guidance from a qualified source has been obtained.

REFERENCES

1. H. R. Roberts, *Fed. Proc.*, 37 (12), 2575 (1978).
2. Food and Nutrition Board, *Recommended Dietary Allowances*, 9th ed., National Academy of Sciences, Washington, D.C., 1980.
3. Office of the Federal Register, *Code of Federal Regulations*, Title 21: Food and Drugs, U.S. Government Printing Office, Washington, D.C., 1979.
4. R. S. Goodhart and M. E. Shils, Eds., *Modern Nutrition in Health and Disease*, 5th ed., Lea and Febiger, Philadelphia, 1973.
5. M. V. Krause and L. K. Mahan, *Food Nutrition and Diet Therapy*, 6th ed., Saunders, Philadelphia, 1979.
6. M. D. Muenter, H. O. Perry, and J. Ludwig, *Am. J. Med.*, 50, 129 (1971).
7. C. Nieman and H. J. K. Obbink, *Vitamins and Hormones*, 12, 69 (1954).
8. C. N. Pease, *J. Am. Med. Assoc.*, 182, 980 (1962).
9. A. G. Knudsen and P. E. Rothman, *Am. J. Dis. Child.*, 85, 316 (1953).
10. T. Moore, "Pharmacology and Toxicity of Vitamin A," in W. H. Sebrell and R. S. Harris, Eds., *The Vitamins*, 2nd ed., Academic, New York, 1967, pp. 280–294.
11. A. Gerber, A. Raab, and A. Sobel, *Am. J. Med.*, 16, 729 (1954).
12. J. Marie and G. See, *Am. J. Dis. Child.*, 87, 731 (1954).
13. W. K. Woodard, L. J. Miller, and O. Legant, *J. Pediatr.*, 59, 260 (1961).
14. T. K. Oliver, Jr., *Am J. Dis. Child.*, 95, 57 (1958).
15. G. Bartolozzi, G. Bernini, L. Marianelli, and E. Corvaglia, *Rev. Clin. Pediatr.*, 80, 231 (1967).
16. E. Rubin, A. L. Florman, T. Degnan, and J. Diaz, *Am. J. Dis. Child.*, 119, 132 (1970).
17. R. J. DiBenedetto, *J. Am. Med. Assoc.*, 201, 700 (1967).
18. J. T. Dingle and J. A. Lucy, *Biol. Rev.*, 40, 422 (1965).
19. S. Bazin and A. Deluanay, *Ann. Inst. Pasteur,* 110, 487 (1966).

20. Federation of American Societies for Experimental Biology, *Evaluation of the Health Aspects of Vitamin A, Vitamin A Acetate, and Vitamin A Palmitate as Food Ingredients (Hearing Draft)*, Bethesda, Maryland, 1979.
21. American Academy of Pediatrics, *Pediatrics*, 48, 655 (1971).
22. H. G. Day, "Vitamin A," in National Nutrition Consortium, *Vitamin-Mineral Safety, Toxicity and Misuse*, American Dietetic Association, Chicago, 1978, pp. 2–4.
23. Federation of American Societies for Experimental Biology, *Evaluation of the Health Aspects of Vitamin* D_2 *and Vitamin* D_3 *as Food Ingredients*, Bethesda, Maryland, 1978.
24. H. I. Chinn, "A Review of the Adverse Effects of Excessive Intakes of Vitamin D," LSRO Report to FDA, Washington, D.C., 1979.
25. J. L. Omdahl and H. F. DeLuca, "Vitamin D," in R. S. Goodhart and M. E. Shils, Eds., *Modern Nutrition in Health and Disease*, Lea and Febiger, Philadelphia, 1973, pp. 158–165.
26. M. S. Seelig, *Ann. N. Y. Acad. Sci.*, 147, 537 (1969).
27. J. Anderson, C. Harper, C. E. Dent, and G. R. Philpot, *Lancet*, 2, 720 (1954).
28. J. E. Howard and R. J. Meyer, *J. Clin. Endocrinol. Metab.*, 8, 895 (1948).
29. Committee on Nutrition, *Pediatrics*, 31, 512 (1963).
30. A. Berio and P. Moscatelli, *Minerva Pediatr.*, 19, 972 (1967).
31. R. H. Folis, *Am. J. Clin. Path.*, 26, 400 (1956).
32. G. DeLuca and M. Cozzi, *Minerva Pediatr.*, 16, 210 (1964).
33. P. Davies, *Ann. Intern. Med.*, 53, 1250 (1960).
34. Editors, *Nutr. Rev.*, 37, 323 (1979).
35. Food and Nutrition Board, *Nutr. Rev.*, 33, 61 (1975).
36. P. M. Farrell and J. G. Bieri, *Am. J. Clin. Nutr.*, 28, 1381 (1975).
37. D. K. Melhorn and S. Gross, *J. Lab. Clin. Med.*, 74, 789 (1969).
38. Federation of American Societies for Experimental Biology, *Evaluation of the Health Aspects of the Tocopherols as Food Ingredients*, Bethesda, Maryland, 1975.
39. J. J. Corrigan, Jr. and F. I. Marcus, *J. Am. Med. Assoc.*, 230, 1300 (1974).
40. A. L. Tappel, *Nutr. Today*, 8 (4), 4 (1973).
41. L. A. Witting, *Am. J. Clin. Nutr.*, 25, 257 (1972).
42. R. E. Olson, *The Fat Soluble Vitamins*, Univ. Wisconsin Press, Madison, Wisc., 1970.
43. R. E. Olson, "Vitamin K," in R. S. Goodhart and M. E. Shils, Eds., *Modern Nutrition in Health and Disease*, 5th ed., Lea and Febiger, Philadelphia, 1973, pp. 166–174.
44. C. A. Owen, Jr., "Vitamin K Group. XI. Pharmacology and Toxicology," in W. H. Sebrell and R. S. Harris, Eds., *The Vitamins*, 2nd ed., Vol. 3, Academic, New York, 1971, pp. 492–509.

45. K. C. Hayes and D. M. Hegsted, "Toxicity of the Vitamins," in National Academy of Sciences, *Toxicants Occurring Naturally in Foods,* 2nd ed., Washington, D.C., 1973, pp. 235–253.

46. H. G. Day, "Vitamin K," in National Nutrition Consortium, *Vitamin-Mineral Safety, Toxicity and Misuse,* American Dietetic Association, Chicago, 1978, pp. 11–12.

47. I. Clark and C. A. L. Bassett, *J. Exp. Med.,* 115, 147 (1962).

48. J. T. Matschiner and E. A. Doisy, Jr., *Proc. Soc. Exp. Biol. Med.,* 109, 139 (1962).

49. L. W. McCuaig and I. Motzok, *Poultry Sci.,* 49, 1050 (1970).

50. D. G. Hazard, C. G. Woelfel, M. C. Calhoun, J. E. Rousseau, Jr., H. D. Eaton, S. W. Nielsen, R. M. Grey, and J. J. Lucas, *J. Dairy Sci.,* 47, 391 (1964).

51. M. Cantin, J. M. Dieudonne, and H. Selye, *Exp. Med. Surg.,* 20, 318 (1962).

52. P. Griminger, *J. Nutr.,* 87, 337 (1965).

53. L. Pauling, *Vitamin C and the Common Cold,* W. H. Freeman, San Francisco, 1970.

54. T. W. Anderson, D. B. W. Reid, and G. H. Beaton, *Canad. Med. Assoc. J.,* 107, 503 (1972).

55. T. W. Anderson, *Ann. N. Y. Acad. Sci.,* 258, 513 (1975).

56. J. L. Coulehan, L. Kapner, S. Eberhard, F. H. Taylor, and K. D. Rogers, *Ann. N. Y. Acad. Sci.,* 258, 513 (1975).

57. E. Greer, *Med. Times,* 83, 1160 (1955).

58. W. F. Korner and F. Weber, *Int. J. Vit. Nutr. Res.,* 42, 528 (1972).

59. M. P. Lamden and G. A. Chrystowski, *Proc. Soc. Exp. Biol. Med.,* 85, 190 (1954).

60. S. N. Gershoff, *Metabolism,* 13, 875 (1964).

61. P. C. Lee, J. R. Ledwich, and D. C. Smith, *Canad. Med. Assoc. J.,* 97, 181 (1967).

62. C. H. Hill and B. Starcher, *J. Nutr.,* 85, 271 (1965).

63. V. Herbert, E. Jacob, and K. J. Wong, *Am. J. Clin. Nutr.,* 30, 297 (1977).

64. C. A. Owen, G. M. Tyce, E. V. Flock, and J. T. McCall, *Mayo Clinic Proc.,* 45, 140 (1970).

65. G. N. Schrauzer and W. J. Rhead, *Int. J. Vit. Nutr. Res.,* 43, 201 (1973).

66. G. N. Schrauzer, D. Ishmael, and G. W. Kiefer, *Ann. N. Y. Acad. Sci.,* 258, 377 (1975).

67. W. A. Cochrane, *Canad. Med. Assoc. J.,* 93, 893 (1965).

68. C. G. King, *Annals N. Y. Acad. Sci.,* 258, 540 (1975).

69. Staff Report, *Nutr. Today,* 14 (5), 6 (1979).

70. H. G. Day, "Vitamin C," in National Nutrition Consortium, *Vitamin-Mineral Safety, Toxicity and Misuse,* American Dietetic Association, Chicago, 1978, pp. 13–18.

71. T. Spies, R. Hillman, S. Cohlan, B. Kramer, and A. Kanof, "Vitamins and Avitaminoses," in G. Duncan, Ed., *Diseases of Metabolism,* Saunders, Philadelphia, 1959, pp. 142–159.
72. M. K. Horwitt, "Niacin," in R. S. Goodhart and M. E. Shils, Eds., *Modern Nutrition in Health and Disease,* 5th ed., Lea and Febiger, Philadelphia, 1973, pp. 198–202.
73. W. J. Darby, K. W. McNutt, and E. N. Todhunter, *Nutr. Rev.*, 33, 289 (1975).
74. W. Parsons and J. Flinn, *J. Am. Med. Assoc.*, 165, 234 (1957).
75. P. Handler and W. Dann, *J. Biol. Chem.*, 146, 357 (1942).
76. V. Herbert, "Drugs Effective in Megaloblastic Anemias: Vitamin B_{12} and Folic Acid," in L. S. Goodman and A. Gilman, Eds., *The Pharmacological Basis of Therapeutics,* 4th ed., Macmillan, New York, 1970, pp. 1414–1423.
77. T. Sheehy, *Am. J. Clin. Nutr.*, 9, 708 (1961).
78. C. Weigand, C. Eckler, and K. K. Chen, *Proc. Soc. Exp. Biol. Med.*, 44, 147 (1940).
79. T. Friedmann, T. Kmieciak, P. Keegan, and B. Sheft, *Gastroenterology,* 11, 100 (1948).
80. A. Tetreault and I. Beck, *Ann. Intern. Med.*, 45, 134 (1956).
81. D. W. Woolley, *A Study of Antimetabolites,* Wiley, New York, 1952.
82. E. Shaw, *Metabolism,* 2, 103 (1953).
83. J. C. Somogyi, "Antivitamins," in National Academy of Sciences, *Toxicants Occurring Naturally in Foods,* 2nd ed., Washington, D.C., 1973, pp 254–275.
84. J. C. Somogyi, "Natural Toxic Substances in Food," in R. Truhaut, Ed., *Toxicology and Nutrition,* Vol. 29 in *World Review of Nutrition and Dietetics* (G. H. Vourne, Series Ed.), Karger, New York, 1978, pp. 237–254.
85. D. C. March, *Handbook: Interactions of Selected Drugs with Nutritional Status in Man,* 2nd ed., The American Dietetic Association, Chicago, 1978.
86. Food and Drug Administration, *Fed. Reg.*, 44, 16126 (1979).
87. P. C. Paul, "Proteins, Enzymes, Collagen and Gelatin," in P. C. Paul and H. H. Palmer, Eds., *Food Theory and Applications,* Wiley, New York, 1972, pp. 115–149.
88. L. E. Holt, Jr., *Curr. Ther. Res.*, 9 (Suppl.), 149 (1967).
89. A. A. Albanese, "The Protein and Amino Acid Requirements of Man," in A. A. Albanese, Ed., *Protein and Amino Acid Requirements of Mammals,* Academic, New York, 1950, pp. 116–134.
90. M. K. Horwitt, *J. Am. Dietet. Assoc.*, 34, 914 (1958).
91. S. Margen and D. H. Calloway, *Fed. Proc.*, 27, 726 (1968).
92. W. H. Martinez, "Other Antinutritional Factors of Practical Importance," in C. E. Bodwell, Ed., *Evaluation of Proteins for Humans,* AVI, Westport, Conn., 1977, pp. 137–146.

93. W. G. Jaffé, "Toxic Proteins and Peptides," in National Academy of Sciences, *Toxicants Occurring Naturally in Foods*, 2nd ed., Washington, D.C., 1973, pp. 106–129.

94. Federation of American Societies for Experimental Biology, *Evaluation of the Health Aspects of Protein Hydrolyzates as Food Ingredients*, Bethesda, Maryland, 1978.

95. Food and Drug Administration, *Fed. Reg.*, 43, 60883 (1978).

96. A. E. Harper, "Amino Acids of Nutritional Importance," in National Academy of Sciences, *Toxicants Occurring Naturally in Foods*, 2nd ed., Washington, D.C., 1973, pp. 130–152.

97. W. L. Nyhan, Ed., *Amino Acid Metabolism and Genetic Variation*, McGraw-Hill, New York, 1967.

98. Federation of American Societies for Experimental Biology, *Evaluation of the Health Aspects of Certain Glutamates as Food Ingredients*, Bethesda, Maryland, 1978.

99. R. Levine, "Carbohydrates," in R. S. Goodhart and M. E. Shils, Eds., *Modern Nutrition in Health and Disease*, 5th ed., Lea and Febiger, Philadelphia, 1973, pp. 99–116.

100. V. N. Patwardhan and J. W. White, Jr., "Problems Associated with Particular Foods," in National Academy of Sciences, *Toxicants Occurring Naturally in Foods*, 2nd ed., Washington, D.C., 1973, pp. 477–507.

101. A. Holzel, V. Schwarz, and K. W. Sutcliffe, *Lancet*, 1, 1126 (1959).

102. Federation of American Societies for Experimental Biology, *Evaluation of the Health Aspects of Sucrose as a Food Ingredient*, Bethesda, Maryland, 1976.

103. Federation of American Societies for Experimental Biology, *Evaluation of the Health Aspects of Corn Sugar (Dextrose), Corn Syrup, and Invert Sugar as Food Ingredients*, Bethesda, Maryland, 1976.

104. G. W. Irving, Jr., letter to Dr. Corbin I. Miles of the Food and Drug Administration, May 23, 1979.

105. H. W. Sherp, *Science*, 173, 11 (1971).

106. J. B. Richmond, *Surgeon General's Report on Health Promotion and Disease Prevention*, Department of Health, Education and Welfare, Washington, D.C., 1979.

107. C. M. Leevy, L. Cardi, O. Frank, R. Gellene, and H. Baker, *Am J. Clin. Nutr.*, 17, 259 (1965).

108. C. H. Halstead, *Nutr. Rev.*, 33, 33 (1975).

109. E. R. Eichner and R. S. Hillman, *Am. J. Med.*, 50, 218 (1971).

110. J. D. Hines and D. H. Cowan, *New Engl. J. Med.*, 281, 333 (1969).

111. E. D. Palmer, *Medicine*, 33, 199 (1954).

112. C. S. Liever, *J. Am. Med. Assoc.*, 233, 1077 (1975).

113. T. B. Van Itallie, *Food Technol.*, 33 (12), 43 (1979).

114. Nutrition and Your Health, *Dietary Guidelines for Americans*, U.S. Department of Agriculture, U.S. Department of Health, Education and Welfare, Washington, D.C., 1980.

115. R. B. Alfin-Slater and L. Aftergood, "Fats and Other Lipids," in R. S. Goodhart and M. E. Shils, Eds., *Modern Nutrition in Health and Disease,* 5th ed., Lea and Febiger, Philadelphia, 1973, pp. 117–141.
116. T. P. Labuza and A. E. Sloan, *Food for Thought,* 2nd ed., AVI, Westport, Conn., 1977.
117. I. S. Scarpa, "An Introduction to the Cholesterol Problem," in I. S. Scarpa and H. C. Kiefer, Eds., *Sourcebook on Food and Nutrition,* Marquis Academic Media, Chicago, 1979, pp. 251–252.
118. A. E. Harper, *Am. J. Clin. Nutr.,* 31, 310 (1978).
119. R. W. Miller, F. R. Earle, and I. A. Wolff, *J. Am. Oil Chem. Soc.,* 42, 817 (1965).
120. U.S. Senate Select Committee on Nutrition and Human Needs, *Dietary Goals for the United States,* U.S. Government Printing Office, Washington, D.C., 1977.
121. W. B. Kannel, *J. Nutr. Educ.,* 10, 10 (1978).
122. R. Carol, *Diet Modification: Can it Reduce the Risk of Heart Disease?,* American Council on Science and Health, New York, 1980.
123. U.S. Senate, Subcommittee on Nutrition, *Heart Disease: Public Health Enemy No. 1,* U.S. Government Printing Office, Washington, D.C. (1979).
124. K. Schwarz, "Essentiality Versus Toxicity of Metals," in S. S. Brown, Ed., *Clinical Chemistry and Chemical Toxicology of Metals,* Elsevier/North-Holland Biomedical Press, New York, 1977, pp. 42–49.
125. R. P. Hamilton, M. R. S. Fox, B. E. Fry, Jr., A. O. L. Jones, and R. M. Jacobs, *J. Food Sci.,* 44, 738 (1979).
126. O. Mickelson, M. G. Yang, and B. S. Goodhart, "Naturally Occurring Toxic Foods," in R. S. Goodhart and M. E. Shils, Eds., *Modern Nutrition in Health and Disease,* 5th ed., Lea and Febiger, Philadelphia, 1973, pp. 412–433.
127. R. R. Cavalieri, "Trace Elements, Section A: Iodine," in R. S. Goodhart and M. E. Shils, Eds., *Modern Nutrition in Health and Disease,* 5th ed., Lea and Febiger, Philadelphia, 1973, pp. 362–371.
128. H. Suzuki, T. Higuchi, K. Sawa, S. Ohtaki, and Y. Horiuchi, *Acta Endocrinol.,* 50, 161 (1965).
129. F. W. Clements, H. B. Gibson, and J. F. Howeler-Coy, *Lancet,* 1, 489 (1970).
130. J. Wolff, *Am. J. Med.,* 47, 101 (1969).
131. Food and Nutrition Board, *Iodine Nutriture in the United States,* National Academy of Sciences, Washington, D.C., 1970.
132. Federation of American Societies for Experimental Biology, *Iodine in Foods: Chemical Methodology and Sources of Iodine in the Human Diet,* Bethesda, Maryland, 1974.
133. L. H. Knott and R. C. Miller, *J. Pediatr. Surg.,* 13, 720 (1978).
134. C. V. Moore, "Iron," in R. S. Goodhart and M. E. Shils, Eds., *Modern Nutrition in Health and Disease,* 5th ed., Lea and Febiger, Philadelphia, 1973, pp. 297–323.

135. R. W. Charlton and T. H. Bothwell, "Hemochromatosis: Dietary and Genetic Aspects," in E. B. Brown and C. V. Moore, Eds., *Progress in Hematology*, Vol. V., Grune and Stratton, New York, 1966, pp. 298–303.
136. R. A. MacDonald, B. J. P. Becker, and G. S. Picket, *Arch. Intern. Med.*, 3, 315 (1963).
137. Editors, *Nutr. Today*, 13 (1), 6 (1978).
138. J. Waddell, H. F. Sassoon, K. D. Fisher, and C. J. Carr, *A Review of the Significance of Dietary Iron on Iron Storage Phenomena*, Federation of American Societies for Experimental Biology, Bethesda, Maryland, 1972.
139. J. Waddell, *The Bioavailability of Iron Sources and Their Utilization in Food Enrichment*, Federation of American Societies for Experimental Biology, Bethesda, Maryland, 1972.
140. C. F. Whitten and A. J. Brough, *Clin. Toxicol.*, 4, 585 (1971).
141. G. K. Davis, "Iron," in National Nutrition Consortium, *Vitamin-Mineral Safety, Toxicity, and Misuse*, American Dietetic Association, Chicago, 1978, pp. 27–28.
142. E. J. Underwood, *Trace Elements in Human and Animal Nutrition*, 4th ed., Academic, New York, 1977.
143. A. S. Prasad, *Zinc in Human Nutrition*, CRC Press, Boca Raton, Florida, 1979.
144. "Outbreaks of Food Poisoning Due to Zinc, 1942–1956," U.K. Ministry of Health Laboratory, *Serv. Mon. Bull.*, July (1957).
145. G. Lawrence, *Br. Med. J.*, 1, 582 (1958).
146. B. L. O'Dell, "Zinc," in National Nutrition Consortium, *Vitamin–Mineral Safety, Toxicity and Misuse*, American Dietetic Association, Chicago, 1978, pp. 34–35.
147. M. R. S. Fox, R. P. Hamilton, A. O. L. Jones, B. E. Fry, Jr., R. M. Jacobs, and J. W. Jones, *Fed. Proc.*, 37, 324 (1978) (abstract).
148. M. R. S. Fox, R. P. Hamilton, A. O. L. Jones, B. E. Fry, Jr. and R. M. Jacobs, *Abstr. XI Internat. Congr. Nutr.*, 140 (1978) (abstr.).
149. C. H. Hill, "Copper," in National Nutrition Consortium, *Vitamin–Mineral Safety, Toxicity and Misuse*, American Dietetic Association, Chicago, 1978, pp. 29–30.
150. P. O. Nicholas, *Lancet*, 11, 40 (1968).
151. C. H. Paine, *Lancet*, 11, 520 (1968).
152. I. H. Scheinberg and I. Sternlieb, *Annu. Rev. Med.*, 16, 119 (1965).
153. H. G. Petering, *Environ. Health Perspect.*, 25, 141 (1978).
154. New York Academy of Medicine Committee on Public Health Relations: Report on Fluoridation of Water Supplies, *Bull. N. Y. Acad. Med.*, 28, 175 (1952).
155. N. C. Leone, *Arch. Indust. Health*, 21, 324 (1960).
156. J. Jowsey, B. L. Riggs, P. J. Kelley, and D. L. Hoffman, *Am. J. Med.*, 53, 43 (1972).
157. National Research Council, *Fluorides*, National Academy of Sciences, Washington, D.C., 1971.

158. H. C. Hodge and F. A. Smith, "Biological Effects of Inorganic Fluorides," in J. H. Simons, Ed., *Fluorine Chemistry,* Vol. IV, Academic, New York, 1965, pp. 247–256.

159. R. H. Hartman, G. Matrone, and G. H. Wise, *J. Nutr.*, 57, 429 (1955).

160. A. B. R. Thompson and L. S. Valberg, *Am. J. Physiol.,* 223, 1327 (1972).

161. G. C. Cotzias, *Physiol. Rev.*, 38, 503 (1958).

162. G. K. Davis, "Manganese Interactions with Other Elements," in National Nutrition Consortium, *Vitamin–Mineral Safety, Toxicity and Misuse,* American Dietetic Association, Chicago, 1978, p. 33.

163. Editors, *Nutr. Rev.*, 26, 173 (1968).

164. C. S. Alexander, *Ann. Intern. Med.*, 70, 411 (1969).

165. H. T. Grinvalsky and D. M. Fitch, *Ann. N. Y. Acad. Sci.*, 156, 544 (1969).

166. I. Rosenfeld and D. A. Beath, *Selenium: Geobotany, Biochemistry, Toxicity and Nutrition,* Academic, New York, 1964.

167. J. T. Rotruck, A. L. Pope, H. E. Ganther, A. B. Swanson, D. G. Hafeman, and W. G. Hoekstra, *Science,* 179, 588 (1973).

168. T. Noguchi, A. H. Cantor, and M. L. Scott, *J. Nutr.*, 103, 1502 (1973).

169. O. H. Muth, Ed., *Selenium in Biomedicine,* AVI, Westport, Conn., 1967.

170. R. J. Shamberger and D. V. Frost, *Can. Med. Assoc. J.*, 100, 682 (1969).

171. K. M. Hambidge, *Am. J. Clin. Nutr.*, 27, 505 (1974).

172. R. A. Levine, D. H. P. Streten, and R. J. Doisy, *Metabolism,* 17, 114 (1968).

173. W. Mertz, *Physiol. Rev.*, 49, 163 (1969).

174. W. Mertz, "Chromium Metabolism: The Glucose Tolerance Factor," in W. Mertz and W. E. Cornatzer, Eds., *Newer Trace Elements in Nutrition,* Dekker, New York, 1971, pp. 123–153.

175. H. P. Brinton, E. S. Fraiser, and A. L. Koven, *U.S. Public Health Rep.*, 67, 835 (1952).

176. E. S. Higgins, D. A. Richert, and W. W. Westerfeld, *J. Nutr.*, 59, 539 (1956).

177. J. B. Nielands, F. M. Strong, and C. A. Elvehjem, *J. Biol. Chem.*, 172, 431 (1948).

178. R. Hill, *Br. Vet. J.*, 133, 365 (1977).

179. V. V. Kovalskii, G. A. Yarovaya, and D. M. Shmavonyan, *Zh. Obshch. Biol.*, 22, 179 (1961); *Biol. Abstr.*, 40, 9498 (1962).

180. W. Layton and J. M. Sutherland, *Med. J. Aust.*, 1, 73 (1975).

181. I. H. Tipton, P. I. Stewart, and P. G. Martin, *Health Phys.,* 12, 1683 (1966).

182. A. L. Forbes, "Dietary Trends and Nutrient Status," presented at "Food Update '79," Food and Drug Law Institute, Washington, D.C., 1979.

183. G. B. Gori, *Food Technol.*, 33 (12), 48 (1979).

184. *Proceedings of Marabou Symposium on Food and Cancer*, Caslon Press, Stockholm, 1978.

185. National Cancer Institute, *"Cancer—A Social Disease?", Proceedings of the Eighteenth Meeting Interagency Collaborative Group on Environmental Carcinogenesis,* Washington, D.C., 1975.

186. F. Coulston, Ed., *Regulatory Aspects of Carcinogenesis and Food Additives: The Delaney Clause,* Academic, New York, 1979.

187. U.S. Senate Committee on Agriculture, Nutrition and Forestry, *Food Safety: Where Are We?* U.S. Government Printing Office, Washington, D.C., 1979.

188. Report of the Task Force on the Evidence Relating Six Dietary Factors to the Nation's Health, *Am. J. Clin. Nutr.,* 32, 2621 (1979).

189. F. J. Stare and E. M. Whelan, *Eat OK, Feel OK!,* Christopher, N. Quincy, Mass., 1978.

190. American Medical Association, statement submitted to U.S. Senate Select Committee on Nutrition and Human Needs Re: Dietary Goals for the United States, Chicago, 1977.

191. R. Passmore, D. F. Hollingsworth, and J. Robertson, *Nutr. Today,* 14 (5), 23 (1979).

192. A. E. Harper, *Nutr. Today,* 14 (5), 23 (1979).

193. E. M. Nelson, *J. Am. Dietet. Assoc.,* 30, 948 (1964).

194. S. T. Coulter and E. L. Thomas, *J. Agr. Food Chem.,* 16 (2), 158 (1968).

195. C. Le Bovit, *J. Agr. Food Chem.,* 16 (2), 153 (1968).

196. M. W. Lamb and M. L. Harden, *The Meaning of Nutrition,* Pergamon Bio-Medical, New York, 1973.

197. National Academy of Sciences, Council on Foods and Nutrition, *J. Am. Med. Assoc.,* 325 (9), 116 (1973).

198. Food and Drug Administration, *Fed. Reg.,* 45, 6314 (1980).

199. W. J. Darby, *Nutr. Rev.,* 36 (3), 65 (1978).

CHAPTER 4

Environmental Contaminants

I. C. MUNRO and S. M. CHARBONNEAU

Of the numerous chemical substances to which humans are exposed, including those added to or naturally found in the food supply, the group of chemicals loosely classified as the environmental contaminants probably presents the greatest potential threat to health. Outbreaks of human disease due to ingestion of certain chemicals, on some occasions involving large numbers of individuals, provide ample evidence that such a threat to human health exists.

Although comprised of a large group of substances of quite diverse chemical structure, the environmental contaminants can be divided into two broad chemical classes: (*a*) trace elements and organometallic compounds and (*b*) organic substances, the most significant of which are the aromatic halogenated hydrocarbons.

Any discussion of the hazards presented by environmental contaminants must take cognizance of those features they may have in common. For example, although differing widely in chemical structures, they possess certain common physical properties that tend to increase their potential hazard to humans. Contaminants that are persistent in the environment resist degradation and are extremely stable in many environmental compartments. Second, they tend to accumulate in the food supply, especially in fish, and it is this accumulation that renders them a potential hazard to humans. The characteristics that make them an actual hazard are their slow rate of elimination and/or metabolism, which results in their accumulation in tissues. Finally, their toxicity is usually greater in higher-order mammals than in species of lower phylogenetic order. For example, fish, seals, and crustaceans can tolerate much higher tissue levels of mercury and arsenic than can humans.

Chemicals released into the environment that have one or more of these properties can be predicted to have a potential to cause human harm. Although not every environmental contaminant possesses all of these properties, there is a surprising similarity among these substances, to the extent that any aromatic halogenated substance or organometallic substance detected in the environment with these properties should be viewed with suspicion.

The importance of environmental contaminants in food safety is illustrated by the 1979 assessment performed by the Office of Technology Assessment at the request of the House Committee on Interstate and Foreign Commerce (1). In a survey of the 50 states and 10 federal agencies, OTA identified 243 incidents involving food contamination during 1968 to 1978. Every region of the United States and every food category were impacted. Major incidents include PCB contamination of the Hudson River, polybrominated biphenyl (PBB) contamination of animal feed in Michigan, and kepone contamination of the James River in Virginia. Other reported incidents involved dieldrin, mercury, pentachlorophenol, pentachloronitrobenzene (PCNB), picloram, chlordane, dichlorodiphenyltrichloroethane (DDT), toxaphene, parathion, diazinon, and pesticides (collectively). Subsequent to the Office of Technology Assessment survey, a damaged transformer led to PCB contamination of animal fats at a packing plant in Montana. The consequent contamination of feeds occurred in 10 states, and ultimately hundreds of thousands of pounds of foods in 17 states were affected.

Environmental contaminants that find their way into food arise from two major sources. The first and most important of these is the inadvertent release of manufactured chemical products into the environment, particularly watersheds. In a limited number of other cases contamination of food may result from misuse of chemical substances or as a result of lack of knowledge regarding the potential for minute residues of certain chemicals to accumulate in the food chain. The second major source of food contamination results from contaminants being released from natural sources such as geological formations.

This chapter addresses itself to the general principles that should be considered in health evaluation of environmental contaminants in food. To facilitate this discussion, the sources of contaminants are dealt with first. Both industrial contaminants and those from natural sources are considered and the ways in which they contaminate certain foods discussed. The latter part of the chapter deals with the toxicology and the metabolism of the major contaminants of current interest. Special emphasis is placed on the toxicology of environmental contaminants to the neonate. The chapter concludes with a discussion of regulatory controls and safety considerations.

INDUSTRIAL CONTAMINANTS

The majority of food contaminants of industrial origin are complex organic substances that are either end-products or by-products of industrial chemical processes. In some instances the contaminant of interest may be an impurity in the final product that arises during the manufacturing process. In other cases inorganic or organometallic substances have been released as a result of human activity and have eventually contaminated the food supply.

One of the complexities involved with industrial contaminants is the multiplicity of substances involved. Over 43,000 chemical substances were listed by the Environmental Protection Agency (EPA) in its initial inventory of chemicals subject to the Toxic Substances Control Act (2). Under most conditions of use, these chemicals do not pose a threat to the safety of the food supply, but incidents such as those involving PCBs and PBBs indicate the potential for hazard. Table 4.1 indicates the sources of some major contaminants of industrial origin that contaminate food. These were selected because of current concern over their potential to produce adverse effects in humans. For reasons to be discussed later, not all foods are contaminated to a similar degree, and in Table 4.1 only those foods considered to be most affected are listed. Each listed contaminant is briefly commented on in the following pages. In addition, a summary of the PBB incident in Michigan is presented. Although the PBB incident represents a localized problem, it is an excellent example of a serious hazard arising from accidental contamination.

Polychlorinated Biphenyls

First synthesized in 1881, PCBs are a complex mixture of chlorinated isomers of biphenyl. An excellent discussion of their chemistry and toxicity is given in a recent report by the New York Academy of Sciences (36) and another by the HEW subcommittee on Health Effects of PCBs and PBBs (3). In the United States and Canada PCBs were sold commercially under the tradename of Aroclor and were used primarily as dielectrics in the electrical industry. A major source of environmental contamination was the destruction of old transformers, capacitors, and similar devices in landfills with subsequent leakage of the contents into subsoil and watersheds (37). Another source of PCBs in food resulted from the use of food packaging materials made from recycled paper containing PCBs (38).

It has been suggested that highly contaminated bottom sediments in sewers and receiving streams may represent a reservoir for the continued release of PCB (39). In this connection, Lawrence and Tosine (40) have demonstrated that the concentration of PCB in sewage sludge samples in six Ontario cities was comparable to results of analysis 2 years prior, in spite of implementation of restrictions in 1971.

In Lake Michigan, although efforts have been made to eliminate point source industrial losses of PCB, recent monitoring revealed that PCB contamination continues to be a problem. Results from a 3-year study of two species of Lake Michigan fish by the U.S. Fish and Wild-

TABLE 4.1 Major Contaminants of Industrial Origin

Chemical	Source	Foods Contaminated
Polychlorinated biphenyls	Electrical industry	Fish (3)
		Human milk (4,5,6)
Dioxins	Impurities in PCP and some other chlorophenols	Fish (7), Cows milk (8), Beef fat (9)
Pentachlorophenol	Wood preservative	Various foods (10,11)
Dibenzofurans	Impurities in PCP and PCB	Fish (12)
Hexachlorobenzene	Fungicide, industrial by-products	Animal fat (6)
		Dairy products (13,14)
		Human milk (15,16,17)
Mirex	Pesticide	Fish (18,19)
		Edible mammals (20)
		Human milk (21)
DDT and related halogenated hydrocarbons	Pesticides	Fish (22)
		Human milk (16)
Alkyl mercury compounds	Manufacture of chlorine soda lye, acetaldehyde,	Fish (23,24)
	seed dressing	Grain (25)
Lead	Automobile exhaust emmission, coal combustion, lead industry;	Vegetables (26)
	Solder in can seams;	Canned milk (27)
		Canned fish (28)
	Lead glazed pottery	Acidic foods (29)

TABLE 4.1 Major Contaminants of Industrial Origin (Continued)

Chemical	Source (Original)	Food Contaminated
Cadmium	Sewage sludge	Grains and vegetables (3)
	Smelter operations	Farmlands, meat products (31)
Arsenic	Smelter operations	Milk (32)
		Vegetables (33)
		Fruits (34)
Tin	Canning industry	Canned foods (35)

life Service Laboratory indicate that PCB levels in lake trout had risen over the study period (41).

Saltwater fish are seldom contaminated by PCBs in the United States as is the case in Japan. Therefore, the major source of PCBs in the American diet is mainly due to freshwater fish instead of saltwater fish (42).

In terms of total exposure, Jelinek and Corneliussen (43) have reported that the estimated PCB intake in the United States has decreased yearly since 1971 because the levels of PCBs in all foods, except fish, have declined drastically. Levels in freshwater fish have remained a problem and have increased in some areas as a result of additional discharge into lakes and rivers. In particular, fish from parts of the Great Lakes and the Hudson River consistently have PCB levels in excess of 5 ppm, prompting state officials in those areas to restrict fishing and/or issue public warnings. Although clams and oysters from Atlantic coastal waters usually do not contain detectable levels of PCBs, certain shellfish can be contaminated. For example, the FDA reported in 1979 that lobster in some areas of the New Bedford, Massachusetts harbor showed PCB levels as high as 68 ppm (44).

Already ubiquitous in the environment, the highly persistent PCBs will have to be monitored for years to come. There is also the distinct possibility that the environmental load will continue to be increased, at least in certain localities. Examples of the latter hazard include the accidental contamination of animal fats from a damaged transformer in Montana and the deliberate dumping of waste transformer fluid along highways in North Carolina.

Recent studies suggest that cooking may significantly reduce the level of PCBs, dieldrin, and DDT in fish. Losses of PCB ranged from 26 to 70%, of dieldrin from 25 to 57%, and of DDT compounds from 30 to 57%. Losses were greatest from broiled fillets and from whole pieces roasted without the skin (45).

Dioxins

In recent years there has been a growing concern over the presence of trace amounts of dioxins in some commercial herbicidal and other chemical formulations. Trace quantities of the 2,3,7,8-tetrachloro-*p*-dioxin isomer, the most toxic form, have been discovered in some commercial formulations of 2,4,5-trichlorophenoxyacetic acid (2,4,5-T) (46) and 2,4, 5,-trichlorophenol (47). This isomer has not been found in 2,4-dichlorophenoxyacetic acid (2,4-D) preparations, nor would it be expected, based on the reaction chemistry involved in synthesis of 2,4-D. Other isomers of dioxins have been detected in commercial formulations of pentachlorophenol (48, 49), as is discussed in the next section.

Limited data such as those from dairy cattle and fish samples indicate the occurrence of dioxins in at least some parts of the food supply at parts per trillion levels. However, the extent of such occurrence is not yet completely characterized.

The presence of 2,3,7,8-tetrachlorodibenzo-*p*-dioxin (TCDD) has been reported in cow's milk in Seveso, Italy, where TCDD escaped from a chemical plant and contaminated the surrounding area. The highest TCDD, levels (≤7 ppb) were found in samples collected from farms close to the chemical plant. In addition, elevated TCDD levels were noted in cows located distant from the plant but that had been fed food harvested in the contaminated area (8).

Analysis of beef fat samples taken from cattle that had grazed on rangelands treated with 2,4,5-T revealed that, out of 85 samples, one had levels of 60 ppt TCDD two had 20 ppt, and five may have had levels of TCDD, within a range of 5 to 10 ppt (9).

Pentachlorophenol

Pentachlorophenol (PCP) is an extremely effective pesticide and wood preservative. Annual production in the United States is estimated to be 46 million pounds (50), whereas world production exceeds 200 million pounds. About 80% of this amount is used for wood preservation. Unlike many other contaminants, PCP does not resist environmental degradation, and thus residues in foods tend generally to be low and of little toxicological significance, except in cases of accidental misuse. Of major concern, however, is the presence of toxic isomers of the dioxin series in commercial PCP. Most PCP preparations are contaminated to some degree with dioxins, mainly the hexa, hepta, and octa isomers (48).

Polychlorinated Dibenzofurans (PCDFs)

The polychlorinated dibenzofurans (PCDFs) have been reported to be present in a variety of polychlorinated biphenyls (51) and in pentachlorophenol (52). Studies have demonstrated the presence of PCDFs in fly ash from municipal incinerators (53). In addition, studies have demonstrated that pyrolysis of technical PCB mixtures may yield up to 60 PCDF isomers, one of the major constituents of which is 2,3,7,8-CDF (54). Therefore, the uncontrolled burning of PCBs could be an important environmental source of PCDFs. Polychlorinated dibenzofurans have recently been identified in fish from the Ohio, Hudson, and Connecticut Rivers and Lake Michigan at the parts per billion level or higher (12). Toxicity studies with PCDFs have demonstrated a pattern of toxic effects similar to that reported in dibenzodioxins (55).

Hexachlorobenzene

Hexachlorobenzene (HCB) is a fungicide used primarily to control bunt in seed grains. It was thought to be responsible for a mass poisoning epidemic in Turkey that resulted from the consumption of treated wheat (56). Its use in recent years for seed dressing has been curtailed or abolished in many countries. Presently, its major source as an environmental contaminant arises from the manufacture of pentachlorophenol (57). Residues in food generally tend to be low, at least in North America, and of limited toxicological significance (58).

Mirex

A chlorinated insecticide that was used to control the fire ant (*Solenopsis* sp.), mirex is also sold as a flame retardant under the tradename Dechlorone. Mirex is structurally related to Kepone (chloroderone). It is extremely stable in the environment and has been detected in human adipose tissue of persons living considerable distances from its point of use or manufacture (59). It is known to be widely distributed in environmental samples (60) and readily accumulates in the food chain.

DDT

Certain chemically related chlorinated hydrocarbon insecticides, including the DDT compounds (consisting mainly of dichlorodiphenyltrichloroethane), tend to be persistent environmental contaminants. The DDT compounds were extensively used in the United States as pesticides prior to the ban on their use in 1973. In 1963 DDT production reached 176 million pounds, and there were registered uses for over 300 commodities (61). Following the ban on its use, environmental exposure levels have declined considerably, judging from current levels in fish and dairy products (58), but DDT and its metabolites continue to be present in substantial concentrations in some human milk (16).

Mercury

The use of mercury compounds in the synthesis of chlorine, soda lye, and acetaldehyde has resulted in considerable environmental contamination. As is discussed in more detail later, both inorganic mercury compounds and elemental mercury are readily converted by aqueous biota to the more toxic alkyl mercury forms. All animal and vegetable tissues contain at least trace amounts of mercury since all living organisms have the ability to concentrate mercury.

Prior to the enactment of regulations prohibiting the release of mercury into the environment, considerable quantities were lost, particularly into watersheds. This resulted in contamination of fish and wildlife (62) and was responsible for several cases of human neurologic disease due to consumption of contaminated fish (24). Alkyl mercury compounds also were used extensively as seed dressings, and the consumption of mercury-treated seed has been responsible for numerous cases of illness and death (25). In recent years this use has been banned. Following the introduction of regulations prohibiting the release of mercury compounds into the environment (62), the levels of mercury have declined considerably in fishery products taken from previously polluted waters (63).

Lead

Health problems have been presented by lead since time immemorial. The natural background occurrence of lead in soil and water results in its presence in all living organisms. Additional lead is contributed to foods through environmental pollution and through food processing activities involving lead. Use of lead compounds as antiknock additives in gasoline has resulted in considerable release of lead into the environment especially near roadsides. Environmental contamination also occurs from lead smelting operations and water runoff from mining operations. Lead-containing pesticides may directly increase lead levels in fruits and vegetables and, where such use has been of sufficient duration, also contribute lead indirectly through the affected soil. Fortunately, however, levels of lead in commercially available fruits and vegetables tend to be low (26, 64) and of only limited toxicological significance.

In food processing the major source of lead is the tin can, which is used to package 10 to 15% of U.S. foods. This contribution of lead comes from the lead solder in the can seams. The FDA has estimated that about 20% of the lead in the average daily diet of persons over 1 year of age is from canned food, with about 13 to 14% from the solder and the remaining 6 to 7% from the food itself (64). More recently, studies have demonstrated that up to 99.5% of the lead present in tuna canned in lead soldered cans originates from the lead solder (28). The use of lead solder in the can seams and as a plug in the vent holes has accounted, in large part, for the presence of lead in canned (condensed) milk preparations (27), but recently levels in such products are declining as new methods of soldering and can seaming are introduced. The presence of lead in canned milk is of concern because many infants and young children consume large quantities of this product.

Drinking water consumed directly and used in food processing also contributes lead to the dietary intake, as does atmospheric lead. In addition to the efforts of FDA to control lead levels in foods, the EPA has also moved to limit lead intake from water and air (65).

Cadmium

One source of cadmium in food has been mining operations. In a well-documented incident in Japan (66), cadmium compounds were transported by the Kakehashi River from a mine upstream to rice fields where the river water was used for irrigation. Examination of the cadmium concentration in rice, paddy soils, and river water performed in 1974 demonstrated that 23 villages were considered to be contaminated. Levels of cadmium in rice in these villages ranged from 0.19 to 0.69 ppm, whereas concentration in rice in nonpolluted areas was almost always less than 0.2 ppm (66).

Recently it has been demonstrated that the increased use of sewage sludge containing elevated levels of cadmium may result in increased cadmium levels in vegetables as well as in edible animal tissues (67).

Arsenic

Like cadmium, arsenic has also been found to result in contamination of food and water as a result of mining operations (32, 33). Apart from this localized heavy contamination, contamination from manmade sources is usually low. While arsenic-containing compounds may be used as insecticides in certain fruit crops (34), exposure to arsenic from this source is low, provided the pesticides are used in accordance with the manufacturer's recommendations.

Tin

The major portion of tin in food is due to canning processes; levels of up to 250 ppm are generally tolerated. These levels are based on investigations in humans in which it was found that gastrointestinal disturbances occurred when foods containing higher levels of tin were consumed (35). Hamilton et al. (68) measured levels of various metals in evaporated milk used for infant formulas. They found that the level of tin was elevated in milk canned in nonlacquered cans. The concentration of tin in canned food depends on the quality of the tin lining, the type of food, and the duration of exposure. Levels were generally higher in tomato paste and vegetables.

Recently, Braman and Tompkins (69) and Hodge et al. (70) have identified the presence of methylated forms of tin as well as inorganic tin in water, air, and human urine. The source of this tin has not been established. It may be due to the environmental migration of organotins used as stabilizers in plastics or to methylation of tin in biological systems.

Polybrominated Biphenyls—A Case Study

The incident involving PBBs in the state of Michigan provides an excellent example of accidental environmental contamination. In 1973 a fire retardant product containing PBBs was mistakenly shipped to farmers' cooperatives in Michigan in place of an intended feed supplement. The PBBs then unknowingly became part of the finished feeds used on farms in the southern part of the state. By the time the resulting illnesses in livestock were traced to the PBBs, there was widespread contamination. On the order of 30,000 livestock (primarily dairy cattle), over 1 million poultry, and thousands of tons of produce had to be destroyed. In a report to the Office of Technology Assessment, the state of Michigan estimated the cost of the PBB contamination at $215 million (1).

When livestock first showed signs of illness from the contaminated feed, analyses of feed and blood samples revealed no clue to the problem since the samples were not specifically tested for PBBs. It was nearly a year after the accidental contamination of animal feed that PBBs were discovered to be the source of the problem.

Ultimately many, if not most, Michigan residents were exposed to food products, contaminated to some degree with PBBs. Those residents of areas near the original accident, in the southern part of the state, had the greatest exposure. In a study of PBBs in human breast milk by the Michigan Department of Public Health, 96% of tested women in the southern peninsula and 43% of tested women in the northern peninsula had detectable levels of PBBs in breast milk (71). Some 2000 Michigan farm families were estimated to have received the heaviest exposures (72).

In a sample of 165 residents of quarantined farms examined by the Michigan Department of Public Health, about half had blood PBB levels above 0.02 ppm, and the highest observed level was 2.26 ppm. A similar examination of residents on nonquarantined farms showed only two individuals with blood PBB levels above 0.02 ppm (73).

Generally, consumers and residents of nonquarantined farms had lower PBB levels than did residents of quarantined farms, although quarantined farm families were advised not to consume meat and milk from their own livestock. Serum PBB was found to be consistently related to exposure.

Wolff et al. (74) studied serum PBB levels in farm residents, consumers, and Michigan Chemical Corporation employees. It was observed that very young children and those living on farms for less than 1 year had lower serum PBB levels than did others. Although there were no consistent trends with age, serum PBB levels tended to occur in decreasing order among young males, young females, older males, and older females.

The accidental contamination of animal feed and subsequently food with PBBs was largely confined to Michigan. However, low PBB concentrations were detected in animal feeds in Indiana and Illinois. Although not officially confirmed, very low PBB levels have also been reported, mostly in poultry products, in Alabama, Indiana, Iowa, Mississippi, New York, Texas, and Wisconsin (75).

Similar to PCBs, the PBBs are quite stable and resist biodegradation. They have become a continuing low-level contaminant of the environment in Michigan. Polybrominated biphenyl residues remaining in pasture lands and farm buildings can continue to contaminate livestock on affected farms (1). The Michigan Department of Agriculture has estimated expenditures of $40 to $60 million for monitoring of PBBs in animals and animal by-products during the next 5 years (76).

The adverse effects on health from PBB exposure are not completely known. Studies of exposed residents of Michigan have indicated some effects on lymphocyte production and an association with fatigue, hypersomnia, irritability, and arthritis-like complaints (1). A recently completed study of the health effects of PBBs by the University of Michigan School of Public Health indicated some association between exposure and mental disorders, diseases of the nervous system and sense organs, malignant neoplasms, disorders of the skin, and subcutaneous tissue and infections. Negative findings included fetal deaths, congenital anomalies, thyroid disorders, benign neoplasms, neoplasms of the breast and endocrine, and nutritional and metabolic disorders. The study concluded, however, that the uncertainty about the health effects of PBBs would persist for some years (77).

CONTAMINANTS FROM NATURAL SOURCES

The ubiquitous natural occurrence of lead in soil and water has already been noted. Other major contaminants of natural origin are shown in Table 4.2. All of these arise from erosion of geological formations or from soils containing naturally high levels of these substances.

Mercury

It has now been established that high mercury levels in certain marine species as well as pelagic species of fish taken from inland lakes, with no known source of industrial contamination, are due entirely to the presence of natural deposits of cinnabar in the affected areas. Levels of mercury in these species may readily approach those found in contaminated areas (84). Similar situations have been found with other naturally occurring contaminants.

The levels of mercury in museum specimens of marine species taken from the seas 95 years ago are similar to those found in these species today. Miller et al. (85) have reported that mercury levels in museum specimens of tuna range from 180 to 640 ppb. These are similar to present levels in tuna of 200 to 1000 ppb. Thus human industrial activities of the past 100 years or so have not substantially influenced the levels of mercury in marine fish. Where human activity has influenced

TABLE 4.2 Major Contaminants of Natural Origin

Chemical	Source (Origin)	Foods Contaminated
Elemental mercury and salts	Geological	Fish (78)
Arsenic - various chemical forms	Geological	Soft drinks (79) Fish (80) Health food supplements (81)
Selenium	Seleniferous soils	Grains (82)
Cadmium	Geological	Fishery products (83)
Tin	Geological	Fish (35)

the level of mercury in fish, the effects have been profound, as demonstrated in Table 4.3. The levels of mercury in pelagic species of fish taken from three inland bodies of water in Canada are shown. Fish taken from Ball Lake, a body of water contaminated through industrial use of mercury adjacent to the lake, contain levels of mercury considerably above the 0.5-ppm Canadian regulatory guideline of mercury in fish (the U.S. action level is 1 ppm). In the center column are listed the mercury levels of fish taken from Lac Seul, where there is no known industrial source of mercury but where certain of the fish still contain levels of mercury slightly above the 0.5-ppm guideline. These data are considered to reflect a high natural background of mercury. The elevated levels of mercury in fish from this lake are due to the presence of geological formations in this area from which mercury is leaked and assimilated by the food chain. Fish taken from Pelican Pouch Lake contain levels of mercury considerably below the 0.5-ppm regulatory guideline. These data are considered to reflect the low end of the natural background level of mercury in inland fish of the species shown.

Cadmium

Elevated levels of cadmium (9.0 to 8.0 ppm) have been reported in two shellfish: sea scallop (*Placopecten magellanicus*) and ocean gauhaug (*Arctica islandica*). These samples were obtained from an area considered relatively unpolluted and a considerable distance from any major source of input (86).

Similar findings were reported by Bull et al. (87), who reported high cadmium concentrations in sea birds sampled from areas distant from environmental pollution.

Oysters are an especially rich source of cadmium, with levels on the order of 3 to 4 ppm. Cadmium levels in lobsters ranging from 2.82 to

TABLE 4.3 Mercury Levels (in Parts Per Million) in Fish from Selected Areas with Natural and Industrial Sources of Mercury

Fish (Common Name)	Industrial Mercury Pollution (Chlor-Alkali Plant)	Natural Mercury Sources	
	Ball Lake	Lac Seul	Pelicanpouch Lake
Pike	2.95	0.84	0.29
Pickerel	2.45	0.60	0.23

Source. G. W. McGregor, Industry Services Branch, Department of Fisheries and Environment, Canada, 1978.

16.73 ppm in the digestive gland, which is used for lobster paste, have been recorded. Muscle levels were less than 1 ppm. Most of the cadmium appeared to be of geologic origin, as the nonindustrialized areas gave the highest values (83).

Arsenic

Contamination of drinking water from artesian wells by arsenic has resulted in numerous cases of chronic toxicity, such as have been documented in Taiwan (88) and in Nova Scotia, Canada (89). The source of this arsenic is arsenopyrite, which is found in certain geological formations. Not only well water but also public drinking water supplies have been found to have elevated levels of arsenic, as in Antofagasta, Chili. In Antofagasta many food commodities, such as soft drinks, had arsenic levels of 0.24 ppm. When the source of municipal drinking water was changed, the arsenic levels in these food commodities fell dramatically. Arsenic levels in soft drinks decreased from 0.24 to 0.060 ppm (79).

The presence of high levels of arsenic in certain species of marine fish and shellfish has been well established (90). In addition, aquatic vegetation such as seaweed has been found to contain elevated arsenic concentrations, ranging from 0.7 to 142 ppm, depending on species and location (90). Health food supplements prepared from kelp may contain up to 20 ppm arsenic and have been reported to cause toxicity in humans ingesting these preparations (91).

Tin

Data on the tin content of food are quite sparse, and there have been few attempts to identify the sources of naturally occurring contamination of foods. Such natural contamination can occur, however, as witnessed by the relatively large amounts of tin (2.57 to 5.43 ppm) found in fish from Moose Lake, Manitoba, a lake free of major industrial development. This was considered to come from tin-bearing minerals in the region (35).

METABOLISM AND TOXICOLOGY OF ENVIRONMENTAL CONTAMINANTS

As noted earlier, the environmental contaminants have quite divergent chemical structures, but many share common physical characteristics that tend to increase their potential hazard. The physical characteristics of major importance include their stability and hence persistence in the environment and their tendency to bioaccumulate, especially in fish. In addition, they have a slow rate of elimination and/or metabolism and tend to be more toxic to the higher mammals, including humans.

In this section these characteristics are elaborated, by example and discussion of certain of the substances listed in Tables 4.1 and 4.2. It is impractical to provide exhaustive coverage of the extensive toxicology literature on environmental contaminants, but references to detailed reviews are included.

Polychlorinated Biphenyls

Some of the many existing reviews of the literature pertaining to the acute and chronic toxicity of PCBs in animals and humans (3, 43, 91–94) suffice to cover PCB toxicology very well. Therefore, this section concentrates on the metabolism and related properties of PCBs. Subsequently, the toxicity of PCBs is briefly reviewed in comparison to other halogenated hydrocarbons.

The tendency of the environmental contaminants to concentrate in the food chain is, for the most part, selective and limited to certain food products such as those shown in Tables 4.1 and 4.2. Fish accumulate PCBs to more than 100,000 times the level present in water (95). In Japan in 1972 more than 1 ppm was found in the edible parts of 16% of seawater fish and in 18% of freshwater fish. Fish from the most highly polluted localities contained more than 3 ppm PCBs (96). There appears to be a selective accumulation of the more highly chlorinated components of the commercial mixtures in biological materials.

Jensen et al. (97) analyzed a wide range of marine organisms from the coastal waters of Sweden. They found that low-PCB members are metabolized or excreted faster than are the higher members, so that there is an increase in the latter as they pass through the food chain. Fish from lower and intermediate levels of the food web have been found (98) to contain lower amounts of hexachloro and higher amounts of tetrachloro and pentachloro compounds than higher trophic level species such as white and silky sharks and aquatic birds, thus suggesting that selectivity in bioaccumulation is also a function of animal species (99). A typical PCB residue from fish resembles the Aroclor 1254 mixture more closely than it does other Aroclors (98).

The ratio of residue levels in food products to levels in the animal's diet are greatest with PCBs of 54% chlorination (100). The residue levels in milk are approximately four times the residue levels in the cow's diet (101). In hens, residue levels in eggs are approximately equal to those in the diet, whereas residue levels in the body tissue (fat basis) are approximately six times the dietary level (102). The ratio of residue levels in products and tissues to intake is lower with PCBs of greater than 54% chlorination. These observations suggest that absorption from the gastrointestinal tract is lower for the more highly chlorinated PCBs (103). Polychlorinated biphenyls with four or fewer chlorines rarely accumulate in animal products when the animals are fed dietary levels less than 0.5 ppm (103). Polychlorinated biphenyls are not excreted to an appreciable extent prior to metabolism to more polar compounds, and long-term PCB storage occurs in the skin and adipose tissue (104).

The initial half-life of 2,4,5,2',4',5'-hexachlorobiphenyl appears to approach infinity (104). This particular hexachlorobiphenyl has been shown to be in the highest concentration of any PCB found in the adipose tissue of the Swedish population (105). These observations have been confirmed in animal studies. Studies of the single PCB homologs of various degrees of chlorination have shown that those with five or less chlorine atoms are more readily metabolized and excreted than are the PCBs with higher chlorination. The position of chlorine substitution also affects retention and elimination of single homologs (100).

Recent work with five symmetrical hexachlorobiphenyl isomers with chicks (106, 107) and mice (108) has demonstrated that separate and distinct differences in isomer toxicity do occur. These differences are related to variations in chemical structure such as the alteration of chlorine substitution on compound lipophilicity and metabolism. Those hexachloro isomers with 4,4'-substitution appear to be more rapidly accumulated and may metabolize more slowly. In addition, the pentachlorobiphenyl isomers with 4,4'-substitution, have been shown to have a higher toxicity in chickens (99).

Cadmium

The bioaccumulation of cadmium in edible plants and its bioavailability from soils treated with sewage sludge is affected by a number of factors including the plant species involved, soil pH, the presence of other trace elements, and the rate of sludge application (62). Leafy plants such as spinach, lettuce, curly cress, and swiss chard are accumulators of cadmium. Based on solution culture experiments, cadmium concentrations as low as 0.05 μg Cd/ml in soil solutions (or saturated extracts of soils) are sufficiently high for a number of crop species to accumulate cadmium in levels that render the plants unsafe for consumption (109).

Similar studies were conducted with livestock fed corn grown on sludge-treated soils. Liquid digested sewage sludge from a Chicago waste treatment plant was applied to experimental corn plots. Corn grain harvested in treated plots was fed to growing swine for 56 days. The sludge-fertilized corn contained higher concentrations of nutrient and toxic elements but did not interfere with swine performance. Minor changes in hepatic microsomal oxidases and red blood cells accompanied significant increases in renal cadmium and decreases in hepatic iron. Swine foraging on these plots ingested considerable amounts of sludge soil and accumulated significantly higher concentrations of renal cadmium than did those fed the corn only (110).

Movement of cadmium through estuarine ecosystems and uptake by commercially important benthic shellfish were found to be influenced by a number of variables of which salinity appears to be of primary importance (67). Kneip and Hazen (111) studied the uptake of marsh-cove ecosystems heavily contaminated by cadmium. Their results indicated that uptake of cadmium occurs in marsh and aquatic plants and animals. Although significantly elevated concentrations were observed in comparison to noncontaminated areas, the edible portions of most fish did not appear to present a hazard. Crabs appeared to present the most likely source of hazard to humans.

The bioaccumulation of cadmium in marine organisms is a potential problem for human health. Because of their high affinity for cadmium, molluscs pose the greatest problem. Significant bioaccumulation of cadmium occurs in the American oyster at aqueous concentrations of cadmium as low as 0.005 ppm. Factors that affect cadmium concentrations in oyster tissues include regional effects, such as locality, seasonal influences, and salinity dependence. Laboratory studies designed to investigate mechanisms of cadmium accumulation in oysters have demonstrated the existence of an inducible cadmium binding protein similar to metallothionein (112).

The extent of human hazard arising from cadmium in foods is not completely clear. In a review by Underwood (113) it was noted that 15 ppm has been concluded to be the concentration required for mild symptoms (nausea, vomiting, and diarrhea) of cadmium toxicity in humans. It is important to note that the toxicity of cadmium is related to dietary levels of zinc. Zinc deficiency symptoms can be enhanced by high intakes of cadmium, and the toxicity of cadmium can be ameliorated by increased intakes of zinc (113).

Arsenic

Arsenic appears to be only a potential problem in the meat, fish, and poultry group of foods; the levels of all other food categories are low

TABLE 4.4 Average Levels of Arsenic (in Parts Per Million), Total Diet Composite[a]

Food Class Composite	1973	1974
Dairy Products	0.003	ND
Meat, Fish and Poultry	0.020	0.059
Grain and Cereal Products	0.003	ND
Potatoes	0.003	ND
Leafy Vegetables	T	ND
Legume Vegetables	T	ND
Root Vegetables	T	T
Garden Fruits	ND	T
Fruits	T	0.017
Oils, Fats and Shortening	T	ND
Sugar and Adjuncts	T	ND
Beverages (Including Water)	ND	ND

Source. Jelinek and Corneliussen, Environmental Health Perspective, 19:83-87 (1977).

[a] ND - Not Detected; T = < 0.001 ppm.

and reasonably consistent among the food groups (114). Fish is a major source of dietary arsenic (Table 4.4), and the high levels reported for fish products are due entirely to its tendency to concentrate in certain marine animals, notably bottom feeding species such as grey sole and shrimp (Table 4.5). The high arsenic levels present in sole and shrimp are confined to areas where the source of arsenic is natural (geological) (115).

The toxicity of arsenic in humans varies a great deal because of both individual susceptibility and the differing toxicity of different forms. Inorganic arsenic is generally considered to be more toxic than the organic forms of arsenic present in fish. Recent studies in humans have demonstrated that excretion of inorganic arsenic may be represented by a triexponential model, with some subjects having a flat, third compartment indicating permanent retention of a fraction of the dose (116). Because retained arsenic has an affinity for keratin, concentrations in human hair and fingernails are higher than in other tissues. This fact makes possible the well-known analysis of hair and fingernails in the diagnosis of suspected arsenic poisoning (113).

Symptoms of acute toxicity from oral ingestion of arsenic (nausea, vomiting, diarrhea, and abdominal pains) are also well known. Chronic toxicity symptoms include weakness, muscular aches and prostration. Changes in skin (pigmentation and wart development) and mucosa together with peripheral neuropathy are also seen. Drowsiness, headaches, confusion, and convulsions can occur from both acute and chronic exposure (113). In addition, there is now strong epidemiological evidence that inorganic arsenic is a skin and lung carcinogen in humans (117).

TABLE 4.5 Typical Mean Arsenic Levels in Canadian Fish and Fishery Products

Fish Product	Arsenic, ppm
Freshwater	
Pickerel	0.4
Pike	0.3
Smelt	0.5
Whitefish	0.2
Marine	
Groundfish	
Cod	2.7
Halibut	4.9
Flounder (nonmineral area)	2.5
Flounder (mineral area)	18.3
Gray sole	56.4
Pelagic	
Herring	0.8
Salmon (Atlantic)	0.3
Tuna (bluefin)	0.6
Crustaceans & Molluscs	
Clams	1.2
Lobster	5.9
Scallops	1.2
Shrimp	10.9

Source. Fish Inspection Branch, Fisheries and Marine Service, Environment Canada.

The chemical form of arsenic in the various species of marine life varies between species (118). Cooney et al. (119) have identified the form of arsenic in algae to be phosphatidyltrimethylarsenium lactic acid, a phospholipid with properties similar to phosphatidylcholine and phosphatidylserine. The arsenophospholipid from algae is thought to be degraded in marine animals since Edmonds et al. (120) have recently identified trimethylarsoniumbetaine in lobsters. Studies in humans have demonstrated that the 71 to 77% of arsenic present in fish appears to be excreted in the urine within 8 to 10 days following consumption (121, 122). Excretion of arsenic through the feces is minimal. The fate of the fish arsenic not eliminated in the urine and feces is unknown. It may be stored in the body indefinitely, eliminated through some other mechanism, such as hair or sweat, or the elimination could take place in the urine and feces at a rate that is too slow to be detected experimentally. Further studies in nonhuman primates indicate age-related differences in the amount retained in the body (Table 4.6).

In this study nonhuman adult and adolescent primates were given a single oral dose of 1 mg As/kg from a fish slurry of gray sole that contained high levels of arsenic. Whereas adult monkeys excreted approximately 80% of the dietary arsenic over a 7-day period, adolescent monkeys excreted a somewhat lesser amount. The data clearly indicate that with fish arsenic, about one-third of the dose was retained. The

TABLE 4.6 Percent of Dose of Fish Arsenic Excreted by Monkeys[a]

	Urine	Feces	Total
Adults	66.6 ± 6.1	10.1 ± 3.5	76.7 ± 4.0
Adolescents	44.7 ± 5.5	17.9 ± 4.5	62.6 ± 5.9

[a] Dose: 1 mg/kg Body Weight.

significance of this retention is currently being investigated in multiple-dose studies using high-arsenic-content fish. However, more definitive studies on the pharmacodynamics and the mechanism of toxicity of fish arsenic in mammalian species must await the isolation and structural identification of the form of arsenic in fish.

In freshwater lakes polluted with arsenic, fish were not found to have elevated arsenic levels (123, 124). In this connection the distribution, accumulation, and degradation of the arsenical herbicide cacodylic acid was investigated in aquatic ecosystems. Bioconcentration ratios obtained suggested that cacodylic acid does not accumulate greatly in the aquatic environments (125). Similarly, arsenic concentrations were measured in aquatic invertebrates, macrophytes, sediments and water of arsenic-contaminated lakes. Arsenic concentration factors were calculated and found to decrease with increasing concentration of arsenic in ecosystem components of the lake (123).

Mercury

The significance of biological methylation on the toxicity of trace elements has not been investigated adequately. Except for methylmercury, which has been extensively studied, there is a relative paucity of information on the toxicity (particularly long-term effects) of other methylated metals and metalloids. On the basis of the electrochemical characteristics of the trace elements, Wood (126) has suggested that metals such as thallium, tin, palladium, and platinum be closely scrutinized for the possibility of biological methylation. Available data on methylmercury would support the need for further studies in this regard.

Methylmercury provides a classic example of the effect that environmental transformations may have on the toxicity and pharmacodynamics of the metals. If we look, for example, at the elimination rates for inorganic and methylmercury from mammals, we find surprising differences. The half-time of elimination of mercuric chloride from the whole body of humans is 40 days, compared to 72 days for methylmercury. Also, the half-time of elimination of methylmercury appears to vary considerably between individuals. The longer half-time of elimination of methylmercury means that a greater amount of methylmercury will accumulate in the body than will mercuric chloride administered at comparable dose levels for extended periods. This fact, coupled with the high CNS susceptibility to methylmercury, accounts in large part for the increased sensitivity of humans and animals to the toxic effects of methylmercury compared to mercuric chloride (24).

The methylated form of mercury, because of its greater lipid solubility, can also cross biological membranes much more readily than can inorganic mercury. Consequently, methylmercury more readily crosses the placenta, resulting in higher exposure of the developing embryo and fetus. In fact, as shown in Table 4.7, methylmercury concentrations in blood samples of newborn infants shortly after birth are consistently higher than those in their mothers. The ratio of methylmercury concentration in newborn blood compared to maternal blood ranges from 1.3 to 2.1, in cases of long-term medium level exposure as in Japan (127), in long-term, low-level exposure as in Canada (128), and in short-term, high-level exposure as in Iraq (129). It has been demonstrated (129) that the elimination rate of methylmercury from infants is slower than from adults; therefore, the mother would excrete methylmercury more rapidly than the fetus, and an abnormally high infant blood: maternal blood ratio could be expected. In all cases the higher level of methylmercury in the blood of newborns is probably due to the greater binding of methylmercury to fetal red blood cells compared to the adult.

The higher blood levels of methylmercury in the fetus, coupled with the vulnerability of the developing nervous system to toxic insult, suggest that the fetus may be at greater risk from methylmercury exposure *in utero* than the newborn after birth or the adult. This possibility has recently attracted considerable interest. Japanese studies (78) have demonstrated that infantile Minamata disease can occur in infants exposed to methylmercury *in utero* whereas their mothers do not show symptoms of methylmercury intoxication. Similar data have been collected from accidental exposure of pregnant women to methylmercury in New Mexico (130) and Iraq (129). These data all indicate that exposure *in utero* to methylmercury may be much more critical than during the postnatal period or adulthood.

Lead

The toxicity of other environmental contaminants to the neonate is also currently a matter of great concern in public health. This is exemplified by the concerns of regulatory agencies regarding the exposure of infants to lead.

Abundant evidence supports the fact that during early life, human infants as well as the young of other animals are uniquely susceptible

TABLE 4.7 Methylmercury Distribution between Mother and Newborn Infant Blood

Mercury concentration (ng/g)		
Newborn Blood	Maternal Blood	Newborn/ Maternal Ratio
115	86	1.3 (100)
26.7	15	1.8 (101)
4,220	2,390	1.8 (101)
3,190	1,505	2.1 (102)
3,190	1,083	1.7 (102)

to lead exposure. This apparent increased sensitivity of the young to lead intoxication relates to differences in the pharmacokinetics of lead in infants as well as the greater susceptibility of developing organ systems to the toxic effects of this metal. For example, it is now well established that infants retain more of an oral lead dose than do adults. Studies in our laboratories have indicated that infant monkeys retained 60 to 70% of an oral lead dose whereas adult monkeys retained about 4%. Similar data are available for other animal species (131, 132) and humans (133, 134). This lower lead retention in adults would be reflected in lower blood lead levels, even at the same level of exposure as children. For example, the blood lead concentration of infant monkeys on milk diets was 25 to 35 μg Pb/dl whole blood with a lead exposure of 100 μg Pb/kg body weight/day. The infant monkeys were weaned onto an adult primate diet at 28 to 29 weeks of age. The blood lead concentrations declined to 15 to 20 μg Pb/dl whole blood even though the lead dose remained constant at 100 μg Pb/kg body weight/day (135). These data indicate that the change in lead retention is related to a change in diet.

A variety of dietary factors are known to influence lead absorption from the gut. Dietary levels of calcium, phosphates, iron, fat, protein, and vitamins D and E are known to affect lead absorption in rodents (136, 137). Also, the type of diet consumed (i.e., milk diet versus solid foods) alters lead absorption by rodents (137–139).

Other factors, such as differences in lead distribution within the body or changes in the excretion of lead from the body due to changes in diet, are currently under investigation in an attempt to explain the toxicological significance of the dietary effect. In this connection it has been established that in addition to differences in lead retention between adults and infants, lead is distributed differently within the body of adults than in infants. Table 4.8 shows the ratio of tissue to

TABLE 4.8 Brain Lead:Blood Lead Ratios in Infant and Adult Monkeys

Tissue	Age Group		
	10 Days	150 Days	Mature/Adult
Frontal cortex	0.22	0.13	0.08*
Temperal cortex	0.22	0.13	0.07*
Occipital cortex	0.21	0.14	0.08*
Cerebellum	0.20	0.11	0.05*
Thalamus	0.22	0.13	0.05*
Hypothalamus	0.25	0.11	0.06*
Pons	0.25	0.15	0.06*
Medulla	0.40	0.17	0.06*

Source. Reference 135

* Significantly different ($P < 0.01$)

blood lead was much greater for bone and brain in infant monkeys than adults. These data indicate that a greater portion of the retained lead is distributed to bone and brain in infants than in adults. This is consistent with the fact that the major symptoms of lead intoxication in infants are of CNS origin, whereas in adults the primary effects occur in the kidney and liver (140, 141).

It is well established that acute or subacute ingestion of lead by children results in encephalopathy, convulsions, and mental retardation (142, 143). Damage to the developing brain by chronic low-level intake of lead is more difficult to determine in the human population, as it must be inferred from behavioral tests in children, the interpretation of which is often subjective. There is also a problem in matching between control and treated subjects (144) since results may be subject to influence by nutritional and socioeconomic factors, age, sex, education, and previous history of lead exposure. Effects on general intelligence and perceptual, visual perceptual, and fine motor skills have been found by some investigators (145–149). Conflicting results have even been reported from the same population of children (150, 151).

In 1972 the World Health Organization (152) convened an Expert Committee to evaluate the lead problem. The committee was able to establish a provisional tolerable weekly intake of 3 mg lead/person or about 7 μg lead/kg body weight/day for adults. This was based on toxicity data in adult humans and the assumption that only 10% of orally ingested lead was absorbed. The committee noted that this figure did not include infants and young children due to uncertainties regarding the degree and the nature of increased risk in this population subset; no guidelines could be established for these groups. This has posed a difficult problem for those of us concerned with regulations relating to the acceptable levels of lead in infants' foods, particularly whole-milk and canned milk products.

To develop a further appreciation of the degree of risk for infants and young children, a study was undertaken by the Health Protection Branch (HPB) in 1975, using data from the 1972 Nutrition Canada Survey of 275 children. The aim of the study was to determine (*a*) the proportion of infants in Canada getting all or part of their milk from canned products, (*b*) estimates of the lead intake of infants in Canada consuming milk products, and (*c*) estimates of the percentages of infants with daily lead intakes above the WHO guidelines for adults. Table 4.9 shows the proportion of Canadian infants getting all or part of their milk intake from canned milk products. The majority of the remaining 60% had as their source of milk either powdered milk or 2% fluid milk. Based on levels of lead in milk and a consumption of 150 ml milk/kg body weight/day, the daily lead intake in infants from fluid milk and canned evaporated milk was estimated. The results revealed that infants fed canned milk had higher lead intake than did those fed whole milk. Furthermore, it was readily apparent that a substantial proportion of infants could exceed the WHO recommended intake for adults. These data indicate that every effort should be made to reduce the lead levels in infant foods to the minimum technologically feasible.

Recognizing the need for an integrated approach to the reduction of lead intake, especially for infants and young children, U.S. regulatory authorities have agreed to coordinate their programs. The EPA has established 50 μg/liter as the maximum permissable level for lead in drinking water, and thus in water used in food processing (153). The EPA has also established a national ambient air quality standard for lead of 1.5 μg/m³ as a monthly average (65). More recently the FDA issued an Advance Notice of Proposed Rulemaking relative to lead in food (64). The Occupational Safety and Health Administration (OSHA) has established a standard for occupational exposure to lead of 50 μg/m³ of air per 8 hours (154). A limitation of 0.06% lead in paint and in surface coatings of toys and furniture has been issued by the Consumer Product Safety Commission (155).

In its notice the FDA announced the intention to establish action levels for lead in evaporated milk products as a first priority followed by action levels for lead in canned infant formulas, canned infant fruit and vegetable juices, and glass-packed infant foods. These plans are based on the initial goal of reducing by at least 50% the contribution of lead intake from lead-soldered cans within the next 5 years. The ultimate goal is to reduce lead intake from all sources (air, water, and food) to less than 100 μg/day for children 1 to 5 years old (64).

Halogenated Hydrocarbons

Unlike the inorganic substances, many of the persistent organic contaminants show an affinity for fatty tissues. The extent to which this

TABLE 4.9 Estimates of the Proportions of Infants in Canada Getting All or Part of Their Milk from Canned Products[a]

Age Category	Proportion Getting All Milk from Canned Products	Proportion Getting Milk All or Partly from Canned Products
0-1 Year	.25	.29
0-5 Months	.37	.44
6-11 Months	.15	.17

Source. Kirkpatrick and Sandi, Bureau of Chemical Safety, Health Protection Branch.

[a] Based on Nutrition Canada Survey - Dietary Recall Data

may be expected to occur can be predicted from the partition coefficient of these materials in *n*-octanol–water mixture. Neely et al. (156) demonstrated that the short chained aliphatic halogenated substances are less likely to persist for long periods in body fat than are the more complex aromatic substances. This should not be interpreted to mean that the former are less toxic, for many of the short-chained substances, in addition to several of the aromatic substances, have been demonstrated to be carcinogenic in animals (157). Generally speaking, the straight-chained and branched and cyclic halogenated compounds are included in this group of carcinogens.

Apart from occupational exposure, humans are exposed to low levels of short-chained halogenated substances. This minor degree of exposure is not considered to be significant in comparison with widespread, low-level, continuous exposure to the more complex aromatic substances that tend to persist in the environment and have slow rates of elimination and accumulate in adipose tissue. At present there is a growing concern over the presence of these substances in the fat of human breast milk. It is well known that lactation results in a mobilization of body fat with a concomitant release of these substances into the milk fat. In Table 4.10 the intake by breast-fed infants of several halogenated hydrocarbons is compared with the current acceptable daily intake (ADI) for adults. The intake data on these chemicals assumes a daily consumption of 150 ml whole milk/kg body weight. In many instances the intake is very close to the ADI and for some substances exceeds it.

Following the marked restrictions placed on the use of the halogenated hydrocarbon pesticides, the levels of most of these substances in human milk appear to be declining. This is particularly true of DDT and its major metabolites, the levels of which have decreased dramatically from 1967 to the present (Table 4.11). Of more pressing concern at the moment is the presence of PCB in human breast milk. Their persistence in the environment, coupled with new knowledge regarding their toxicity, has heightened concern regarding the potential for adverse effects on the health of infants from these compounds.

In 1968 a disease known as *Yusho* was reported in Japan. The symptoms of this disease include chloracne, pigmentation of the skin and nails, swelling of eyelids with loss of eyelashes and eye discharge, and low birth weights. This disease was caused by the consumption of rice oil contaminated with PCBs from a leaking heat exchanger (158). It should be noted that the PCBs may have contained other contaminants, including the chlorinated dibenzofurans, which may have accounted, at least in part, for some of the observed symptoms (158). Table 4.12 shows that the smallest estimated dose of PCB causing toxic symptoms in the Yusho patients was estimated to be 0.10 mg/kg/day consumed over a period of 50 days.

TABLE 4.10 Comparison of Intake of Halogenated Hydrocarbons and ADIs (WHO)[a]

Compound	Levels in Whole Milk, μg/kg		Intake, μg/kg		ADI, μg/kg
	Mean	Max	Mean	Max	
PCB (Aroclor 1260)	12	68	1.8	10.2	1[b]
HCB	2	21	0.3	3.2	0.6
Heptachlorepoxide	1	3	0.15	0.45	0.5
Oxychlordane	1	2	0.3	0.6	1
Trans Nonachlor	1	2			
Dieldrin	2	6	0.3	0.9	0.1
DDT	44	213	6.6	32.2	5

Source. Health Protection Branch

a ADI, Acceptable Daily Intake, established by the World Health Organization.
b Canadian conditional ADI.

TABLE 4.11 Daily Intake of DDT from Human Milk by Nursing Infants Expressed as Percent of the ADI

	1 Month Infant			6 Month Infant		
	1967	1970	1975	1967	1970	1975
% of ADI	486	272	154	270	152	85.8

Source. Health Protection Branch

The PCB levels in the blood plasma of babies born to Yusho patients were considerably higher than those born to mothers without a history of PCB exposure. These data, coupled with symptomology in the infants, indicate that PCBs readily cross the placenta and accumulate in the fetus. Animal studies (159, 160) using monkeys further demonstrated the increased sensitivity of the fetus to PCB exposure. In these studies PCB was fed in the diet to monkeys at dose levels of 0.1 to 0.2 mg/kg body weight/day. Within 2 months, female monkeys showed clinical signs of toxicity similar to those observed in humans. Following 6 months of exposure, the monkeys were bred. The conception rate for the 0.1- and 0.2-mg/kg dose groups was 8/8 and 6/8, respectively. However, only five monkeys from the 0.1-mg/kg group and one from the 0.2-mg/kg group were born alive. These infant monkeys had a decreased body weight and showed focal areas of hyperpigmentation of the skin. Within 2 months of birth, during which time the infants suckled their mothers, the clinical signs worsened and three of the six infants died, including the one born to the dam receiving 0.2 mg/kg. Analysis of milk samples from the lactating monkeys revealed levels ranging from 150- to 400-ppb PCB. The estimated intake of PCB from milk was 0.03 to 0.075 mg/kg body weight/day.

After 4 months the remaining five infants were placed on a synthetic milk replacer, and their condition improved decidely over the next 4 months. It was reported, however, that these monkeys displayed learning difficulties and hyperkinesis following cessation of exposure (161).

The Yusho incident and the data on monkeys (159, 160) demonstrate clearly the transplacental toxicity of PCBs and raise concern over the health of human infants born to women with high PCB exposure, particularly if the mother breast feeds the infant. In 1975 the HPB undertook a survey of PCBs in human breast milk in Canada. In this survey of 100 samples the maximum PCB level was found to be 68 ppb and 98% of the samples had detectable levels. The mean level of PCB was 12 ppb. For the average human infant, this would provide an average intake of 0.001 mg/kg body weight/day (Table 4.13). The average daily intake of PCB by the infant monkeys was 0.03 to 0.07 mg/kg/day, which is roughly 10 to 80 times higher than the amount to which

TABLE 4.12 Lowest Estimated Doses of PCB Producing Toxicity in Adults

	Daily Dose	Duration
Adult Humans (Yusho Disease)	0.10 mg/kg/day	50 days
Adult Female Monkeys	0.10 mg/kg/day	60 days

Source. References 158-160

human infants are exposed. It should also be remembered that the infant monkeys were exposed to substantial amounts of PCB *in utero*.

Samples of human breast milk collected by EPA throughout the United States in 1974 and 1975 for the analysis of pesticide content have also been analyzed for PCBs (162). For 1038 of these samples, 309 i.e., 29.8% were positive (i.e., 50 ppb or greater on a whole-milk basis). In addition, 720 samples were classified as having trace amounts of PCBs and only 9 (i.e., <1%) were reported as zero. The maximum observed value was 563 ppb. Of the positive samples (50 ppb or greater), approximately 50 percent were less than 70 ppb.

Although the margin of safety for human infants exposed to PCB through their mother's milk is not definitively known at present, there have been no reports in the United States or Canada of the symptoms resembling those seen in PCB poisoning. The HPB has initiated a program to analyze breast milk for PCB in instances where the family physican feels, on the basis of detailed examination of the infant and the mother (including a history of her PCB exposure), that this would be desirable. Considering the many benefits of breast feeding, it is considered unlikely that Canadian mothers having elevated levels of PCB in their breast milk would be advised to discontinue breast feeding except in those instances where, in the view of the attending physican, this would be considered necessary. United States public health officials have arrived at a similar position, advising that decisions about breast feeding should be made by mothers in consultation with the attending physician.

REGULATORY CONTROL

Public health concern regarding contaminants in the food supply stems from the fact that the margin of safety between the actual exposure level in foods and the levels that will produce adverse effects may be quite narrow. For the problem contaminants, safety factors rarely exceed one order of magnitude and in some instances may not exist at all. Furthermore, many of these chemical have been detected in the environment in the last few years, and it is only recently that legisla-

TABLE 4.13 Estimated Intake of PCB through Breast Milk

	Daily Dose
Human Infants	0.002 mg/kg/day[a]
Infant Monkeys	0.03-0.07 mg/kg/day

[a] Based on data by J. Mes and D. J. Davies; Bull. Environ. Contam. Toxicol., 21:381-387 (1979).

tion has been enacted to control these materials in environmental media.

Regulatory choices for dealing with contaminants are limited. Banning of foodstuffs containing potentially hazardous levels of environmental contaminants is not usually an acceptable alternative since this will restrict the availability of otherwise nutritious food, place severe hardships on the producers of those foods, and increase food costs in the long run. Removing the offending substance from commerce may be acceptable in some instances but has the disadvantage that it deprives the economy of potentially useful chemical substances and will not, in any event, produce the desired early results because of the persistence of most contaminants in environmental media. Additionally, in many instances natural background levels contribute significantly to the total exposure, and obviously this cannot be controlled through regulatory actions. The regulatory option selected in most cases has been to establish some sort of limit for acceptable levels of contaminants in food commodities and to restrict their use in commerce.

Legal Considerations

Section 406 of the FFD&C Act provides the legal basis for controlling the levels of environmental contaminants in foods. It specifically states:

> Any poisonous or deleterious substance added to any food, except where such substance is required in the production thereof or cannot be avoided by good manufacturing practice shall be deemed to be unsafe for purposes of the application of clause (2) (A) of section 402 (*a*); but when such substance is so required or cannot be so avoided, the Secretary shall promulgate regulations limiting the quantity therein or thereon to such extent as he finds necessary for the protection of public health, and any quantity exceeding the limits so fixed shall also be deemed to be unsafe for purposes of the application of clause (2) (A) of section 402 (*a*). While such a regulation is in effect limiting the quantity of any such substance in the case of any food, such food shall not, by reason of bearing or containing any added amount of such substance, be considered to be adultered within the meaning of clause (1) of section 402 (a). In determining the quantity of such added substance to be tolerated in or on dif-

> ferent articles of food the Secretary shall take into account the extent to which the use of such substance is required or cannot be avoided in the production of each such article, and the other ways in which the consumer may be affected by the same or other poisonous or deleterious substances.

In addition, Section 408 of the FFD&C Act provides for "tolerances for pesticide chemicals in or on raw agricultural commodities" in a similar fashion.

The FDA has published procedural regulations in Part 109, Title 21 of the Code of Federal Regulations for the establishment of "tolerances" and "action levels," both of which set limits on unavoidable contaminants in foods. In both cases, if deemed necessary to protect the public health, these limits may prohibit any detectable amounts of the substance in question. Basically, tolerances, which entail formal rule-making procedures, are established when changes in permitted levels are not anticipated in the near future. Action levels can be established quickly and more informally and are used when changes in permitted levels may occur in the forseeable future.

Formal tolerances have been established by the FDA only for PCBs. These limits have been set at 5 ppm for fish, 1.5 ppm (on a fat basis) for milk and dairy products, 3 ppm (on a fat basis) for poultry, and 0.3 ppm for eggs (85). A tolerance of 0.3 ppm was proposed for lead in evaporated milk and evaporated skim milk. However, that proposal will be withdrawn and action levels will be set for lead in evaporated milk, infant and toddler foods, and adult foods consumed by young children as a first priority for controlling lead exposure.

Among the action levels established by the FDA are those for mercury in fish (changed in 1978 from 0.5 ppm to 1.0 ppm) and leachable lead from ceramic dinnerware, enamelware, pewter, and silver-plated hollowware (7 ppm for shallow items and 0.5 ppm for silver-plated cups intended for use by infants). A variety of action levels for pesticide residues have also been established jointly by the FDA and the EPA. Some examples are shown in Table 4.14. Canada and other countries throughout the world also utilize similar limits or guidelines for regulatory control of environmental contaminants, although the specific numerical limits may vary somewhat from country to country at any given point in time. For example, guidelines have been established in Canada that prohibit the sale of fishery products containing in excess of 0.5 ppm of mercury or 2 ppm of PCBs.

SAFETY CONSIDERATIONS

In developing limits for environmental contaminants, regulators must balance the risk from consumption of contaminated products against the economic and other losses to the affected industries.

TABLE 4.14 Action Levels for Pesticide Residues in Foods

Residue/Commodity	Action Level (ppm)
Aldrin and Dieldrin	
Artichokes, figs, small fruits	0.05
Eggs	0.03
Fish and shellfish	0.3
Melons	0.15
Milk and dairy products	0.3 (fat basis)
DDT, DDE and TDE	
Grains: barley, corn, milo, oats, rice, rye, wheat	0.5
Cocoa beans, whole, raw	2.0
Eggs	1.5
Fish	5.0
Kepone	
Fish, shellfish	0.3
Crabmeat	0.4
Mirex	
Fish	0.1
Dibromochloropropane	
Raw agricultural commodities	0.05
Milk	1.5

Source. Food and Drug Administration, Adulteration - Pesticide Residues; Washington, D.C., 1978.

Further study of this problem leads to the conclusion that the concept of establishing guidelines or tolerable levels of contaminants in food requires some reexamination. The guideline approach has several disadvantages that require consideration. Toxicity is a function of the dose, and with environmental contaminants the dose is the product of the level of contaminant in the food and the amount of food consumed per day. Thus it is easily reckoned that without information on tolerable consumption patterns, guidelines do not provide the ultimate in the way of health protection. On the contrary, announcements by governments of guidelines on levels of contaminants in food is usually interpreted by the public to mean that the product is now safe for consumption. The fallacy in this approach becomes readily apparent when one considers that consumption of as little as 60 g of fish containing 0.5 ppm of mercury would result in an individual reaching the World Health Organization provisional tolerable daily intake of 30 μg/day. Admittedly, the average person would not consume that much fish containing 0.5 ppm mercury on a daily basis, but it has been well established that certain fish consumers could easily exceed this quantity. Moreover, regulatory use of guidelines implies that all individuals in the population are of equal senstitivity to contaminants. In other words, the special sensitivity of pregnant mothers, infants, and the very old are not necessarily reflected in the guideline approach. Further, establishment of across-the-board guidelines, say, for all fish products, have had marked economic impact on certain segments of the food industry. For example, in the United States and Canada consumption of swordfish essentially ceased following pronouncement of the 0.5-ppm guideline.

In addition to economic hardship, the guideline has decreased the availability of important sources of dietary protein, not only from swordfish, but from several other marine and freshwater species, many of which exceed the guideline depending on their size, food supply, and geographical location.

It will become increasingly important in the future for regulatory agencies to deal effectively with this problem. One means of accomplishing this would be to couple the guideline approach with programs aimed at increasing consumer awareness of toxicological concerns much the same as we are doing through nutrition labeling. The position that the consumer has the "right to know" must be met through education and labeling programs designed to educate the consumer regarding the hazards of environmental contaminants. It would be useful to provide the consumer with information on suggested maximum daily or weekly consumption of contaminants, as has been done in Scandinavia. In this way the consumer has the ultimate decision, and the special sensitivity of certain population subsets would be taken care of. Follow-up studies would have to be conducted to determine whether such a program was successful. This approach has the additional advantage of reducing economic hardship on affected segments of the food industry and at the same time optimizing the availability of otherwise nutritious food.

Another concern with respect to environmental contaminants and food safety is that of the unregulated contaminants. For those contaminants that have guideline or action level limits in certain foods, regulatory programs, coupled with public education, can be relatively effective in protecting against food hazards. However, that type of regulatory effort by itself does not address those potential contaminants for which guidelines do not currently exist. To effectively reduce food hazards arising from environmental contaminants, attention must also be paid to these latter contaminants.

The technology clearly exists for improved investigatory monitoring to identify potential environmental contaminants of food (1). What is needed initially is a recognition of the importance of this area and the assignment of resources commensurate with the priorities of food safety. With such priority assignment, integrated regulatory-investigatory programs could be implemented involving coordination among federal agencies and between the federal and state agencies. Only in this way can we reduce or prevent the entry of environmental contaminants into the food supply or, in the case of existing and accidentally introduced contaminants, effectively control the safety of food.

REFERENCES

1. Office of Technology Assessment, *Environmental Contaminants in Food,* U.S. Government Printing Office, Washington, D.C., 1979.
2. *Report to the President by The Toxic Substances Strategy Committee* (public review draft), Washington, D.C., 1979.
3. F. Cordle, P. Corneliussen, C. Jelinek, B. Hackley, R. Lehman, J. McLaughlin, R. Rhoden, and R. Shapiro, *Environ. Health, Perspec.,* 24, 157 (1978).
4. E. P. Savage, J. D. Tessari, J. W. Malberg, H. W. Wheeler, and J. R. Bagby, *Pestic. Monitoring J.,* 7, 1 (1973).
5. J. Mes and D. J. Davies, *Chemosphere,* No. 9, 699 (1978).
6. M. N. Brady and D. S. Siyali, *Med. J. Aust.,* 1, 158 (1972).
7. E. Homberger, G. Reggiani, J. Sambeth, and H. Wipf, draft document issued by Givaudan Research Company Ltd. and R. Hoffman-LaRoche Company Ltd. (1976).
8. R. Fanelli, M. P. Bertoni, M. Bonfanti, M. G. Castelli, C. Chiabrando, G. P. Martelli, M.A. Noè, A. Noseda, S. Garranttini, C. Binaghi, V. Marazza, F. Pezza, D. Pozzoli, and G. Cicognetti, *Bull. Environ. Contam. Toxicol.,* 24, 634 (1980).
9. J. D. McKinney, *Ecol. Bull.* (Stockholm) 27, 55 (1978)
10. V. Zitko, O. Hutzinger, and P. M. K. Choi, *Bull. Environ. Contam. Toxicol.,* 12, 649 (1974).
11. R. C. Dougherty and K. Piotrowsak, *Proc. Nat. Acad. Sci.,* 73, 1777 (1976).
12. R. C. Dougherty, L. Smith, D. Stalling, C. Raffe, and D. W. Kuehl, Abstr. No. 38, 178th American Chemical Society Conference, Washington, D.C., September 1979.
13. D. D. Manske and P. E. Corneliussen, *Pestic. Monitoring J.,* 8, 110 (1974).
14. D. D. Manske and R. D. Johnson, *Pestic. Monitoring J.,* 9, 94 (1975).
15. I. Graca, A. M. S. Silva Fernandes, and H. C. Mourao, *Pestic. Monitoring J.,* 8, 148 (1974).
16. J. Mes and D. J. Davies, *Bull. Environ. Contam. Toxicol.,* 21, 381 (1979).
17. C. I. Stacey and B. W. Thomas, *Pestic. Monitoring J.,* 9, 64 (1975).
18. Fisheries and Environment Canada and Health and Welfare Canada, Technical Report 77-1 (1977).
19. K. S. Khera, in R. A. Goyer and M. A. Mehlman, Eds., *Advances in Modern Toxicology,* Vol. 1, Wiley, New York, 1976, Part 1, Chapter 12.

20. K. M. Hyde, J. B. Graves, J. B. Fowler, F. L. Bonner, J. W. Impson, J. D. Newsom, and J. Haygood, *La. Agr. Exp. Sta. Bull.*, 17 (1), 10 (1973.)

21. J. Mes, D. J. Davies, and W. Miles, *Bull. Environ. Contam. Toxicol.*, 19, 564 (1978).

22. C. A. Edwards, *Persistent Pesticides in the Environment*, 2nd ed., CRC Press, Cleveland, Ohio, 1973.

23. Studies on the Health Effects of Alkylmercury in Japan, Environment Agency, Japan (1975).

24. Report from an Expert Group, *Nordisk Hygienisk Tidskrift*, Suppl. 4 (1971).

25. *Bulletin of the World Health Organization*, Supplement to Vol. 53 (1976).

26. *Lead in the Canadian Environment*, National Research Council Canada, NRCC, No. 13682 (1973).

27. D. J. Snodin, *J. Assoc. Pub. Anal.*, 11, 47 (1973).

28. D. M. Settle and C. C. Patterson, *Science*, 207, 1167 (1980).

29. A. C. Kolbye, K. R. Mahaffey, J. A. Fiorina, P. C. Corneliussen, and C. F. Jelinek, *Environ. Health Perspec.*, 7, 65 (1974).

30. M. Webb, *Br. Med. Bull.*, 31 (3), 246 (1975).

31. M. Fleischer, A. Sorafim, D. C. Fasset, P. Hammond, H. T. Shacklette, I. C. T. Nixbet, and S. Epstein, *Environ. Health Perspec.*, Exp. Issue, No. 7, 253 (1974).

32. R. M. Orheim, L. Lippman, D. J. Johnson, and H. H. Bovee, *Environ. Lett.*, 7 (3), 229 (1974).

33. R. A. Goyer and M. A. Mehlman, Eds., *Advances in Modern Toxicology*, Vol. 2, Wiley, New York, 1977.

34. C. F. Jelinek and P. E. Corneliussen, *Environ. Health Perspec.*, 19, 83 (1977).

35. L. Fishbein, *Environmental Health Aspects of Germanium, Titanium and Tin*, National Center of Toxicological Research, Jefferson, Arkansas (1976).

36. W. J. Nicholson and J. A. Moore, "Health Effects of Halogenated Aromatic Hydrocarbons," *Ann. N. Y. Acad. Sci.*, 320 (1979).

37. G. D. Vieth, *Environ. Health Perspec.* 1, 51 (1972).

38. Interdepartment Task Force on PCBs, Department of Health, Education and Welfare, Washington, D. C. (1972).

39. J. Sarkka, *Arch. Environ. Contam. Toxicol.*, 8 (2), 161 (1979).

40. J. Lawrence and H. M. Tosine, *Ontario Bull. Environ. Contam. Toxicol.*, 17, 49 (1977).

41. H. E. B. Humphrey, H. A. Price, and M. L. Budd, Final Report of FDA Contract No. 233-73-2209 (1976).

42. I. Watanabe, T. Yakushiji, K. Kuwabara, S. Yoshida, K. Maeda, T. Kashimoto, K. Koyama, and N. Kunita, *Arch. Environ. Contam. Toxicol.*, 8, 67 (1979).

43. C. F. Jelinek and P. E. Corneliussen, *Proceedings of National Conference on PCB's*, Washington, D. C. (1976).
44. Anonymous, *Food Chem. News*, 22 (2), 23 (1980).
45. M. E. Zabik, P. Hoojjat, and C. M. Weaver, *Bull. Environ. Contam. Toxicol.*, 21, 136 (1979).
46. Advisory Committee on 2, 4, 5-T, Report to the Administrator of the Environmental Protection Agency, Washington, D.C. (1971).
47. R. J. Kociba, D. G. Keyes, J. E. Beyer, R. M. Carreon, C. E. Wade, D. A. Dittenber, R. P. Kalnins, L. E. Frauson, C. N. Park, S. D. Barnard, R. A. Hummel, and C. G. Humiston, *Toxicol. Appl. Pharmacol.*, 46, 279 (1978).
48. R. H. Shehl, R. R. Papenfuss, R. A. Bredeweg, and R. W. Roberts, *Advances in Chemistry*, American Chemical Society, Washington, D.C., 1973.
49. B. A. Schwetz, P. A. Keeler, and P. J. Gehring, *Toxicol. Appl. Pharmacol.*, 28, 151 (1974).
50. National Academy of Sciences—National Research Council, the Report of the Executive Committee, Washington, D. C. (1975).
51. G. W. Bowes, M. J. Mulvihill, B. R. T. Simonet, A. L. Burlingame, and R. W. Risebrough, *Nature*, 256, 305 (1975).
52. J. A. Goldstein, M. Friesen, R. E. Linder, P. Hickman, J. R. Hass, and H. Bergman, *Biochem. Pharmacol.*, 26, 1549 (1977).
53. K. Olie, P. L. Vermuelen, and O. Hutzinger, *Chemosphere*, 8, 455 (1977).
54. C. Rappe, H. R. Buser, and H. Bosshardt, *Ann. N. Y. Acad. Sci.*, 320, 1 (1979).
55. J. A. Moore, E. E. McConnell, D. W. Dalgard, and M. W. Harris, *Ann. N. Y. Acad. Sci.*, 320, 151 (1979).
56. C. Cam and G. Higogosyan, *J. Am. Med. Assoc.*, 183, 88 (1963).
57. N. H. Booth and J. R. McDowell, *J. Am. Vet. Med. Assoc.*, 166, 591 (1975).
58. *Summary Report of Data Received from Collaborating Centres for Food Contamination Monitoring Stage 1—1977*, World Health Organization (1979).
59. F. W. Kutz, A. R. Yobs, W. G. Johson, and G. B. Wiersma, *Environ. Entimol.*, 3 (5), 882 (1974).
60. E. M. Waters, J. E. Huff, and H. B. Gerstner, *Environ. Res.*, 14, 212 (1977).
61. The National Research Council, *Drinking Water and Health*, Washington, D.C. (1977).
62. R. A. Wallace, W. Fulkerson, W. D. Shutls, and W. S. Lyon, ORNL NSF-EP-1, Oak Ridge, Tenn. (1971).
63. Anonymous "The Decline in Mercury Concentration in Fish from Lake St. Clair 1970–76," Ontario Ministry of the Environment, report No. AQS 77-3 (1977).
64. Food and Drug Administration, *Fed. Reg.*, 44, 51233 (1979).

65. Environmental Protection Agency, *Fed. Reg.* 43, 46246 (1978).
66. K. Nogawa, A. Ishizaki, and S. Kawano, *Environ. Res.*, 15 (2), 185 (1978).
67. B. Fowler, *Environ. Health Perspec.*, 28, 297 (1979).
68. E. I. Hamilton, M. J. Minski, J. J. Cleary, and V. S. Halsey, *Sci. Total Environ.*, 1, 205 (1972).
69. R. S. Braman and M. A. Tompkins, *Anal. Chem.*, 51, 12 (1979).
70. V. F. Hodge, S. L. Seidel, and E. D. Goldberg, *Anal. Chem.*, 51, 1256 (1979.).
71. R. Lillis, H. A. Anderson, J. A. Valcinkas, S. Freedman, and I. J. Selikoff, *Environ. Health Perspec.*, 23, 105 (1978).
72. W. D. Meester and D. J. McCoy, *Clin. Toxicol.*, 10 (4), 474 (1977).
73. K. Kay, *Environ. Res.*, 13, 74 (1977).
74. M. S. Wolff, B. Aubrey, F. Camper, and N. Hames, *Environ. Health Perspec.*, 23, 177 (1978).
75. Environmental Protection Agency, "Assessment of the Hazards of Polybrominated Biphenyls (PBB's)," draft report of The Special Actions Group, Office of Toxic Substances, Washington, D.C., 1977.
76. G. Whitehead, Michigan Department of Agriculture, January 16, 1978.
77. Anonymous, *Food Chem. News*, 21 (33), 48 (1979).
78. *Environmental Health Criteria 1*, World Health Organization, Geneva (1976).
79. J. M. Borgono, P. Vincent, H. Venturino, and A. Infante, *Environ. Health Perspec.*, 19, 103 (1977).
80. G. Lunde, *Environ. Health Perspec.* 19, 47 (1977).
81. O. Walkiw and D. R. Douglas, *Clin. Toxicol.*, 8, 325 (1975).
82. H. L. Cannon, *Symposium on Environmental Quality in Food Supply*, Futura Publishers, Mount Kisco, New York, (1974).
83. H. C. Freeman and J. F. Uthe, International Council for the Exploration of the Sea, Copenhagen, 1974.
84. I. C. Munro and S. M. Charbonneau, *Fed. Proc.*, 37, 2582 (1978).
85. G. E. Miller, P. M. Grant, R. Kishore, F. J. Steinkruger, F. S. Rowland, and V. P. Guin, *Science*, 175, 114 (1972).
86. J. B. Palmer and G. M. Rand, *Bull. Environ. Contam. Toxicol.*, 18 (4) 512 (1977).
87. K. R. Bull, R. K. Murton, D. Osborn, P. Ward, and L. Cheng, *Nature*, 269, 507 (1977).
88. W. Tseng, *Environ. Health Perspec.*, 19, 109 (1977).
89. J. T. Hindmarsh, O. R. McLetchie, L. P. M. Hefferman, O. A. Hayne, H. A. Ellenberg, R. F. McCurdy, and H. J. Thiebaux, *J. Anal. Toxicol.*, 1, 270 (1976).
90. G. Lunde, *J. Sci. Food Agr.*, 24, 1021 (1973).
91. L. Fishbein, *Ann. Rev. Pharmacol.*, 14, 139 (1974).
92. R. D. Kimbrough, *CRC Crit. Rev. Toxicol.*, 2, 445 (1974).
93. Food and Drug Administration, *Fed. Reg.*, 42, 17487 (1977).

94. Food and Drug Administration, *Fed. Reg.*, 44, 38330 (1979).

95. T. Taksukawa, *PCB Poisoning and Pollution*, Kodansha Ltd, Tokyo, 1976.

96. K. Fujiwara, *Sci. Total Environ.*, 4, 219 (1975).

97. S. Jensen, A. G. Johnels, M. Olsson, and G. Otterlind, *Nature*, 224, 247 (1969).

98. V. Zitko, O. Hutzinger, and M. K. Choi, *Environ. Health Perspec.*, 1, 47 (1972).

99. I. Pomerantz, J. Burke, D. Firestone, J. McKinney, J. Roach, and W. Trotter, *Environ. Health Perspec.*, 25, 133 (1978).

100. H. Matthews, G. Fries, A. Gardner, L. Garthoff, J. Goldstein, Y. Ku, and J. Moore, *Environ. Health Perspec.*, 24, 147 (1978).

101. G. F. Fries, G. S. Marrow, and C. H. Gorden, *J. Agr. Food Chem.*, 21, 117 (1973).

102. G. F. Fries, R. J. Lillie, H. C. Cecil, and J. Bitman, Paper presented at 165th Meeting, American Chemical Society (abstr.) (1973).

103. L. W. Smith, G. F. Fries, and B. T. Weinland, *J. Dairy Sci.*, 59, (1976).

104. H. B. Matthews and M. W. Anderson, *Drug Metals Disp.*, 3, 371 (1975).

105. S. Jensen and G, Sundström, *Ambio*, 3, 70 (1974).

106. J. D. McKinney, K. Chae, B. N. Gupta, J. A. Moore, and J. A. Goldstein, *Toxicol. Appl. Pharmacol.*, 36, 65 (1976).

107. J. A. Goldstein, J. D. McKinney, G. W. Lucier, P. Hickman, H. Bergman, and J. A. Moore, *Toxicol. Appl. Pharmacol.*, 36, 81 (1976).

108. M. Blocca, paper presented at National Conference on Polychlorinated Biphenyls, Chicago (1975).

109. F. Bingham *Environ. Health Perspec.*, 28, 39 (1979).

110. L. G. Hansen and T. D. Hinesly, *Environ. Health Perspec.*, 28, 51 (1979)

111. T. J. Kneip and R. E. Hazen, *Environ. Health Perspec.*, 28, 67 (1973).

112. J. M. Frazier, *Environ. Health Perspec.*, 28, 75 (1979).

113. E. J. Underwood, "Trace Elements," in National Academy of Sciences, *Toxicants Occurring Naturally in Foods*, 2nd ed., Washington, D.C., 1973, Pp. 43–87.

114. C. F. Jelinek and P. E. Corneliussen, *Environ. Health Perspec.*, 19, 83 (1977).

115. I. C. Munro, *Clin. Toxicol.*, 9, 647 (1976).

116. C. Pomroy, S. M. Charbonneau, R. S. McCullough, and G. K. H. Tam, *Toxicol, Appl. Pharmacol.*, 53, 550 (1980).

117. Committee on Medical and Biological Effects of Environmental Pollutants, *Arsenic*, National Academy of Sciences, Washington, D.C.

118. J. S. Edmonds and K. A. Francesoni, *Nature*, 265, 436 (1977).

119. R. V. Cooney, R. O. Mumma, and A. A. Benson, *Proc. Nat. Acad. Sci.*, 75, 4262 (1978).

120. J. S. Edmonds, K. A. Francesoni, J. R. Cannon, C. L. Raston, B. W. Skelton, and A. H. White, *Tetrahedron Lett.*, 18, 1543, (1977).

121. H. C. Freeman, H. F. Uthe, R. B. Fleming, P. H. Odense, R. G. Ackman, G. Landry, and C. Musial, *Bull. Environ. Contam. Toxicol.*, 22, 224, (1979).

122. S. M. Charbonneau, G. K. H. Tam, F. Bryce, and E. Sandi, abstract No. 362, 19th Annual Meeting of the Society of Toxicology, Washington, D.C., March 1980.

123. R. Wagermann, N. B. Snow, D. M. Rosenberg, and A. Lutz, *Arch. Environ. Contam. Toxicol.*, 7, 169 (1978).

124. *Task Force on Arsenic, Yellowknife, Northwest Territories,* Canadian Public Health Association, Ottawa, Canada (1977).

125. C. K. Schuth, A. R. Isensee, E. A. Woolson, and P. C. Kearney, *J. Agr. Food Chem.*, 22 (6), 999 (1974).

126. J. M. Wood, *Science*, 183, 1049 (1974).

127. T. Suzuki, T. Miyama, and H. Katsunuma, *Bull. Environ. Contam. Toxicol.*, 5, 502 (1972)

128. C. A. R. Dennis and F. Fehr, *Sci. Total Environ.*, 3, 275 (1975).

129. L. Amin-Zaki, S. Elhassani, M. A. Majeed, T. W. Clarkson, R. A. Dohery, and M. R. Greenwood, *J. Pediatr.*, 85, 81 (1974).

130. R. D. Synder, *New Engl. J. Med.*, 284, 1014 (1971).

131. G. B. Forbes and J. C. Reina, *J. Nutr.*, 102, 647 (1972).

132. K. Kostial, I. Simonovic, and M. Pisonic, *Nature*, 233, 564 (1971).

133. F. W. Alexander, *Environ. Health Perspec.*, 7, 155 (1971).

134. R. A. Keohe, *J. R. Inst. Public Health Hyg.*, 24, 1 (1960).

135. R. F. Willes, D. Rice, and J. Truelove, in R. Singal, Ed., *Lead Toxicity*, Arbon and Swansenberg, Baltimore (in press), Chapter 9.

136. D. Barltrop and H. E. Khoo, *Postgrad. Med. J.*, 51, 795 (1975).

137. R. Stephens and H. A. Waldron, *Food Cosmet. Toxicol.*, 13, 555 (1955).

138. D. Kello and K. Kostial, *Environ. Res.*, 6, 355 (1973).

139. K. Kostial, D. Kello, S. Jugo, I. Rabar, and T. Maljkovic, *Environ. Health Perspec.*, **25**, 81 (1978).

140. J. S. Lin Fu, *New Engl. J. Med.*, 286, 702 (1972).

141. H. A. Waldron, *Prev. Med.*, **4**, 135 (1975).

142. D. Thurston, J. Middlekamp, and E. Mason, *J. Pediat.*, 47, 413 (1955).

143. M. Perlstein and R. Attala, *Clin. Pediat.*, 5, 292 (1966).

144. T. Kamstra, *Environ, Health Perspec.*, 19, 297 (1977).

145. A. Beattie, M. Moore, O. Golberg, M. Findlayson, J. Graham, E. Mackie, J. Maui, D. McLaren, R. Murdock, and G. Stewart, *The Lancet*, 3, 589 (1975).

146. B. De la Burdé and M. Choate, *J. Pediat.*, 87, 638 (1975).

147. P. Landrigan, R. Whitworth, R. Baloh, N. Stachling, W. Barthel, and B. Rosenblum, *The Lancet*, 1, 708 (1975).

148. R. Baloh, R. Sturon, B. Green, and G. Glesi, *Arch. Neurol.*, 32, 326 (1975).

149. R. Lansdown, J. Shephard, B. Clayton, H. Delves, P. Graham, and W. Turner, *The Lancet*, 3, 538 (1974).

150. J. McNeil, J. Ptasnik, and B. Croft, *Arch. za Higijenu Rada in Toksikilogijie*, 26 (Suppl.), 97 (1975).

151. R. Whitworth, B. Rosenblum, M. Dickerson, and R. Baloh, *CDC Morbid. and Mortal. Weekly Rep.*, 23, 157 (1974).

152. FAO/WHO Expert Committee on Food Additives, *16th Report of the Joint FAO/WHO Expert Committee on Food Additives*, World Health Organization Technical Report Series, 505, FAO Nutrition Meeting Report Series, 51 (1972).

153. *Code of Federal Regulations, Title 40*, Section 141.11, U.S. Government Printing Office, Washington, D.C., 1979.

154. Occupational Safety and Health Administration, *Fed. Reg.*, 43, 52952, 1978.

155. Consumer Product Safety Commission, *Fed. Reg.*, 42, 44193 (1977).

156. W. B. Neely, D. R. Branson, and G. E. Blau, *Environ. Sci. Technol.*, 8, 113 (1974).

157. L. Tomatis, C. Agthe, H. Batsch, J. Huff, R. Montesano, R. Saracci, E. Walker, and J. Wilbourn, *Cancer Res.*, 38, 877 (1978).

158. *Environmental Health Criteria—2*, World Health Organization, Geneva (1976).

159. J. R. Allen and D. A. Barsotti, *Toxicology*, 6, 331 (1976).

160. D. A. Barsotti, R. J. Marlar, and J. R. Allen, *Food Cosmet. Toxicol.*, 14, 99 (1976).

161. R. E. Bowman, M. P. Heironimus, and J. R. Allen, *Pharmacol. Biochem. Behav.*, 9, 49 (1978).

162. B. W. Miller, *J. Pediat.*, 90 (3), 510 (1977).

CHAPTER 5

Food Hazards of Natural Origin

J. V. RODRICKS and A. E. POHLAND

Food is by far the most chemically complex substance to which human beings are continuously and directly exposed. The various plant, animal, and marine products that humankind has found acceptable for the dinner plate naturally contain a vast number of chemical compounds, most of which have not been identified. These products contain not only the substances that have nutritional value or that supply calories, but also a host of compounds that impart color or flavor. Numerous other compounds, having no known nutritional role and possessing no characteristics that may have otherwise influenced the human decision to accept the products in which they are present, can also be found in food, especially those of plant origin. No accurate estimate of the number of compounds naturally present in food can be made, but surely the range is in the tens if not hundreds of thousands. The small fraction of these compounds that have been identified display the diversity of structural characteristics that can be found in any comprehensive organic chemistry textbook. (Interestingly, the only exception to the last statement appears to be the class of compounds containing the carbon–halogen bond. Such compounds are relatively rare in nature but are among the classes of synthetic compounds that have frequently been implicated as potentially hazardous environmental pollutants.)

Of course, men and women have never been content simply to accept what nature had to offer and, for a variety of practical, cultural, or aesthetic reasons, have found ways to modify nature's produce. The discovery of fire brought cooking. Because cooking results in numerous chemical changes in the natural components of food, this early technological advance brought with it a multitude of new chemical additions to the human diet. The way in which food is cooked (over an open fire, roasting, frying, boiling, etc.) and the temperature and the duration of cooking influence the course of chemical change and the nature of the products introduced, so that any close examination of the chemistry of cooking requires that the conditions of treatment be specified.

Spices and condiments were introduced in ancient times primarily for purposes of food preservation and added a huge number of new compounds to the human dietary. Certain horticultural specialties came into use when they were found pleasing to the palate (e.g., coffee and tea) and added their share of chemicals. The practices of drying, fermenting, pickling, salting, and smoking food, introduced over a period of several centuries, further increased the number and the type of compounds to which humans are exposed through food. Of course, the nineteenth century revolution in synthetic chemistry yielded many substances that have proved useful in production, processing, and packaging of food (pesticides, drugs used in food animal production, colors, sweeteners, preservatives, etc.). Although as yet poorly understood, genetic manipulation of food plants and animals to improve their characteristics (i.e., the introduction of new varieties and breeds) have no doubt resulted in further additions to the natural background of human exposure to chemicals.

A final source of chemicals in the human diet includes those substances of either synthetic or natural origin that *contaminate* food. Contamination may result from a deliberate act of adulteration, the growth of toxin-producing microorganisms on food, an accident such as a chemical spill, or general environmental pollution.

Even this superficial survey of the origins and the types of food chemicals should be sufficient to document the truth of the opening statement of this chapter. No doubt chemists will continue to probe food to add to our knowledge of the number and the types of compounds we are ingesting every day of our lives. One important reason why such research activities should be supported derives from our knowledge that exposure to some chemicals can threaten human health.

HEALTH HAZARDS IN FOOD

During the past decade, and apparently with increasing intensity, a significant share of public health resources has been devoted to increasing our knowledge of the health effects of chemicals. Most of this effort has been devoted to synthetic chemicals contaminating food, air, or water or deliberately added to food (intentional food additives). This is no doubt an important way to spend resources, but it may have obscured the fact that the chemical exposures humans have experienced since the advent of the industrial revolution simply add to an already huge burden resulting from human exposure to natural materials, mostly foods. It is a fundamental precept of toxicological science that the health risks associated with chemicals are a function of their inherent capacities to damage the human organism (i.e., their toxicities) and the level and the frequency of human exposure. Whether a chemical is natural in origin or is synthetic is not relevant to the question of the health risks it may pose. The natural chemical constituents of food provide a huge reservoir of chemicals for toxicological study.

It has been argued, and in some respects reasonably so, that the fruits of the synthetic chemical industry contain some chemical moieties unknown in the natural world and that may render them far more hazardous than anything nature may yield. Such a statement might make sense if our knowledge of the natural constituents of food were not so meager and if we were far more sure than we are about the relationships between chemical structure and toxicity. The statement might also be credible if it could be assumed that exposure to the natural constituents of food over many generations has resulted in some form of human "adaptation" that permits us to ingest such substances *ad libitum* with no attendant risk. Unfortunately, this assumption has no foundation in experience, especially insofar as it deals with chronic and injurious effects that cannot be detected in ordinary experience. It further suffers from our knowledge, to be discussed in later sections of this chapter, that a good many of the natural constituents of food have not been present in the human diet for more than a few generations.

Of course, a number of naturally occurring constituents of the human diet are known to pose health risks, and they are the subjects of this chapter. The theme that should underlie the survey to be made is that what we now know about the health effects of natural food constituents is but a small fraction of what remains to learned.

Substances to Be Examined

Food hazards of natural origin (chemical rather than biological) are surveyed in this chapter, but, as might be apparent from the opening discussion, it is not easy to offer a strict definition of the universe of substances to be covered. Many questions arise. For example, are the substances introduced by charcoal broiling of hamburgers to be considered "natural"? If so, is it also reasonable to include chemical alteration products that might be formed when food is cooked in a microwave oven? Are certain metals present in some plant foods to be considered natural components, even if their accumulation has resulted in part from the use of inorganic fertilizers on the soil in which they are grown? The picture is further complicated by the requirements of federal food laws in which any substance intentionally added to food is to be considered a food additive, even if the subtance is natural in origin (as many food additives are).

For purposes of this chapter, a rather broad view of natual hazards is taken. Synthetic chemicals are not discussed. Naturally occurring substances whose prime uses are covered by the food additive laws are avoided, although not completely. (Food additives are discussed in Chapter 6.) Toxic metals are given only scant attention because the important ones are covered in Chapter 4 of this book. The same is true of bacterial toxins, a subject of such importance and breadth that it is separately covered in Chapter 2. The nutrients are excluded as they, too, are treated separately in Chapter 3. All other food constituents are discussed in the present chapter.

Organization of the Chapter

The substances to be examined in this chapter could be organized according to chemical type or biological (toxicological or pharmacological) property. Both of these modes of organization are impractical because of the great diversity of chemical and biological properties displayed by these substances.

A useful and practical form for presentation of this material is one that rests on grouping the substances according to the manner in which they become components of food. Such a scheme results in the following five broad categories: (*a*) intrinsic components of foods of plant origin, (*b*) natural constituents of soil and water that accumulate in foods, (*c*) metabolites of microorganisms that grow on food, (*d*) compounds of natural origin contaminating edible animal products, and (*e*) compounds produced by chemical reactions occurring during food storage, processing, or preparation.

Organization of food hazards of natural origin according to the preceding scheme was based on the notion that the ease with which any identified risk can be reduced, eliminated, or otherwise managed is dependent largely on the mode of entry of the hazardous substance into the food supply (1, 2). Regulation of food hazards of natural origin also takes into account the relative difficulties involved in controlling such substances. Regulation and other means of controlling food hazards of natural origin are reviewed in the closing section of the chapter.

INTRINSIC COMPONENTS OF FOODS OF PLANT ORIGIN

Compounds in this group constitute the largest number and greatest variety of chemicals present in food; most have not been chemically or biologically characterized.

Although a number of the substances contained in this group display relatively high acute toxicity, most do not present an acute human health hazard unless the foods in which they are present are consumed in extraordinarily large amounts. A few of the intrinsic components of food are known to be injurious after chronic or subchronic exposure in experimental animals, but there has been little done to assess their potential health risks to humans at the (usually) much lower levels of human exposure. Thus, with few exceptions, little can now be said about how serious a public health problem the intrinsic components of food may represent. The evidence is certainly suggestive, however, and in many cases more extensive study is clearly indicated. Some of the better-known substances comprising this group are described in the following.

Oxalates

Salts of oxalic acid are widespread in foods of plant origin and can be found in unusual abundance in rhubarb, tea, cocoa, spinach, and beet leaves (3, 4). The oxalate content of these plant foods has been reported to range from approximately 0.2 to 1.3%. Lesser concentrations, of the order of 10% as much, can be found in a large number of other vegetables (e.g., lettuce, turnips, carrots, peas, and beans) and in a few varieties of fruit (e.g., berries). In most of the plants in which oxalate occurs, the highest levels are associated with the leaves rather than the stalks (as in rhubarb) or the root (as in beets and carrots). It appears that oxalate, which is a simple two-carbon carboxylic acid, is an end-product of carbohydrate metabolism in the plant.

The acute toxicity of oxalate is marked by the appearance of corrosive effects in the mouth and the gastrointestinal tract, sometimes resulting in severe hemorrhaging. Oxalate has a pronounced effect on calcium metabolism in animals, first expressed as lowering of plasma calcium levels. Kidney damage and convulsions can accompany oxalate poisoning (5, 6).

There is some doubt as to whether ingestion of food containing oxalate has been responsible for human intoxication. Although there have been over 20 reports of human poisoning in the United States resulting from the ingestion of foods high in oxalate content (most frequently rhubarb leaves or stalks), little conclusive evidence has been produced to show that oxalate was the sole or even principal cause of the symptoms observed. Fassett (6) has provided a critical analysis of the available literature on human poisonings allegedly linked to oxalate-containing foods. Other authors, however, have found the evidence for human poisoning by food oxalate more compelling and should be consulted for details (5, 7).

Another aspect of oxalate that requires further study is its effect on calcium metabolism; again, Fassett (6) finds no evidence that oxalate interference with calcium absorption results in calcium deprivation in the United States, but in view of the lack of knowledge of the total dietary intake of oxalate, especially in children, continued investigation of this feature of oxalate biochemistry appears warranted.

Oxalate is an end-product of metabolism in mammals and is excreted in the urine. Although quantitative data are lacking, it appears that food oxalate, known to be poorly absorbed from the gut, contributes only a minor fraction of the total amount of oxalate excreted. The largest share of excreted oxalate arises from glyoxylic acid (a product of carbohydrate and protein metabolism) and from ascorbic acid (5). Interestingly, certain foreign compounds (e.g., ethylene glycol) are known to increase urinary output of oxalate or, in some instances, to produce stones (5). Fassett (6) has recommended serious examination of the possibility that excessive ascorbic acid intake, apparently now much in favor among a large segment of the population, may result in undesirably high levels of urinary oxalate (see also Chapter 3).

The hypothesis that chronic ingestion of food oxalate may contribute to kidney disease or to stone formation seems worthy of test, but little work of a definitive nature has been done to date. In 1947 Fitzhugh and Nelson (8) reported no significant health effects in rats administered oxalic acid at dietary levels ranging from 0.1 to 1.2%. The extent of pathological examination of the test animals is not completely documented, and it appears that further investigations along these lines are desirable. A variety of interesting studies of the effects in human subjects of subchronic exposure to oxalate, particularly in regard to its possible interference in calcium metabolism, have been reviewed in detail by Fassett (6). This author generally finds little evidence that food oxalate presents a significant health hazard but, for reasons described previously, it appears that a number of questions could be profitably explored in more detail.

Glycoalkaloids in White Potato (*Solanum tuberosum*)

Some have estimated the number of compounds present in the white potato at about 200, most of which have not been characterized (5). The structural variety of the compounds present in the white potato is wide: there are steroids, terpenes, phenols, and notably some steroidal glycoalkaloids. The last are significant because some are known to be toxic.

The major glycoalkaloids in potatoes are solanine and the closely related chaconine (Figure 5.1). Several other members of this class of compounds are also present. All green parts of the plant contain these compounds, and so do the tubers, especially when the latter have been exposed to light and thereby begin to "green."

The concentrations of glycoalkaloids in potato tubers vary considerably, especially among different varieties (9). Environment also affects the glycoalkaloid content of potatoes. The solanine content can be increased, for example, by any of several environmental factors that can retard the maturation process. Exposure of tubers to some wavelengths of light has been shown to increase glycoalkaloid content. "Netted gem" cultivars, for example, when exposed to infrared light for 4 days, undergo a doubling of glycoalkaloid concentration. Glycoalkaloid synthesis in tubers can be induced by simple mechanical injury (9).

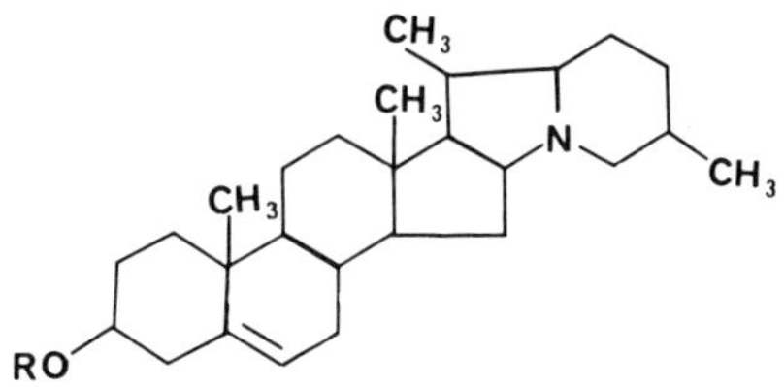

FIGURE 5.1 Solanine, a glycoalkaloid found in the white potato (*Solanum tuberosum*) R = trisaccharide.

Emphasis has been placed on the glycoalkaloid content of potatoes because of knowledge that one of more of these substances, singly or in combination, have been shown to produce toxicity, and even death, in both humans and animals. These compounds possess anticholinesterase activity, although the relative potencies of the various glycoalkaloids are poorly defined. They can produce gastrointestinal and certain neurological disorders (10). Human deaths attributed to consumption of potatoes containing excessive amounts of glycoalkaloids have been reported, although the last such report in the United States dates from 1931 (11). In humans an oral dose of solanine of approximately 2.8 mg/kg body weight can lead to drowsiness, dyspnea, and hyperesthesia, and somewhat higher doses can produce vomiting and diarrhea (12). Although solanine and chaconine have been rather well studied for short-term pharmacologic effects, their potential for producing damage on prolonged low-dose exposure is unknown. This last statement also holds for the huge number of other components of this common food.

There are several other common foods that are members of the Solanaceae plant family and for which similar problems of potential toxicity are known or suspected. Included are eggplant or aubergine (*Solanum melongena*) and tomato (*Lycopersicon esculentum*). Tobacco is also a solanaceous plant. Several poisonous plants, among them Jimson weed and deadly nightshade, are in the same botanical family, and the common biological origin of the poisonous plants and the ones used for food probably accounts for the residual toxicity that has been discussed in this section.

Cyanogenetic Glycosides

Cyanide, a substance whose biological effects are known by almost everyone, can be found in a very large number of plants, a few of which are used as food. Free cyanide, however, does not occur in these plant foods; rather, the HCN is present as an aldehyde or ketone adduct (a cyanohydrin), which is, in turn, bound to a sugar (hence the name "cyanogenetic glycosides"). Release of certain enzymes present in the plant food, a process that can be brought on during food preparation or when plant tissue is damaged (as in chewing) results in cleavage of the cyanohydrin adduct from the sugar molecule and

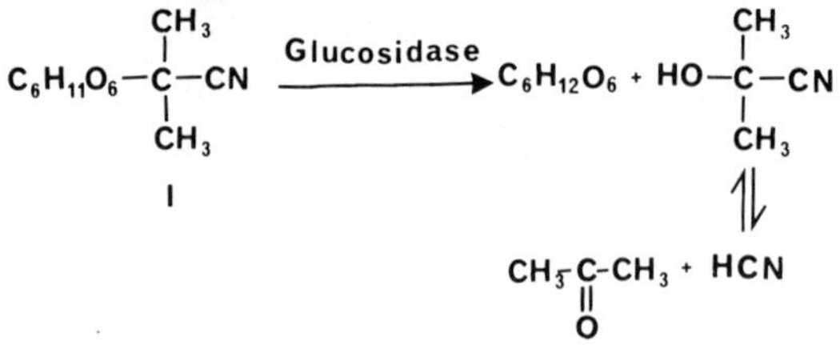

$$C_6H_{11}O_6-\underset{\underset{CH_3}{|}}{\overset{\overset{CH_3}{|}}{C}}-CN \xrightarrow{\text{Glucosidase}} C_6H_{12}O_6 + HO-\underset{\underset{CH_3}{|}}{\overset{\overset{CH_3}{|}}{C}}-CN \rightleftharpoons CH_3-\underset{\underset{O}{||}}{C}-CH_3 + HCN$$

I

FIGURE 5.2 Linamarin (I), a cyanogenetic glycoside, and its enzymatic cleavage to yield HCN.

subsequent collapse of the adduct to yield the aldehyde or ketone and the poisonous HCN (Figure 5.2).

Cyanogenetic glycosides found in food plants include linamarin (Figure 5.2), a constituent of lima beans; amygdalin, a component of the kernels of several stone fruits, including peach and apricot; and dhurrin, a substance present in sorghum. None of these foods, save lima beans, plays a significant role in the human diet in the United States. The potential hazards associated with lima beans have been avoided, at least in most western nations, by the use of varieties of lima beans known to be low in cyanide (9). However, several instances of human cyanide poisoning have been recorded; they have usually been associated with the consumption of large quantities of peach or apricot kernels by people apparently unaware of the possible risks (13).

Cassava root, an important carbohydrate source in some parts of West Africa, notably Nigeria, is another source of human exposure to cyanide. Although the mode of food preparation used results in removal of the cyanogenetic glycosides present or inactivation of the enzymes necessary to release cyanide, it appears that low-level cyanide residues remain in the cassava as consumed (9, 14). Tropical atonic neuropathy, a degenerative disease of the nervous system, has been tentatively linked to the chronic consumption of cyanide in a group of Nigerians having a high cassava consumption rate (14, 15). The effects of chronic cyanide consumption are not yet fully understood. Conn (16) has speculated that the widespread incidence of goiter in Nigeria might be attributable to thiocyanate, an ion known to be goitrogenic and produced as a detoxification product of cyanide.

Hemagglutinins

Not all proteins are nutritionally beneficial. Such is the case with a group of plant proteins that can cause red blood cells to clump (or agglutinate). These substances can be found in kidney beans and numerous other leguminous plants. Ricin, a highly toxic component of the castor bean, is a member of this class of compounds.

Fortunately, the plant hemagglutinins are rendered inactive during cooking, so that these substances, under normal conditions of food preparation, are not a threat to human health. Their interesting and unique biological properties have, however, made them objects of considerable study (17, 18).

Quercetin and Related Plant Phenolics

Phenols are present in virtually all plants and occur in a large variety of structural arrangements, among them the class known as *flavonols*. Quercetin (Figure 5.3) is a prominent member of the flavonol class. This substance has been the subject of increased interest in recent years following reports that it possesses mutagenic activity in bacteria (19, 20). Whether it also possesses carcinogenic properties in animals, as might be predicted from its mutagenic properties, is yet to be determined.

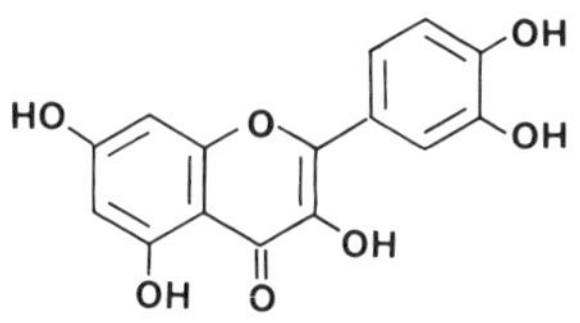

FIGURE 5.3 Quercetin, a flavonal widely distributed in the plant kingdom, including some plants used as foods.

If quercetin proves to be carcinogenic, then a host of related substances that are found in plants may become suspect. Such a finding may reawaken interest in this class of natural products that, based on previous experience, had been thought to be toxicologically innocuous (21). Additional information on plant phenolics can be found in Singleton and Kratzer (21).

Carotatoxin

Crosby and Ahranson (22) isolated an unusual diacetylenic allyl alcohol from carrots and celery and named it *carotatoxin* (Figure 5.4). The name they attached to this substance was not a frivolous one, because the compound is a neurotoxic agent of moderately high toxicity. A related compound, aethusin, is a constituent of a plant called "fool's parsley"—a poisonous plant known to be responsible for human intoxication (13). The effects of chronic ingestion of carotatoxin are unknown. The FDA and the USDA are jointly developing analytical methodology to begin identifying the levels of human exposure to this substance.

Phytoalexins

A subgroup of the genetically determined, intrinsic components of plant foods are the phytoalexins. These substances, representing a diverse collection of chemical types, are produced as metabolites in

$$CH_2{=}CH{-}\underset{|}{\underset{OH}{CH}}{-}(C{\equiv}C)_2{-}CH_2{-}CH{=}CH{-}(CH_2)_6{-}CH_3$$

FIGURE 5.4 Carotatoxin (Falcarinol).

plants from mechanical damage or, more commonly, fungal attack (23).

Examples of phytoalexins include the terpenoid compounds rishitin and lubminin produced in the white potato, phaseolein in the green bean, and pisatin in the garden pea (9). Several substances produced in the sweet potato in response to stress are toxic in experimental animals. One of these, ipomeamarone (Figure 5.5), is a hepatotoxic agent. Two other, more toxic substances, 4-ipomeanol and 1-ipomeanol (Figure 5.5), have been shown to produce lung edema in mice (24). The extent of contamination of marketplace sweet potatoes with these substances is not known. In some areas of the world where sweet potato consumption is high, it may prove profitable to determine whether a high incidence of lung disease (e.g., emphysema) is present in the consuming population.

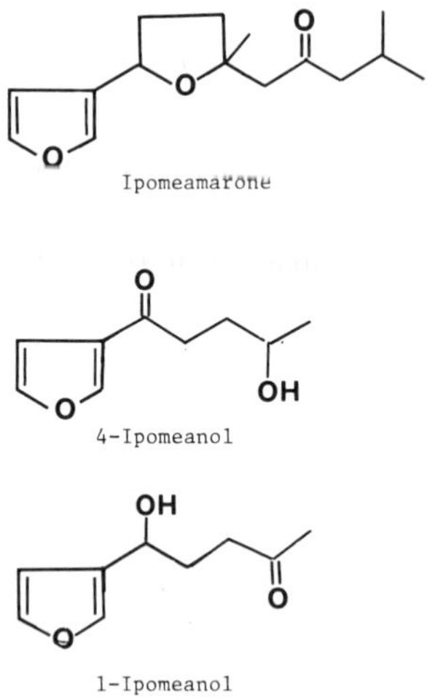

Ipomeamarone

4-Ipomeanol

1-Ipomeanol

FIGURE 5.5 Toxic metabolites of mold-damaged sweet potatoes.

Mushroom Poisoning

There are about 5000 species of fleshy mushrooms in the United States that have a history of food use; about 100 of these species have been implicated in food poisonings, and about 12 species have been identified as containing lethal toxins (the death cap mushroom, *Amanita phalloides,* the closely related destroying angel, and the false morel, *Gyrometra esculenta).* For several reasons, no estimate can be made concerning the severity of the problem of mushroom poisoning in the United States because (*a*) the great majority of cases are not reported to health authorities; (*b*) many cases of intestinal disorders are possibly misdiagnosed or not attributed to mushroom ingestion; (*c*) mushroom poisoning is not on the U.S. Public Health list of reportable diseases; and (*d*) death certificates do not appear to be a reliable source for such information. Nevertheless, the National Clearinghouse for Poison Control estimates about 350 cases of mushroom poisoning annually, about 70% of those in children under 5 years of age. The USDA Poison Fungus Center in Beltsville, Maryland, recently revealed data obtained from only eight states listing 105 cases of mushroom poisoning with two fatalities during 1973; success at arriving at even this limited statistic was attributed to the combined efforts of local mushroom clubs, health authorities, and state poison control centers. The great majority of cases resulted from misidentification of the mushroom eaten (25).

It should be pointed out that the most common type of mushroom poisoning is an adverse allergic reaction probably due to unidentified proteins, peptides, or alkaloids. Some people are even sensitive to the cultivated mushroom *Agaricus bisporus*.

Extensive studies of the toxins in mushrooms has led to a classification, proposed by Lincoff and Mitchell (25), of the poisonings into four categories based on the physical effects and time required after ingestion for onset of poisoning symptoms. This classification is described in the following list:

Class A: Toxins that cause cellular damage (liver and kidney) and death with onset of symptoms in about 10 hours.

Group I: This group of mushrooms, the most common of which are *Amanita* and *Galerina*, contain deadly cyclopeptides. The action of these toxins is insidious, since the individual poisoned may not realize the problem for 10 hours and may wait too long for treatment. An example of these toxins is shown in Figure 5.6.

Group II: This group of mushrooms is characterized by its ability to produce some extremely toxic hydrazine derivatives. It includes the false morels (lorels, lorchels, and brain fungi). Mortality in those individuals reacting adversely ranges from 15 to 50%. The principal offender is *Gyrometra esculenta*; which is known to produce gyromitrin (Figure 5.7) that on hydrolysis releases the toxic monomethylhydrazine. There is some evidence that these hydrazine derivatives are carcinogenic.

Class B: Toxins that affect the autonomic nervous system with onset of symptoms in 0.5 to 2 hours.

Group III: Ingestion of this group of mushrooms *(Coprinus* and *Claviceps)* produces toxins that act much like Antabuse, a drug used

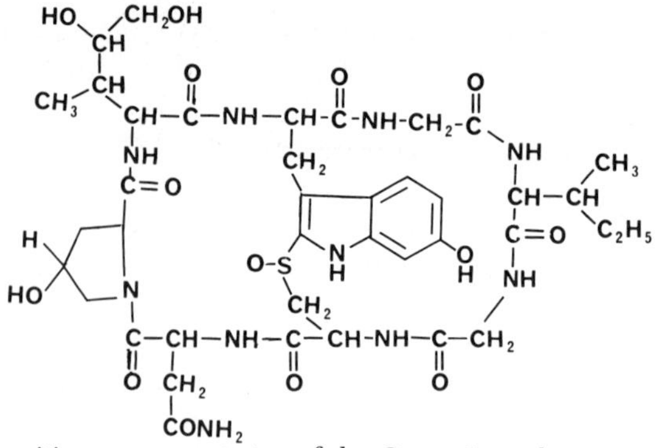

FIGURE 5.6 α-Amanitin, a representative of the Group I mushroom toxins (see text).

$$CH_3CH{=}N{-}N(CH_3)(CHO) + H_2O \longrightarrow CH_3CHO + CO + CH_3NH{-}NH_2$$

FIGURE 5.7 Gyromitrin, a Group II mushroom toxin (see text) and its hydrolysis to methylhydrazine.

to help alcoholics shed the drinking habit. The symptoms of poisoning include flushing, metallic taste, prickling and tingling of the extremities, and eventual nausea and vomiting. It is not known to be fatal. This type of poisoning occurs if the mushroom is eaten in conjunction with an alcoholic beverage. The active toxin, coprine (from *Coprinus atramentarius*), is shown in Figure 5.8.

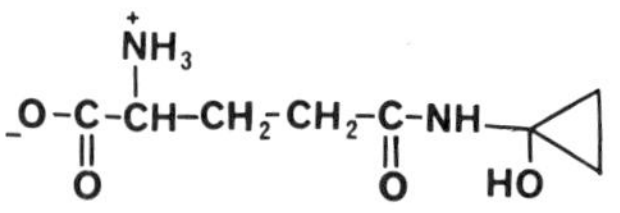

FIGURE 5.8 Coprine, a Group III mushroom toxin (see text).

Group IV: This group of mushrooms produces compounds that, if ingested, lead to symptoms typical of muscarine poisoning (profuse perspiration, salivation, lachrymation, muscle spasms, decreased pulse rate and blood pressure, etc.). Species of *Amanita, Clitocybe, Inocybe,* and *Boletus* have all been implicated; some of these produce small quantities of muscarine as well as other toxins (Figure 5.9).

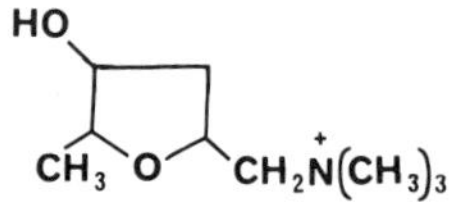

FIGURE 5.9 Muscarine, a Group IV mushroom toxin (see text).

Class C: Toxins that affect the CNS with onset of symptoms in 0.5 to 2 hours.

Group V: This group of mushrooms contains some potent isoxazole derivatives; it includes mainly species of *Amanita,* and in particular *A. muscaria* (fly agaric). Symptoms of intoxication include inebriation, manic behavior, delirium, and deep sleep. Some of the toxins implicated in this type of poisoning are shown in Figure 5.10.

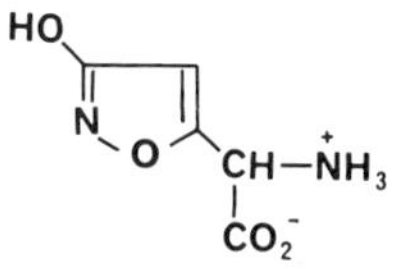

Ibotenic Acid

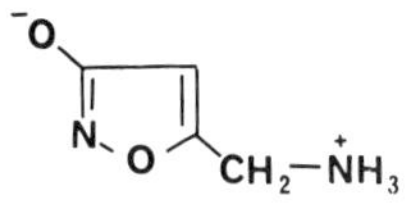

Muscimol

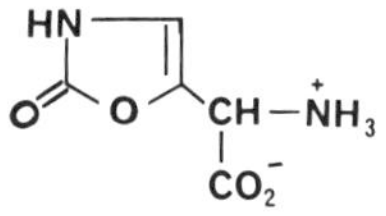

Muscazone

FIGURE 5.10 Some Group V mushroom toxins (see text).

Group VI: This group of mushrooms contains powerful hallucinogens; species of *Psilocybe* and *Panaeolus* are the major producers of these hallucinogens, the majority of which are indole derivatives (Figure 5.11).

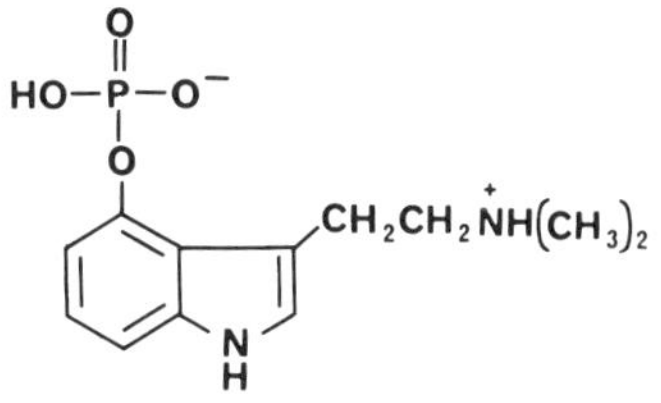

I

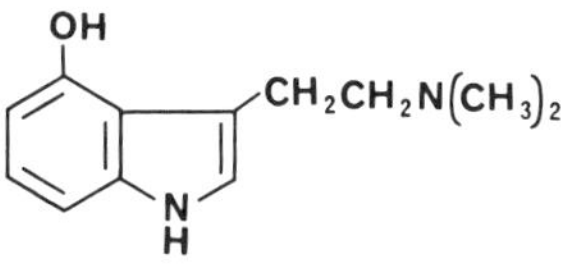

II

FIGURE 5.11 Psilocybin (I) and Psilocin (II), two Group VI mushroom toxins (see text).

Class D: Toxins responsible for gastrointestinal distress with onset of symptoms in 0.5 to 3 hours.

Group VII: Many genera of mushrooms when ingested cause vomiting and diarrhea or both; these include *Agaricus, Amanita, Bo-*

letus, Chlorophyllurm, and many others. In most cases the chemical structure of the causative agent is unknown.

Goitrogens in Foods

Foods contain a large variety of sulfur-containing compounds, some of which are known to cause undesirable effects on ingestion. One such group of compounds includes the glucosinolates (thioglucosides). Early folklore ascribes human goiter (hypothyroidism) to some food plants now known to contain these compounds (26). This early proposal was supported by a report in 1928 that rabbits, fed cabbage as the major portion of their diet, exhibited greatly enlarged thyroids. Other experiments using a variety of different animal species seemed to substantiate, but not conclusively prove, these early findings. There is, however, little epidemiological or experimental data to prove that endemic goiter is a direct result of a diet rich in glucosinolates (27).

The basic glucosinolate structure is shown in Figure 5.12. These compounds invariably are associated with enzymes capable of hydrolyzing the glucosinolate to glucose, bisulfate, and organic sulfur-containing compounds. The latter rearrange quickly, yielding varying amounts of isothiocyanate, thiocyanate, nitrile, and sulfur. These hydrolysis reactions occur readily on crushing of the wet, unheated plant material. The pungent odor of many of the plants of the *Cruciferae* family (horse radish, mustard, etc.) is due to the formation of these hydrolysis products; the formation of the isothiocyanates seems to be the preferred hydrolytic pathway.

$$\mathrm{R{-}C}\begin{matrix}{/\!\!/ NOSO_2^-}\\{\backslash S{-}C_6H_{11}O_5}\end{matrix} \xrightarrow{H_2O} \left[\mathrm{R{-}C}\begin{matrix}{/\!\!/ NH}\\{\backslash SH}\end{matrix}\right] + HSO_4^- + C_6H_{12}O_6$$

$$\downarrow$$

$$S + RCN + R{-}NCS + R{-}SCN$$

FIGURE 5.12 Hydrolysis of a typical glucosinolate to yield organic nitriles, thiocyanates, and isothiocyanates.

About 300 of the 1500 known species of *Cruciferae* have been investigated; several different glucosinolates have been found in each. Cabbage, cauliflower, Brussels sprouts, kale, turnips, and broccoli are the major contributors to human exposure to these substances (5, 28). Generally larger quantities are found in the seed than in other parts of the plant, and in some cases at fairly high concentrations. For example, turnip seeds have been found to contain 0.7 to 2% gluconapin and 0.1% progoitrin. In comparison, fresh cabbage leaves contain about 350 ppm sinigrin, 0.5 ppm progoitrin, and 60 to 800 ppm glucobrassicin. The goitrogenic activity associated with ingestion of plants that contain these glucosinolates is further complicated by the fact that thiocyanate ion, an end-product in the hydrolysis of glucosinolates, is itself goitrogenic, as are some of the thiocyanates, isothiocyanates, and nitriles. These compounds apparently act by inhibiting the incorporation of iodine into thyroxine and related compounds. It has been estimated that a minimum of 20 mg of goitrin or 200 to 1000 mg of thiocyanate would be required to produce an observable goitrogenic effect in humans (28). Consequently, given the levels of such compounds normally present and the amounts of each plant normally consumed, it is unlikely that an acute goitrogenic effect would be observed in an adult; the effect of continuous, low-level exposure over long periods of time is unknown.

Conclusions

The substances chosen for review in this section represent a random selection of food plant constituents about which safety questions have been raised or that are known to be hazardous under some conditions of exposure. The compounds surveyed represent a broad spectrum of chemical types and biological activities. Some possess characteristics that require the imposition of some kind of control on human exposure; these aspects of the natural food hazard problem are examined more fully in the closing sections of this chapter.

The preceding survey is by no means exhaustive, but it does attempt to be representative of the kinds of hazard that have been found in plant foods. For the most part, the substances that have been most closely studied are those that possess moderately pronounced acute toxicities and have a relatively long history as human toxicants (e.g., solanine, cyanogenetic glycosides, oxalate, and mushroom poison).

In more recent years a few of these substances have been examined by some of the techniques of modern toxicology and epidemiology, and suggestive evidence that chronic, low-level exposure may result in adverse effects has been uncovered (e.g., ipomeamarone and quercetin). As was emphasized at the outset, it appears that continued examination of the natural components of plant foods using current techniques will likely reveal a surprisingly high number of deleterious substances. In most instances the extent of human risk will not be known, but, as with so many synthetic chemicals, it will be necessary to rely on animal data to reach conclusions about the possible impact on human health.

NATURAL CONSTITUENTS OF SOIL AND WATER THAT ACCUMULATE IN FOODS

Some of the more important inorganic substances present in food are treated in Chapters 3 and 4. Examples of chemicals in this class are a number of inorganic elements, some of which are well-known nutrients and others, potentially hazardous. (Most of the elements essential for nutrition are themselves hazardous under some conditions; see Chapter 3.) The specific elements present in plant foods and their chemical concentrations and chemical forms depend in part on the characteristics of the soil in which they are grown. The accumulation of such substances in foods derived from both land and marine animals is a result of the ingestion by these animals of plant foods and water containing these elements. The natural background levels of these elements in food can be modified as a result of various types of human activity.

Accumulation of nitrates from soil is a characteristic of several important food plants. The process of accumulation is sometimes very complex and can be influenced by many factors, such as the variety and species of plant, its stage of maturity, its state of health, its moisture content, and the type of soil and fertilizer in which it is grown. Excessive levels of nitrate in some plants can give rise to methemoglobinemia in children who consume such plants; this condition has been reported in children who had eaten spinach puree (29). The spinach consumed by the children contained 218 mg of nitrite per 100 g wet weight and only a trace of nitrate; this finding is consistent with knowledge that nitrate–nitrite reduction can occur in stored vegetables (30). Nitrites can, under some conditions, react with naturally occurring amines to produce the class of carcinogens known as nitrosamines; these compounds are considered in another section of this chapter.

METABOLITES OF MICROORGANISMS THAT GROW ON FOODS

Bacterial toxins are reviewed in Chapter 2 and are not discussed here. The metabolic products of fungi that display toxic properties (mycotoxins) and that can contaminate food are the subjects of this section.

Background—Mycotoxins

Molds are capable of producing a wide variety of secondary metabolites, many of which possess biological activity. Humankind has benefited greatly from the many fungal metabolites that have proved useful as antibiotics but throughout history has been plagued by a number of disease states resulting from inadvertent exposure to toxic fungal products that can contaminate food. Those fungal metabolites capable of producing toxic effects in humans and animals are called *mycotoxins* (from the Greek word *mykes*, meaning fungus). The disease states resulting from exposure to mycotoxins are *mycotoxicoses*. The problem of mycotoxicoses in humans (and in livestock) is not only a historical one; it is still with us and, over the past 15 years, has received increased attention on a worldwide basis. The discovery in the early

1960s of the highly toxic and carcinogenic mycotoxins known as the *aflatoxins* provided the impetus to current research and control activities.

The veterinary and medical literature documents numerous outbreaks of mycotoxicoses in humans and livestock, the best known example of which is ergotism (31). This bizarre disease, characterized by its alarming effects of convulsion, hallucinations, and gangrene of the extremities, was first recognized in the Middle Ages. It is brought on by the ingestion of breads and other products prepared from flour derived from grains contaminated with the chemical by-products of the ergot fungus, *Claviceps purpurea.* Other less widespread but equally severe disease outbreaks associated with moldy foods and feeds include *alimentary toxic aleukia,* a hemorrhagic disease that recurred in various regions of the Soviet Union throughout the first half of the twentieth century (32); *yellow-rice disease,* a form of intoxication characterized by paralysis, convulsions, and respiratory arrest that occurred in Japan (33); and another hemorrhagic disease, *stachybotryotoxicosis,* that has affected both humans and livestock in the Soviet Union and in Eastern Europe (34).

These and the many other known mycotoxicoses, which usually occurred shortly after ingestion of contaminated foods or feeds, resulted from highly moldy products consumed because of food shortages or because of ignorance of the possible consequences. Such practices continue in many areas of the world, and public health officials are just beginning to measure the toll on human and animal health caused by the use of moldy foods and feeds. Unfortunately, this problem is not limited to the acute mycotoxicoses arising from the consumption of highly moldy products; experience gained in attempting to deal with the aflatoxins has revealed an even more insidious aspect to this problem.

Aflatoxins

Definition

The term "aflatoxins" is applied to a group of closely related compounds derived from the common fungi *Aspergillus flavus* and *A. parasiticus.* The term is also used to designate metabolites of these compounds that are produced in animals fed the aflatoxins. Individual members of the group are further identified by the addition of a letter and a number relating to some molecular or biological characteristic of the compound. The major fungal metabolites are two compounds that emit a *blue* fluorescence when exposed to ultraviolet light (these are called aflatoxins "B_1 and B_2"), and two that emit a *green* fluorescence (termed, thus, aflatoxins "G_1 and G_2"). These four aflatoxins are the group commonly occurring in foods and are also the toxins for which assay methods are available. Most reports citing the occurrence of aflatoxins in foods refer to this collection of four compounds (35).

The only other aflatoxin currently of interest as a food contaminant is aflatoxin M_1, a metabolite of B_1 that is excreted in the milk of lactating animals after ingestion of contaminated feed (36).

Chemical Characteristics and Analytical Methodology

The aflatoxins and nonproteinic, low-molecular-weight (e.g., aflatoxin B_1 has a molecular weight of 312) organic compounds containing the unusual difurofuran moiety attached to a substituted coumarin nucleus (Figure 5.13). They are generally stable to heat and can survive most types of food processing. However, some important exceptions to the last statement are mentioned in later sections. Analytical methods are available to measure aflatoxins B_1, B_2, G_1, G_2, and M_1 in all the foods known to be susceptible to contamination. Most of the available methods involve extraction of the toxins from foods, some form of chromatographic separation of the toxins from interfering substances, quantitative measurement of the toxins, and qualitative measurements to confirm their chemical identities. The quantitative measurement step usually involves comparison of the fluorescence intensities of the suspect toxins with standard toxins on thin-layer chromatoplates. Methods for confirming the identity of the toxins are available and should always be applied to obtain unequivocal evidence that the substances under measurement are indeed aflatoxins. In recent years a variety of qualitative methods useful for rapid screening of suspect lots of food have also been made available. A useful compilation of these methods is contained in a publication of the Association of Analytical Chemists (37). The methods published in this chapter have undergone rigorous interlaboratory validation and have proved reliable in many laboratories around the world.

A major problem in obtaining an accurate estimation of the aflatoxin content of a lot of food derives from the fact the contamination tends to be highly localized. Obtaining a representative lot sample is, with the exception of fluid products such as milk, the major difficulty to be overcome if analytical data are to be considered meaningful. Guidance for coping with problems of food sampling and analysis is available (37, 38).

It should be noted that identification of the presence of the producing fungi on foods is not a reliable indicator of the presence of aflatoxins. Conversely, aflatoxins can be found in foods exhibiting no apparent mold growth (39).

Health Effects of Aflatoxins

Foods contaminated with aflatoxins are a public health problem of as yet unknown dimensions. The aflatoxins, particularly B_1, are highly

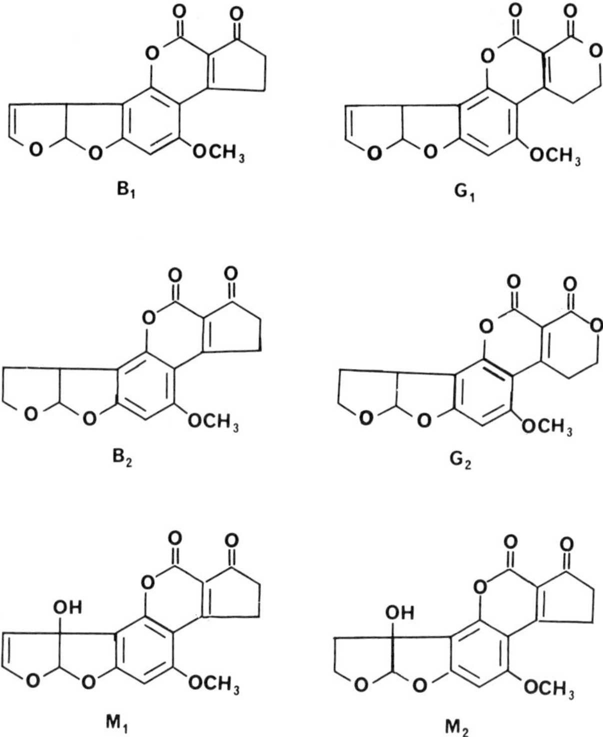

FIGURE 5.13 The aflatoxins commonly occurring as food and feed contaminants; M_1 and M_2 are hydroxylation products excreted in the milk of animals ingesting B_1 and B_2, respectively.

toxic to animals (the toxicity is manifested primarily by the appearance of liver injury), and several episodes of apparent acute aflatoxicoses in humans have been reported. The most important episode of apparent aflatoxicosis in humans occurred in several villages in India in the autumn of 1974. About 400 people were affected, and over 100 succumbed to fatal hepatic disease (40). Maize, a major dietary constituent, was found to be contaminated with aflatoxins at levels ranging from 0.25 to 15.6 mg/kg.

Encephalopathy and fatty degeneration of the viscera (EFDV), a fatal childhood disease, has been associated with aflatoxin consumption in Thailand (41). Reye's syndrome, a disease resembling EFDV and fast becoming recognized as an important contributor to childhood mortality in the United States and elsewhere, has been suggested to be due, at least in part, to aflatoxin ingestion (42). The disease appears to be of complex etiology, however, and no definitive associations have been described.

It is only in recent years that the capability for diagnosing acute aflatoxicoses in humans has become available. As investigators acquire this capability, it is likely that other examples of acute aflatoxicoses in humans will be uncovered. No animal species has proved resistant to the acute toxic effects of aflatoxins (43); therefore, it is reasonable to assume that humans are similarly affected. The important question of relative susceptibilities remains unanswered.

It is likely that acute aflatoxicosis requires a level of food contamination that is, fortunately, not common in most areas of the world, although such a statement may be unduly optimistic in the face of our serious lack of information on the occurrence of aflatoxins. There is, however, another aspect to the aflatoxin problem that is perhaps more significant. Soon after the discovery of aflatoxins in England in the early 1960s, it was discovered that these substances were potent primary liver carcinogens in experimental animals (35, 44). This observation has been repeated many times in test animals. In some strains of rats, aflatoxin B_1 is the most potent carcinogen yet identified (39). Interestingly, mice appear to be rather resistant to the cancer-causing effects of aflatoxins (44). The relative susceptibilities of humans to aflatoxins are not known, although epidemiology studies carried out in parts of Southeast Asia and Africa, where there is a high incidence of hepatoma, have revealed an association between cancer incidence and the aflatoxin content of the diet. These studies have not proved a cause–effect relationship, but the evidence of an association is highly suggestive (45). The International Agency for Research on Cancer (IARC), a body of the World Health Organization, has listed aflatoxin among the known causes of human cancer (46). The IARC Monograph on which this list was based (47) should be consulted for a full understanding of the uncertainties attaching to the classification of aflatoxin as a "known" human carcinogen.

It appears that a major effort should be devoted to minimizing human exposure to aflatoxins. Aflatoxins can cause a variety of diseases in livestock and poultry; the extent to which aflatoxins contribute to the loss of food animals around the world is as yet unknown, but the economic cost is probably appreciable. Concern about aflatoxin contamination of animal feed also arises because residues of the toxins can be found in meat, milk, and eggs (36). Of these three foods, milk appears to be by far the most susceptible to contamination. The specific aflatoxin excreted in milk, M_1, is also carcinogenic (47) and, although the levels that have been reported are rather low (i.e, ordinarily $1 \mu g/kg$), there may still be a health risk, especially for infants. This conclusion is based on evidence that the young are more susceptible to the toxic effects of aflatoxins than are adults and that their diets can consist almost entirely of milk.

For these reasons, many countries have instituted regulatory limits on the aflatoxin content of susceptible foods. Enforcement of such controls has already adversely affected international trade, and increased regulatory activity is expected (35, 48). Other aspects of aflatoxin regulation and control are discussed in a subsequent section.

Susceptible Commodities

Although aflatoxins were first discovered as contaminants of groundnuts (peanuts) and derived products, it now appears that, on a worldwide basis, maize is the most important source of aflatoxin exposure. This last conclusion holds true for the United States but is largely limited to that portion of the maize crop that is produced in the Southeastern states. Peanuts, of course, remain a problem; although the average incidence of peanut contamination is now much lower than it was when aflatoxin was discovered, the size of the exposed population is larger than that limited population in the Southeast exposed to locally produced corn. It is important to note that sweet corn (freshly cooked for direct consumption, freezing, or canning) is not know to be a susceptible commodity (49). Cottonseed, copra, and certain tree nuts are also susceptible to aflatoxin contamination. Surveys of a variety of commodities have been carried out in many parts of the world, and a compilation of the results of these surveys has been prepared (49).

Aflatoxin contamination of foods can occur whenever environmental conditions are suitable for growth of the producing fungi. At present all the factors contributing to contamination have not been clearly identified. It has been established, however, that crops can be contaminated in the field during growth, especially under condition of plant stress (e.g., drought, and insect damage). Inadequate drying of crops before storage and failure to provide storage conditions that will protect crops from moisture can also lead to contamination. The most serious problems of contamination occur in tropical and semitropical regions, where climate favors growth of the producing fungi, but contamination can also occur in temperate regions (35, 39, 49).

Patulin

A mycotoxin that has generated considerable interest over the past few years is patulin (Figure 5.14). This compound was discovered in a long, intensive search for new antibiotics; it was first isolated in crystalline form by Chain et al. in 1942 from *Penicillium claviforme* and given the name "clavicin" (50). Other investigators isolated the identical material from other mold species, each one applying a name descriptive of the mold source; the literature, therefore, contains many synonyms for patulin including clavicin, expansin, mycoin, penicidin, leukopin, and tercinin. The name "patulin" was finally accepted with elucidation of the correct chemical structure by Woodward in 1949 (51). Patulin is an excellent antibiotic against both gram-positive and gram-negative bacteria. This led immediately to the testing of its possible use as an antibiotic against the common cold; however, it was not found to be efficacious and the idea was dropped. Later attempts to use the drug as an antibiotic cream for topical application were also discontinued since it led to dermal irritation; given orally, it caused vomiting and nausea (52).

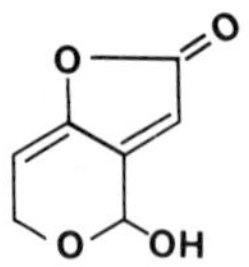

FIGURE 5.14 Patulin.

In 1961 Dickens and Jones (53) reported that subcutaneous injection of 0.2 mg of patulin twice weekly in male rats for 61 to 64 weeks produced sarcomas at the site of injection in six out of eight animals surviving after 64 to 69 weeks. Since that time several other subacute (chronic) toxicity studies have been carried out; none of these studies establishes patulin as a chemical carcinogen, nor has it been found to exhibit mutagenic activity.

Because Dickens and Jones (53) had implicated patulin as a possible carcinogen, and since molds known to elaborate patulin were commonly found on foods, an extensive effort was made to develop and apply sensitive analytical methods for detection of patulin in such foods. Some of these methods depend on the formation of colored reaction products; for example, two of the thin-layer chromatographic (TLC) methods involve reaction of the nascent aldehyde function (Figure 5.14) with phenylhydrazine hydrochloride or 3-methyl-2-benzothiazolinone to yield a brightly colored product suitable for visual or densitometric estimations. Several gas–liquid chromatographic (GLC) methods have been developed based on formation of acetate, trifluoroacetate, or silylether derivatives; these are especially convenient for conducting GC-mass spectrometric analyses. The most recently developed methods are based on detection of the strong patulin ultraviolet absorption near 270 nm after normal or reverse-phase high-pressure liquid chromatographic (HPLC) separation. These HPLC methods are capable of detecting patulin in apple juice at levels approaching 15 ng/ml (54).

Patulin has been associated primarily with the apple rotting fungus *(Penicillium expansum)* and with *P. urticae*, a fungus found commonly in soils and on plant stubble. The analytical methods described previously were used to examine fruit products and grains for the possible occurrence of patulin. These studies showed that natural apple rots can contain as much as 136 mg patulin/kg fruit. It was not surprising, therefore, that when the FDA surveyed commercial apple juice, 37% of 136 samples were found to contain, on the average, 69 ng patulin/ml (range, 40 to 440 ng/ml) (55). Other fruits (peaches, pears, apricots, cherries), have been shown to contain patulin, at least when partially rotted. Patulin has also been found in 5 of 21 samples of baked goods, in soil samples, and in straw. Although *P. expansum* has been used to ripen certain types of sausage, no patulin has been detected in the sausage, presumably because of the high reactivity of patulin toward sulfhydryl compounds (cysteine, glutathione, etc.) (56). The reaction products, although not completely characterized at this time, are apparently nontoxic. It has also been found that fermentation processes degrade or remove patulin.

Zearalenone

In 1927 Buxton and Legenhausen (57) reported the occurrence of severe hyperestrogenism in certain herds of swine in Iowa. The malady was found to be related to severely molded corn used for feed; removal of the pigs from the feed alleviated the condition and the pigs recovered completely. There have been many similar reports of hyperestrogenism in pigs. The syndrome has been attributed to the ingestion of metabolites of *Fusarium tricinctum*, *F. oxysporum*, and *F. moniliforme* growing particularly on corn and barley (58).

The primary symptoms of the estrogenic syndrome in swine are swollen, edematous vulvas, rectal and vaginal prolapse, enlarged uterus, and atrophy of the ovaries. The result is infertility, reduced litter size, and weight loss. Other animals (mice, rabbits, turkey poults, and sheep) are similarly affected. The responsible mycotoxin, zearalenone, isolated and characterized in 1962 by Stob, et al., is a derivative of resorcylic acid lactone; its structure is shown in Figure 5.15 (59). It has been referred to by various authors as FES (fermentation estrogenic substance), F-2, and RAL (resorcylic acid lactone). It occurs naturally along with lesser amounts of a number of closely related derivatives. Chemical reduction of the ketone and exocyclic double-bond functions leads to a pair of epimeric alcohols, both of which occur naturally and one of which (the higher melting isomer) is about four times more active than zearalenone itself. This last isomer is currently produced and marketed as a growth promoter for use in cattle. Over 100 other derivatives (reaction products) of zearalenone have been synthesized and tested for biological activity (58).

A number of reliable analytical methods have been developed for detection of zearalenone in foods and feeds. These include TLC procedures that depend on the natural fluorescence of zearalenone or on the

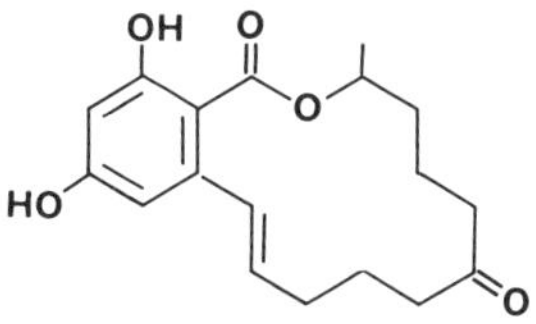

FIGURE 5.15 Zearalenone.

enhancement of fluoresence intensity with the use of an aluminum chloride spray. Other TLC visualization procedures are based on the use of spray reagents (H_2SO_4 and heat; $K_3Fe(CN)_6$–$FeCl_3$–HCl). These methods are simple and inexpensive and are capable of detecting zearalenone in grains at the 50-ng/g level (60).

After derivatization, zearalenone can also be detected by GLC; confirmation of identity is accomplished by preparation of other zearalenone derivatives (dimethyl ether, methyloxime, etc.) followed by GLC or mass spectrometry. These procedures are capable of detecting zearalenone at the 100-ng/g level (60).

A more recent, excellent analytical method is based on the strong ultraviolet absorption of zearalenone at 273 nm; the method uses HPLC and has a detection limit of 10 ng/g (61).

Zearalenone is frequently found together with other mycotoxins, including aflatoxin, ochratoxin, T-2 toxin, and other trichothecene toxins. The realization of this fact led to the development of several screening procedures for the detection of more than one mycotoxin. Most of these procedures are based on TLC and, although efficient, tend to lack low detection limits (60).

The development of reliable analytical methods for detection of zearalenone led to survey studies aimed at the determination of its natural occurrence. The data developed to date indicate that zearalenone is occasionally present in grains, particularly in corn that has suffered "Gibb" ear rot damage, usually at levels of 0.1 to 200 μg/g. In a USDA survey of the 1967 corn crop and 1968/69 export crop, zearalenone was found in 6 of 576 samples at levels of 450 to 800 ng/g. However, after extensive "Gibb" damage of corn was encountered in the Midwest in 1972, the FDA found that 17% of 223 samples taken from terminal elevators contained zearalenone at levels of 0.1 to 5.0 ng/g. Zearalenone has also been occasionally found in wheat, barley, oats, sorghum, sesame, hay, corn silage, corn oil, and starch made from zearalenone-contaminated corn (58).

On the farm zearalenone can be a significant and costly contaminant of animal feed. Of more concern is the fact that zearalenone has been found in human food (corn) and has been found to give a positive response in presumptive screening tests for mutagenic activity (62). Its structural similarity to diethylstilbestrol (DES) and its estrogenic activity raise questions about the health significance of the presence of this compound as well as other naturally occurring estrogenic substances in foods. Answers, however, are not yet available.

Trichothecenes

Nowhere is the ability of the fungi to produce large numbers of secondary metabolites so strikingly illustrated as in the case of the trichothecenes. This class of mycotoxins, all of which contain the trichothecene nucleus (Figure 5.16), is produced by various species of *Fusarium, Myrothecium, Trichoderma, Cephalosprium, Verticimonosporium, Cylindrocarpon,* and *Stachybotrys.* Over 40 trichothecene metabolites are known; many of these are biologically active and some are extremely potent toxins (58, 63). Some of these metabolites have been implicated as the causative agents in several severe outbreaks of mycotoxicoses (see Table 5.1). The toxicoses listed are characterized by some common symptoms, and these same symptoms are elicited in experimental animals exposed to specific trichothecenes. These symptoms include emesis, edema, hemorrhage, skin necrotization, nervous disorders, destruction of the bone marrow, leukopenia, and feed refusal. Frequently observed symptoms of exposure in humans are dermatitis, cough, rhinitis, and hemorrhage in the nose and throat.

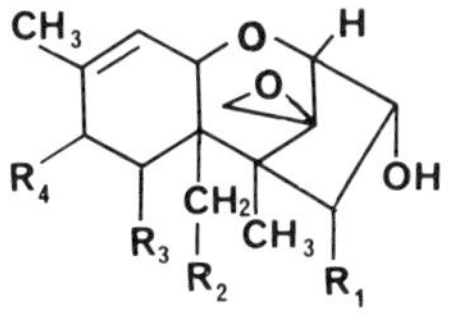

	R_1	R_2	R_3	R_4
T-2 Toxin	OAc	OAc	H	$(CH_3)_2CHCH_2CO_2$
HT-2 Toxin	OH	OAc	H	$(CH_3)_2CHCH_2CO_2$
DAS	OAc	OAc	H	H
DON	H	OH	OH	O

FIGURE 5.16 Some representative Trichothecenes.

During 1942 to 1947 severe outbreaks of alimentary toxic aleukia occurred in Russia, presumably from the ingestion of bread made from mold-damaged (overwintered) wheat; in 1944 over 10% of the exposed population of certain areas were fatally affected (63). The causative agents are now believed to have been trichothecene mycotoxins (63, 64). Of perhaps more concern than acute exposure are the implications of recent reports that certain trichothecenes, particularly T-2 toxin, are carcinogens in certain animal species and, in general, can inhibit DNA and protein synthesis (63, 65).

The naturally occurring trichothecenes have been classified (63) into four groups: (*a*) those having a functional group other than a ketone at C-8; (*b*) those having a carbonyl at C-8; (*c*) the macrocylic derivatives; and (*d*) those possessing an epoxide function at C-7, 8 (Figure 5.16). In general these compounds are colorless, crystalline, optically active solids, only slightly soluble in water. The ester functions are readily hydrolyzed in acid or base to the corresponding alcohols; the epoxide

TABLE 5.1 Trichothecene Mycotoxicoses

Mycotoxicoses	Species Affected	Responsible Organism
Alimentary Toxic Aleukia (ATA)	Man, horse, pig	*F. sporotrichioides*
Stachybotryotoxicosis	Man, horse. pig, poultry	*Stachybotrys atra*
Moldy Corn Toxicosis	Pig,cow	*F. tricinctum*
Dendrochiotoxicosis	Horse, sheep, pig	*Dendrochium toxicum*
Bean-hull Toxicosis	Horse	*F. solani*
Red-mold Toxicosis	Man, horse	*F. graminearum*

Source. Y. Ueno, "Trichothecenes Overview Address" in J. V. Rodricks, C. W. Hesseltine, M. Mehlman, Eds., *Mycotoxins in Human and Animal Health*, Pathotox Publishers, Inc., Park Forest South, IL, 1977, pp. 189-207.

function is structurally protected and stable (58). The development of reliable analytical methods for detection of this class of compounds has been slow because of the large number of closely related derivatives, the ease of hydrolysis of the various ester functions, and the lack of strong ultraviolet absorption at wavelengths greater than 210 nm. The first methods developed were based on TLC using a H_2SO_4 or a *p*-anisaldehyde spray for visualization. Later methods involved formation of suitable derivatives (acetates, silanes, etc.) followed by GLC for identification and quantitation, and gas chromatography–mass spectrometry (GC–MS) for confirmation of identity. These procedures had limits of detection in the 10- to 50-ng/g range (66). More recently an HPLC procedure has been developed, requiring initial preparation of benzoates that may then be detected (lower detection limit for T-2 toxin is about 10 ng/g) and quantitated (67). Radioimmunoassay techniques have been explored for T-2 toxin and show promise for the future (68). Finally, a commonly used biological screen for detecting the trichothecene mycotoxins has been developed, based on the sensitivity of brine shrimp (*Artemia salina*) toward most members of this group of compounds. There are no methods for the identification of all members of the trichothecene class.

The development of reliable analytical methods was the key to the development of data on the natural occurrence of the trichothecenes. The first documentation of the natural occurrence of trichothecenes appeared in 1971 and arose out of a case of mycotoxicosis in a herd of cattle in Wisconsin. Thirty animals died with the typical hemorrhagic syndrome previously associated with moldy corn toxicoses in other animals. The mycotoxicosis was soon identified as being due to the feed that was badly infested with *F. tricinctum*. Extensive investigation of this feed led to the discovery of T-2 toxin; the feed was contaminated at a level of 2 μg/g (69).

Although the fungal species capable of producing the trichothecene metabolites are frequently found on foods and feeds and the trichothecenes are often suspected to be the causative factors in both animal and human mycotoxicoses, the trichothecenes have only rarely been isolated and identified. The few documented cases include the identification of the trichothecenes, nivalenol and deoxynivalenol (DON), in barley produced in Japan in 1972. Vesonder additionally identified DON in corn produced in Ohio at levels of 20 ng/g in 1973; this corn had been implicated in a mycotoxicosis in swine and was apparently responsible for the observed symptoms of feed refusal and vomiting. In 1976 Mirocha reported the natural occurrence of diacetoxyscirpenol (DAS) in swine feed (380 to 500 ng/g) apparently responsible for hemorrhaging in the swine. He also reported finding DON in corn and mixed feeds (50 to 1800 ng/g), which was apparently responsible for feed refusal and vomiting in pigs and dogs in the Midwest and for bloody stools in Nebraska cows (66).

An outbreak of *Fusarium* contamination occurred in 1972 in corn produced in the Midwest. The FDA analyzed 223 samples of the implicated corn; 93 of 173 samples contained a dermal irritant as measured using a rabbit skin test. Chemical analysis of some of these samples showed the presence of T-2 toxin and DON (66, 70).

Although there is a wealth of suggestive evidence that the trichothecene mycotoxins may be a significant hazard to human and animal health, the unavailability of rapid and reliable analytical methods and of standards of the pure mycotoxins make it impossible to pinpoint the extent of the potential hazard. The whole trichothecene problem is complicated by the fact that it is not known which of the many members of this class are likely contaminants of food and feed.

Ochratoxin

The ochratoxins are a group of seven closely related fungal metabolites produced by various species of *Aspergillus ochraceus* isolated from cereals and legumes (57). The structures of the major metabolites are shown in Figure 5.17. The ochratoxins are colorless, crystalline compounds whose structures have been confirmed by total synthesis. On acid or enzyme hydrolysis ochratoxin A yields L-phenylalanine and an isocoumarin residue. This reaction forms the basis for a colorimetric method for detection of ochratoxin A in blood. The ochratoxins are readily detected on TLC because of their intense fluorescence. Ochratoxin A, under long-wave ultraviolet light, appears as a green fluorescent spot; treatment with base yields a more intense blue fluorescence. Reliable methods for the detection of ochratoxins in food and feeds by GLC and HPLC are available (71).

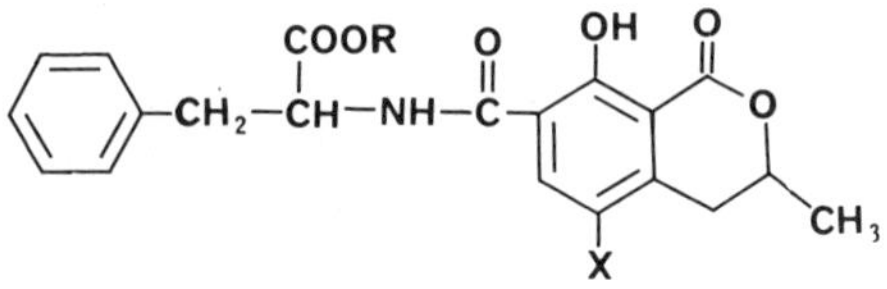

	X	R
Ochratoxin A	Cl	H
Ochratoxin B	H	H
Ochratoxin C	Cl	C_2H_5

FIGURE 5.17 The Ochratoxins commonly occurring in foods.

The ochratoxins are acutely toxic in many animal species; the LD_{50} (oral lethal dose, rat) is about 22 mg/kg for ochratoxin A. Ochratoxin A has also been reported to be teratogenic in mice, rats, and chick embryo (72). Recently Japanese investigators have reported the finding of renal carcinomas in mice after oral administration of ochratoxin A (73). There is also a report that ochratoxin A and sterculic acid act synergistically to produce hepatoma in rainbow trout (74).

Ochratoxin A has been found to be nephrotoxic in all animal species tested to date; studies in rats revealed that at high dose levels liver injury occurs as well (75). Swine are particularly sensitive to the nephrotoxic action of ochratoxin A. The ochratoxin induced nephropathy in swine is comparable to chronic interstitial nephritis in humans, a malady known to occur throughout the world. The etiology of the latter is, however, unknown at this time. One particular form of the disease is Balkan endemic nephropathy, a severe human disease found in certain Balkan countries (75). It has been proposed that ochratoxin A may be one cause of the disease. This proposal is supported by recent findings of ochratoxin A in the foods from certain endemic areas at higher incidence and level than in other areas studied; definitive studies have not yet been conducted.

The acute and subacute toxicological properties of the ochratoxins prompted an examination of foods and feeds for the possible presence of these mycotoxins. Surveys conducted in the United States indicated both a low incidence and a low level of contamination (4 of 1600 corn samples contained 83 to 166 ng/g, 9 of 848 wheat samples contained 20 to 114 ng/g; 22 of 159 barley samples had 10 to 29 ng/g; and none of 138 malt barley or 200 sorghum samples contained ochratoxin) (76). On the other hand, in Denmark a high incidence and level of contamination was observed; in one survey 58% of 33 feed samples (mostly barley) contained ochratoxin at levels of 0.03 to 28 μg/g (75). Further study showed that such contaminated feed resulted in ochratoxin transmission to the meat and blood of pigs; levels as high as 67 μg/g were observed in the kidneys (75). Denmark subsequently instituted a control program in which the kidneys of pigs are examined at slaughter for evidence of nephrosis. Nephrotic kidneys are analyzed for ochratoxin, and evidence of ochratoxin contamination is the basis for disposal of the carcass (75).

Ochratoxin has also been found in beans, peanuts, moldy green coffee beans, and hay. Transmission to edible poultry tissue has also been observed. The significance to human health of the natural occurrence of ochratoxin in foods has yet to be determined (71).

Sterigmatocystin

Sterigmatocystin was first isolated in 1957 as a yellow, high-melting crystalline compound from mycelial mats of *Aspergillus versicolor.* Its complex and unusual structure was elucidated in 1963; it was the first naturally occurring compound identified to contain the furfuran ring system (77). Its structure is shown in Figure 5.18. The elucidation of the structure of sterigmatocystin simplified greatly the later structural work on the chemically related aflatoxins, the versicolorins, and the austocystins (77).

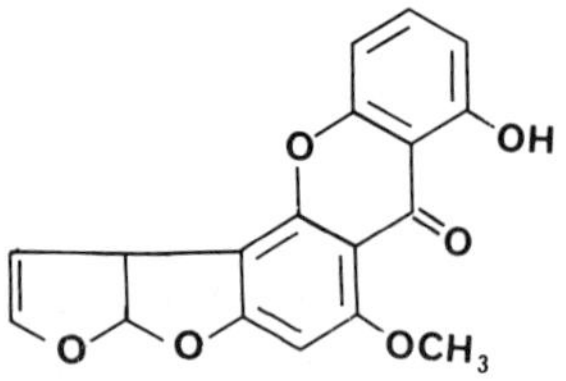

FIGURE 5.18 Sterigmatocystin.

Sterigmatocystin is also produced in lesser quantity by other molds frequently found on foods, including *A. nidulans*, *A. rugulosus*, and *A. flavus*. These mold species are prolific producers of colored anthroquinone and xanthone derivatives, including at least seven methylated, demethylated, methoxylated, hydroxylated, and dihydro derivatives of sterigmatocystin. There is good evidence to show that sterigmatocystin is an intermediate in the biosynthesis of aflatoxin (77).

Sterigmatocystin has been found to be less acutely toxic than aflatoxin (LD_{50} of 166 mg/kg in rats). On the other hand, like aflatoxin, it has been found to induce hepatomas in rats and both pulmonary and liver tumors in mice. Unlike aflatoxin, it produces neoplastic skin lesions in mice. It has been shown to possess both mutagenic and teratogenic activity. (77).

The finding of carcinogenic activity and the frequent occurrence of molds capable of producing sterigmatocystin in foods posed the question of possible human exposure to this carcinogen. Analytical methods based on TLC were developed to detect sterigmatocystin in grains at the 50-ng/g level (78). The use of these methods in extensive surveys has failed to identify a significant natural occurrence of this carcinogen in food.

Other Mycotoxins

A large variety of molds may be found naturally on foods. Mold profile studies are commonly used to identify the most prevalent species. A great deal of effort has been applied to determine what factors (environmental, cultural, etc.) influence the type of mold found in foods. The most commonly found species are usually examined for toxin-producing capability. The toxins are then isolated and identified chemically and their toxicological properties determined. Extensive studies of the type described previously have led to the realization that many mycotoxins have a potential to contaminate foods and feeds; the full significance from a public health point of view of the presence of such toxins in foods and feeds is unknown. Table 5.2 lists some of the mycotoxins found in foods and feed, the organism first identified

TABLE 5.2 Some Important Mycotoxins

Name	Origin	Susceptible Commodities	Effects
Brevianamides	*P. viridicatum*	Corn, grains	--
Citreoviridin	*P. citreo-viride*	Rice	Neurotoxins
Citrinin	*P. citrinin*	Corn, grains	Nephrotoxin
Cyclochlorotine	*P. islandicum*	Grains, rice	Hepatotoxin
Cyclopiazonic acid	*P. cyclopium*	Grains	Neurotoxins
Cytochalasins	*Helminthosporium*	Grains, corn	Cytochemical
Ergot alkaloids	*Claviceps*	Grains	Neurotoxins hallucigens
Luteoskyrin	*P. islandicum*	Rice, grains	Hepatotoxin
Penitrems	*P. cyclopium*	Corn, grains	Tremorgen
Penicillic acid	*P. puberulum*	Corn, flour, beans	Antibacterial
PR Toxin	*P. roqueforti*	Cheese, grain, rice	Tremorgen
Roquefortine	*P. roqueforti*	Cheese, grain, rice	Neurotoxin
Rubratoxin B	*P. rubrum*	Corn, grains	Hepatotoxin, Nephrotoxin
Rugulosin	*P. rugolosum*	Grains, rice	Hepatotoxin

as producing the toxin, and the major toxicological properties of the toxins. The list is by no means all inclusive; over 100 mycotoxins have been isolated and identified, and undoubtedly many more have yet to be discovered.

COMPOUNDS OF NATURAL ORIGIN CONTAMINATING EDIBLE ANIMAL PRODUCTS

In the earlier section on aflatoxin, mention was made of a metabolite called aflatoxin M_1 that is excreted in the milk of lactating animals ingesting aflatoxin B_1. This is but one of several examples of plant or microbial metabolites that can accumulate in edible animal products and thus pose a potential risk to human beings.

Little is known about the overall incidence of this type of food contamination, but, as with many of the other food hazards of natural origin, there are compelling reasons to believe that the area of the unknown greatly exceeds the area of the known. A few of the better-known examples of this phenomenon are cited in this section.

Coniine

Hall (79) has identified a passage from the Bible that must be not only one of the oldest but surely one of the most reliable references in the toxicological literature. It is found in the eleventh chapter of the Book of Numbers (verses 31 through 33) and is worth quoting:

> Then a wind from the Lord sprang up; it drove quails in from the west, and they were flying all around the camp for a day's journey, three feet above the ground. The people were busy gathering quails all that day, all night, and all next day, and even the man who got least gathered ten omers. They spread them out to dry all about the camp. But the meat was scarcely between their teeth, and they had not so much as bitten it, when the Lord's anger broke out against the people and he struck them with a deadly plague.

It may be suspected that the author exaggerated the speed with which the poisonous tissue acted ("not so much as bitten it"), but it seems reasonable to conclude that the acute toxicity of the substance ingested was more than moderately high.

Hall speculates that the quail wintered in Africa, as they do today, and ingested poisonous berries, including those from hemlock (*Conium maculatum*). The latter contains a neurotoxic alkaloid called *coniine* (Figure 5.19). The quail are able to ingest and accumulate this substance in their tissues without apparent problem, but human beings are, in this regard, not so fortunate. (Of course, for other reasons, the quail did not survive the incident either.)

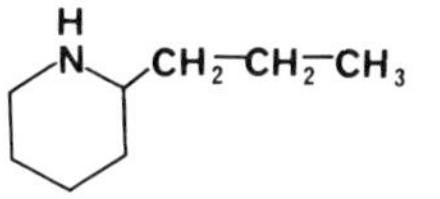

FIGURE 5.19 Coniine.

Pyrrolizidine Alkaloids

On a more serious plane are the possible threats posed by the class of substances known as *pyrrolizidine alkaloids.* More than 100 of these plant products have been characterized, and a large fraction of those that have been studied are hepatotoxic. At least four are well-characterized mutagens and animal carcinogens, and it is likely that more of the substances are similarly active (80).

These alkaloids all possess a common heterocyclic ring system, exemplified by retronecine (Figure 5.20) and differ from one another in the manner in which the ring system is substituted. A broad subclass also contains macrocyclic ring systems of the type found in the hepatocarcinogenic retrorsine (Figure 5.20). These substances are produced by a number of plant species, of which *Senecio, Heliotropium,* and *Crotolaria* are the important members (80). Livestock grazing on plants in the preceding families suffer liver and lung lesions, but human health is also of concern. Two possible routes of human exposure are (*a*) direct consumption of plants containing the pyrrolizidine alkaloids for medicinal purposes and (*b*) consumption of edible products derived from livestock grazing on these plant sources.

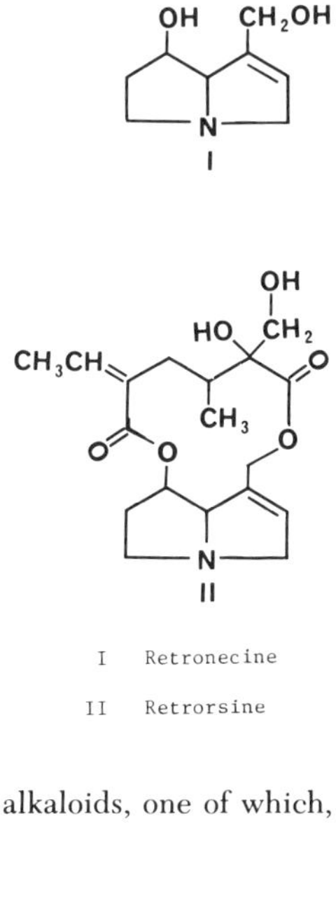

I Retronecine

II Retrorsine

FIGURE 5.20 Two pyrrolizidine alkaloids, one of which, retrorsine, is a known hepatocarcinogen.

Outbreaks of vascular occlusive disease of the liver, apparently associated with consumption of pyrrolizidine alkaloids contained in herbal teas or medicines, have been reported in Jamaica, India, and Afghanistan (81) and recently from the United States (82). The latter case involved two infants in Arizona, both under 1 year of age, who were administered "Gordolobo yerba" tea as a medicine. Both children exhibited acute hepatic disease and severe hypertension; the disease state was characterized as hepatic venoocclusive disease. The teas were found to contain pyrrolizidine alkaloids, at least one of which was probably retrorsine. One of the children died (82).

The possible role in human liver cancer (and even mutagenesis) of consumption of herbal medicines and beverages containing these substances remains largely unexplored. Because these plants are widely used in heavily populated areas of the world, an intensive effort to assess population risks from these substances would seem to be well justified (83).

Far less documentation exists on the occurrence of pyrrolizidine alkaloids in edible animal products. Because they occur in common rangeland plants (tansy ragwort, groundsel, etc.) in the United States (particularly the Southwest), the potential for meat contamination might be high. The USDA and the FDA are pooling resources to survey rangeland *Senecio* species to ascertain which of the large number of pyrrolizidine alkaloids are likely to enter the food supply. Methods for the detection of these alkaloids in animal tissues and milk need to be developed, but the direction this effort should take largely depends on the survey results and experimental studies in cows, sheep, and other animals to identify the expected tissue residues (84). This is another area that does not appear to be receiving the effort it deserves. Of course, any such effort is limited by the extreme difficulty encountered in preparing sufficient amounts of the suspect alkaloids for analytical, metabolic, and toxicological studies.

Several of the pyrrolizidine alkaloids present in *Senecio jacobae* (tansy ragwort) have been found as contaminants of honey produced from nectar of this species (85). This is not the only example of bees accumulating toxic substances from plants and excreting them in honey, and some areas of the world have been closed to bee keeping for this reason (79).

Marine toxins

It has long been recognized that the ocean is a bountiful and relatively inexpensive source of nutritious food. The ever-increasing world population demands that maximum use be made of the rich protein sources found in fish and shellfish. Many species of fish and shellfish, however, are known to contain toxins that, if ingested, can be harmful and even deadly to humans (86). Poisonings related to ingestion of such seafoods have been recorded throughout human history. Over the years humans have learned to avoid eating certain types of seafood; on the other hand, many fish and shellfish are normally safe to eat and only occasionally are toxic. It is the ingestion of these normally "safe" seafood products that leads to the major poisonings noted today. The large majority of such human poisonings may be categorized as follows: (*a*) paralytic shellfish poisoning; (*b*) ciguatera intoxication; (*c*) tetrodotoxin poisoning; and (*d*) scombroid poisoning. Each of these types of poisoning has been known for many years and continues to represent a serious human health problem.

Paralytic Shellfish Poison (PSP)

It has been known for several centuries that shellfish occasionally become toxic. In the early 1600s the Indians of Port Royal, Nova Scotia, were observed to eat their dogs or the bark from trees rather than the plentiful, but deadly, supply of mussels available to them. On the west coast of North America the Indians would not eat shellfish during those periods when the ocean exhibited bioluminescence at night; they had somehow correlated, probably through bitter experience, the presence of an eerily glowing sea with the danger of toxic shellfish (87). The historical record documents many examples of severe illness and death resulting from the ingestion of toxic shellfish by the unwary or careless consumer. It has been estimated that from 1689 to 1977 about 1440 people were affected by PSP worldwide with at least 232 deaths. Undoubtedly the true incidence of poisoning is much greater. Today, as doctors are becoming more familiar with the symptoms of the poisoning and are correctly diagnosing the malady, and for other reasons as well, we are beginning to realize the magnitude of the public health problem resulting from PSP (86).

It is now believed that shellfish become toxic when feeding on benthic algae, particularly the dinoflagellates. These organisms are, along with other phytoplankton, the backbone of the marine food chain. Over 1200 species of dinoflagellates are known; luckily, only a few species seem to elaborate toxins (88). Under certain conditions of growth these organisms undergo a period of rapid growth (a bloom), causing a phenomenon known descriptively as a "red tide." The large numbers of organisms in the water ($\leq$1,000,000/ml) often color the water various shades of red; in Puget Sound, for example, it is not uncommon to observe, particularly during the summer months, water the color of tomato soup due to the dinoflagellate, *Noctiluca,* meaning night light. This particular organism, (which is, fortunately, not toxic) gives rise to a bright bioluminescence when the water is disturbed.

Neither the presence nor the absence of a red tide guarantees that the shellfish growing in such an area will be nontoxic. There are areas, such as Alaska, where the shellfish are toxic, but in which no red tide has ever been observed. It is now suspected that this may be due to ingestion by the shellfish of the resting cysts rather than the motile cells present in the waters of the red tide (88).

Paralytic shellfish poison will concentrate in any marine organism feeding on the toxin-containing dinoflagellates. The shellfish seem to be impervious to the action of the toxins. Other marine organisms, however, are affected; consequently, if an unusually large number of dead fish, crabs, and similar organisms are found on the beach, a red tide may be suspected. On the West Coast of the United States the major red tide organism apparently is *Gonyaulax catenella;* on the East Coast it is *G. tamarensis.* It has been estimated that when the concentration of these organisms in the water reaches 200 cells/ml the bivalves in that area become too toxic for human consumption (88).

The shellfish exhibit marked differences in their ability to assimilate the toxin and eliminate it. For example, the Bay mussel (*Mya edulis*) concentrates PSP mainly in the digestive gland and retains the toxin for about 2 weeks; it accumulates and releases the toxin so rapidly that its toxicity parallels the presence of the red tide. The soft shell clam (*Mya arenaria*) accumulates and releases the toxin slowly; the toxin is concentrated in the digestive glands during summer months and in the gills during the autumn and winter months. The Alaskan butter clam accumulates the toxin primarily in the siphon and only slowly releases the toxin (86, 88).

Although paralytic shellfish poisoning has been known for many hundreds of years, it was not until 1957 that Schantz et al. (89) isolated a pure toxin from the California mussel and the Alaskan butter clam as a highly hydroscopic, colorless powder. The chemical structure, shown in Figure 5.21, was finally elucidated in 1975 by Schantz et al. (90). The compound, having a molecular weight of 372 and a molecular formula of $C_{10}H_{19}Cl_2\ N_7O_4$, is a complex alkaloid containing the very uncommon guanidine moiety. Further research quickly revealed that PSP was not a single toxin, but a mixture of saxitoxin and saxitoxin analogues (gonyautoxins). The relative proportion of the various analogues was dependent on the dinoflagellate source (88).

These compounds are potent neurotoxins; they inhibit the sodium-potassium pump that controls electrical conduction. Ingestion of PSP results, in mild cases, in a tingling sensation or numbness around the lips, face, and neck; headache; dizziness; and nausea. In more severe cases there occurs a stiffness or a numbness of the limbs, a feeling of lightness, general weakness, rapid pulse, and slight respiratory difficulty. Severe poisoning results in muscular paralysis, pronounced respiratory difficulty, and death within 24 hours. The malady is frequently misdiagnosed since the symptoms are often misinterpreted as drunkenness. There is some evidence that an individual can develop a limited immunity to the poison (86, 87). There is no known antidote; this may

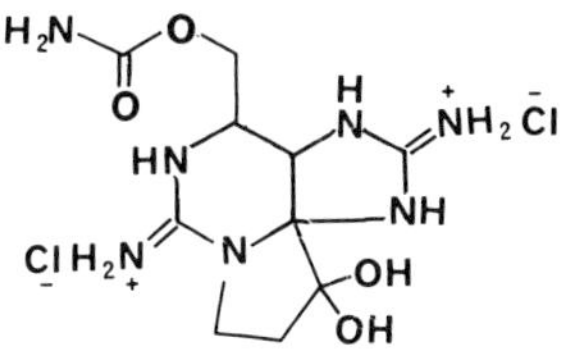

FIGURE 5.21 Saxitoxin.

explain why U2 pilot Gary Powers is said to have carried a suicide vial of saxitoxin in 1960 when he was shot down over Russia.

The control of PSP is costly. To control the problem, public health officials have taken two approaches: (1) monitoring of the shellfish for the presence of PSP during harvest or (2) prohibiting shellfish harvesting from those sections of the coast where evidence of PSP or red tide has been obtained. As early as 1925 the U.S. government attempted to establish such a control program in conjunction with the shellfish-producing states (86). In 1937 Sommer and Meyer devised the first practical assay for PSP, a mouse bioassay, which measured toxicity in terms of mouse units (MU) (86). One MU was defined as the amount of toxin given intraperitoneally, that will kill a 20-g mouse in 15 minutes (0.18 μg of saxitoxin). Toxicity is then expressed in terms of ng/100 g shellfish. Unfortunately, because of animal variation, the variability of toxicity among different saxitoxin analogues, and so on, the mouse bioassay may underestimate the amount present by as much as 60%. Nevertheless, before the structure of saxitoxin had been determined, no other assay was available and control programs were established on the basis of the mouse bioassay.

In 1947 the entire coastline of Alaska was closed to the harvest of the Alaskan butter clam and mussels. In 1951 the FDA promulgated a regulation establishing the "safe" level of shellfish at 400 MU/100 g shellfish meat. This limit was set at 80 ng/100 g in 1958 as a result of improvements in the assay method and an epidemiological survey in Canada. When shellfish exceed that level of toxicity the area is closed to shellfish harvest (86). It has been estimated that levels of 1000 ng/100 g will produce moderate symptoms of poisoning in the adult; 10,000 ng/100 g is apparently a lethal dose. It is not unusual for shellfish to have such extremely high levels of PSP; in 1975 mussels from Maine contained about 22,000 ng/100 g.

Today a great deal of effort is being applied to the development of a chemical assay procedure for PSP. Some methods have already been developed. In one such method saxitoxin is cleaved to guanidine by treatment with acidic H_2O_2; the guanidine is then detected colorimetrically (91). The method has an inadequate lower detection limit of 2,000 ng/100 g. Another method uses a basic H_2O_2 solution to produce a strongly fluorescent derivative; as little as 0.2 ng/100 g can be detected by this method (91). Unfortunately, none of these chemical methods of analysis has been developed for routine and widespread use. Recently, Shimizu of the University of Rhode Island has developed a "PSP analyzer" (92) consisting of a high-performance liquid chromatograph using Bio-Rex-70 ion-exchange resins, a mixing coil for adding the H_2O_2 solution, and a fluorometric detector. The advantage of this system appears to be that it can be used to detect and quantitate not only saxitoxin, but also 8 analogues, some of which are of the same range of toxicity as saxitoxin itself.

Ciguatera

Ingestion of ciguatoxic fish results in a malady known as *ciguatera* (86). In a sense the term is a misnomer; the early Spanish settlers in Cuba were aware that a freshwater snail, locally known as *cigua*, frequently caused digestive and nervous disorders on ingestion. All other fish poisonings were thought to be the same as that caused by the snail. The term "ciguatera" first appeared in a technical paper published in Havana in 1787. We now know that over 400 species of fish may become ciguatoxic; each year numerous people are adversely affected after ingestion of such fish. Halstead reports that at least 4497 persons had been poisoned by ingestion of ciguatoxic fish (worldwide) and that at least 542 deaths resulted. A mortality rate of about two dozen cases is noted each year, virtually all of which are attributable to ingestion of barracuda. The true incidence is unknown because ciguatera is not a reportable disease and many doctors do not recognize the symptoms (which are much like those due to organophosphate poisoning). In the Virgin Islands it has been estimated that only 10 to 15% of the yearly cases are reported; even so, health authorities estimate a minimum of 30 cases a week. It is commonly stated in the Virgin Islands that "virtually everybody" has been poisoned at least once (some as many as 15 times), but medical advice is almost never sought.

The typical symptoms of ciguatera include an initial period of gastrointestinal disorders (abdominal pain, nausea, vomiting, and diarrhea), followed by extended periods of neurological disorders (tingling and numbness in lips, tongue, and extremities; metallic taste, headache, dizziness, visual disturbances, painful teeth, itching, convulsions, reversal of feelings of hot and cold, etc.). In most instances such symptoms are transitory, lasting for a few hours to several weeks; in cases of more severe intoxication the symptoms may be debilitating, lasting for as long as 25 years. In extreme cases death may occur within 10 minutes of ingestion of the toxic fish; normally, however, death does not occur for several days. It should be noted that repetitive exposure to nontoxic levels of the toxin may result in ciguatera (there is some evidence that the toxin accumulates or concentrates in humans as well as fish).

Only tropical, saltwater fish have been found to be ciguatoxic; neither the source nor the structure of the toxin(s) are known. Ciguatera has been described as a particularly insidious disease because it is usually caused by ingestion of fish that are normally safe to eat. Over 400 species of fish have been identified as being at times ciguatoxic; these are usually confined to shallow coastal areas, lagoons, and reefs.

There have been various proposals regarding the origin of the toxin(s). Some have suggested that the toxin is produced by a benthic, blue–green algae. This proposal is based on the fact that most ciguatoxic fish are bottom dwellers (many eating only plants) or predatory fish that feed on the bottom dwellers. This would explain why larger (older) predatory fish are often more toxic than smaller fish of the same variety and also why outbreaks of ciguatoxicity occur following disturbances of coral reefs (storms, dredging, shipwrecks, dumping, etc.); the blue–green algae are early colonizers of such denuded or disturbed areas (86).

A second, frequently postulated theory is that fish become ciguatoxic following the ingestion of dinoflagellates (particularly *Gymnodinium breve* and *G. toxicus*). The dinoflagellates are well-known producers of neurotoxins and are frequently found to colonize coastal waters. The third most prevalent theory is that ciguatoxicity originates with bacteria, perphaps living in the gut of the fish or in close association with dense populations of dinoflagellates or blue–green algae (86).

Comparison of the pharmacological properties of the toxin(s) responsible for ciguatera reported in different, widely separated locations leads to the conclusion that ciguatera is caused by more than one compound (86). Several toxic substances have been isolated, including a lipid-soluble toxin (ciguatoxin), a water-soluble toxin (ciguaterin), and a high-molecular-weight toxin (maitotoxin, molecular weight ca. 300,000). The structures of none of these materials have been elucidated, and none appear to be the principal toxin responsible for ciguatera. Scheuer of the University of Hawaii has recently purified a new toxin that shows an LD_{50} in mice of 10 ng/kg. The toxin (500 mg) was obtained from 2200 lb of moray eel as a water-insoluble, noncrystalline material with an approximate molecular weight of 1500 and a suggested empirical formula of $(C_{35}H_{65}NO_8)_n$. Hydrolysis of the toxin yields glycerol and a nonhydroxylic fatty acid (86).

Although the structure(s) of the toxin(s) is unknown, methods have been developed for identifying ciguatoxic fish. These include a mouse bioassay, a brine shrimp bioassay capable of detecting 1 to 25 μg/ml, and most recently, a radioimmunoassay procedure (93). None of these methods is completely satisfactory or suitable for quality control purposes. In view of this situation, and because the toxin(s) are stable to freezing and cooking, public health officials have provided certain guidelines for the public designed to minimize the problems of (1) knowing which species of fish are dangerous in a particular locale, (2) avoiding the ingestion of internal organs, especially the liver, and (3) avoiding the ingestion of the larger and older fish known to become ciguatoxic (94).

Tetrodotoxin Poisoning

Tetrodon or puffer poisoning is another disease associated with ingestion of toxic fish. In this case the puffer fish, of which about 80 species are known to be toxic, is the culprit. According to Hirata, human intoxications resulting from the consumption of puffer fish were reported 2000 years ago in the early Chinese medical literature (95). The puffer fish is considered to be a delicacy in Japan; consequently, tetrodon poisoning has been a continuing problem in Japan. Public health officials in that country have attempted to control the problem by licensing individuals trained to remove the toxic portions from the fish (liver, ovaries, gonads, intestines, and skin).

The causative agent of tetrodon poisoning is tetrodotoxin. The toxin was first obtained from the ovaries of the puffer fish as a water-insoluble, heat-stable material that has no definable melting point or ultraviolet absorption and an empirical formula of $C_{11}H_{17}N_3O_8$. The structure (Figure 5.22) was finally elucidated in 1966 by x-ray diffraction studies on a crystalline tetrodoic acid hydrobromide (96). The magnitude of the challenge in resolving the structure of this toxin may be inferred from the observation that there are as many hetero atoms as carbons. The total synthesis of tetrodotoxin was reported in 1972 (97).

The mouse bioassay has been used for many years to identify toxic fish. No reliable chemical assay procedures are currently available, although a procedure based on TLC and having a limit of detection of about 3 μg has been reported recently (98).

Tetrodotoxin, one of the most powerful toxicants known, has an LD_{50} in mice of about 12 ng/kg. Worldwide, a mortality rate of 61% has been estimated for tetrodotoxin. This neurotoxin causes convulsions and death in humans in 1.5 to 8 hours as a result of respiratory paralysis. There is no known antidote. The toxin is not inactivated by cooking (99).

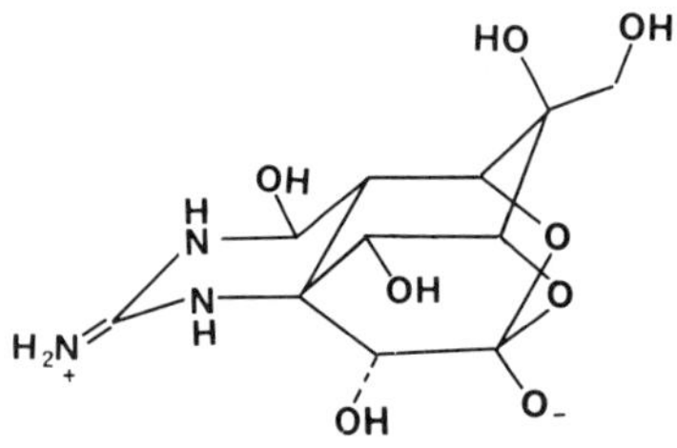

FIGURE 5.22 Tetrodotoxin.

Scombroid Poisoning

The greatest proportion of cases of seafood poisoning have undoubtedly been a result of bacterial decomposition; such bacterial decomposition, resulting from improper preservation of fish, is called *scombroid poisoning* because it frequently results from eating sauries (family Scombroidea) and tunas, mackerels, sardines, anchovies, and wahoo (family Scombroidea). Documented cases of scombroid poisoning have been reported in the United States since the late 1930s. The symptoms of scombroid poisoning resemble strongly an allergic response to histamine (next section) and include facial flushing, intense headache, vomiting, and gastrointestinal pain. The disease is rarely fatal (100).

It is generally believed that bacterial decarboxylation of the histidine present in the muscle of dark-meated fish yields histamine that, in the presence of a synergist, leads to the condition known as *scombroid poisoning*. There is no doubt that bacterial decomposition of such fish leads to high histamine levels (2000 to 5000 μg/g) even before the fish appear to be spoiled or are organoleptically unsuitable. On the other hand, it has been found that humans can tolerate large quantities of pure histamine ($\leq$ 180 ng) with no ill effect. A great deal of research is still under way to unravel the mystery of scombroid poisoning (100).

Miscellaneous Substances

Dairy cattle excrete a carcinogen derived from braken fern when they graze on this plant. Estrogenic substances present in numerous plants can possibly contaminate edible animal products. The number of poisonous plants in rangelands that are used as forage is substantial. Mycotoxins, which are far more likely to occur in animal feeds than in human foods, at least in the temperate zones of the earth, are likely contaminants of meat, milk, and eggs. Public health scientists have barely begun to explore the naturally occurring toxicants that may accumulate in the edible portions of land and marine animals used as food (1).

COMPOUNDS PRODUCED DURING FOOD STORAGE, PROCESSING, AND PREPARATION

The chemical reactions that can take place during food storage, processing, and preparation are probably infinite. Variations within the same food material occur under different conditions of preparation. Reaction products also depend on the types of food combined during preparation. Any attempt to characterize the total number and types of products formed during food processing and preparation is probably hopeless, even excluding the additional sets of reactions involving substances intentionally introduced into foods. There are, however, some examples of compounds in this class that are noteworthy because of their potential to adversely affect human health.

Biologically Active Amines in Foods

Foods naturally contain a large number of physiologically active amines. Most of them are low-molecular-weight organic bases that do not appear to represent a hazard to individuals unless abnormally large amounts are ingested or the individual's natural catabolic mechanisms are genetically deficient or medicinally inhibited (e.g., with monoamine oxidase inhibitors, drugs used for treatment of depression, hypertension, etc.). Many of these amines act directly to increase blood pressure; others are psychoactive (101).

Among the vasoactive compounds are tyramine, histamine, and phenethylamine. Histamine, a capillary dilator, produces hypotensive effects whereas tyramine and phenethylamine cause a rise in blood pressure. Phenethylamine has been implicated in the onset of migraine headache. Tyramine has been found in bananas, broad beans, cheeses of all kinds, wines, beer and ale, beef and chicken livers, sauerkraut, yeast extracts, and fermented and pickled fish products. Histamine has been found in wines (especially red wines), yeast extracts, fish, sausage, sauerkraut, and cheese of all types. Phenethylamine occurs in chocolate (101).

These compounds are described in this section because they appear to become important health problems only in foods that are fermented or decomposed. For example, many bacteria produce enzymes capable of catalyzing the decarboxylation of amino acid components of foods. Bacteria known to promote the formation of histamine and tyramine include *Lactobacillus, Salmonella, Clostridium, Proteus, Betabacterium,* and *Escherichia*. It is not surprising, therefore, that aged, fermented, or decomposed foods often contain considerable quantities of such amines. Fish products are especially susceptible; the muscle tissue normally contains large amounts of free histidine, which in the presence of bacteria is converted to histamine. High levels of histamine can be reached before the fish appears spoiled or is organoleptically unacceptable. One way of controlling fish quality is to use the histamine content as an indicator of decomposition. Many other physiologically active derivatives of histamine, tyramine, and phenethylamine are present in foods, including tryptamine and its hydroxylated derivative serotonin, which can be found in tomato, banana, fruits, and fruit juices. A great deal remains to be learned concerning the incidence and the levels of such compounds in foods (101).

Although the substances listed in the preceding paragraphs are known to serve as neurotransmitters, their ingestion through food appears to have little effect on the CNS because they do not appear to penetrate the blood–brain barrier. On the other hand, some naturally occurring amines at sufficient intake levels can have a marked effect on the CNS (acting as stimulants, such as the caffeine, theophylline, and theobromine in coffee and tea; depressants, such as ethyl alcohol and discorine; the tropane alkaloid found in yams; and hallucinogens such as myristicin in nutmeg; the ergot alkaloids; and some mushroom toxins). These amines will provide a fertile area for research for many years to come.

Nitrosamines

Nitrites, which can appear in the diet because of natural occurrence or deliberate addition, can under some conditions react with secondary amines to form nitrosamines (Figure 5.23). The amines necessary for the nitrosation reaction occur widely in the human diet and are also present in some cosmetics and drugs. The nitrosation reaction can take place during the frying of nitrite-cured bacon and also occurs in the digestive tract (102). Some nitrosamines are carcinogens in a wide range of animal species, and it is probable that at least some of them are similarly active in humans.

$$NO_2 + H^+ \rightleftharpoons HONO \rightleftharpoons NO^+ + H_2O$$

$$NO^+ \xrightarrow{R_2NH} H^+ + R_2N{-}NO$$

FIGURE 5.23 *N*-Nitrosamine formation.

Clearly, the ease of reducing risk from nitrosamines is related to the ease of removal of the insulting agents from the environment. However, it may well be that those sources easiest to control may not represent the greatest risks. Much more information is needed on the relative risks associated with the various routes of human exposure to nitrosamines before it can be decided how best to invest public health resources.

Magee et al. (102) list the following food sources of nitrosamines: smoked sausages, fried bacon, ham, salami, dried sausages, smoked herring and other fish, cheese, milk, flour, wheat, and mushrooms. Recently nitrosamines have been identified in beer and whiskey also (103).

Oxidized Fats

Chemical reactions occurring during the heating of fats and oils can result in the production of a variety of hydroxy, epoxy, and peroxy compounds, some of which are toxicologically suspect because of their high reactivity toward cell constituents. Similar oxidation of certain steroids (e.g., cholesterol) may occur. Animal feeding studies conducted thus far have revealed no carcinogenic action among the compounds examined, but the issue of a human hazard remains open because of the incompleteness of our knowledge of the chemical and biological properties of many of these lipid oxidation products (104).

Polynuclear Aromatic Hydrocarbons

These carcinogenic substances occur widely in the human environment and arise from many sources. It appears likely that the major sources are those found in certain occupational settings, although some of these compounds can be found in foods.

Compounds in this class are among the most widely studied of all carcinogens. Any attempt to summarize the wealth of information available on these substances will necessarily fail to be even close to adequate. One or, more frequently, combinations of the polynuclear aromatic hydrocarbons, which include compounds such as anthracene, benzanthracene, phenanthrene, fluorene, pyrene, benzpyrene, chrysene, and substituted derivatives (all fused-ring polyaromatics), can be found in water, air, tobacco smoke, curing smoke, foods, gasoline, and diesel exhausts and wherever incomplete combustion of fuels occurs (105–107). Although smoking appears to be their prime route of entry into foods, it has also been reported that severe heat treatment of meats can introduce some of these substances (107).

Some of the polynuclear aromatic hydrocarbons have been identified in surface sediments, presumably arising during forest and prairie fires, and may thus accumulate in plants grown in such sediment. Plant biosynthesis has also been reported (106). The relative contributions of these various dietary sources are unknown.

Development of methods for the analysis of compounds in this class has been the subject of strenuous effort, but lack of rapid methods of high specificity still remains a hindrance to the acquisition of complete data on human exposure to these substances through food (105, 108). The polynuclear aromatic hydrocarbons are of interest because of their well-known and long-established carcinogenicity (106). Not all members of the class are equally threatening because of differences in the levels at which they reach humans and variations in (including lack of) carcinogenic effect. These substances represent some of the most thoroughly studied carcinogenic agents, yet questions still remain regarding the manner of their carcinogenic action. It is known that they require metabolic activation to produce cellular damage and that the form and the rate of metabolic activation depend on chemical structure (106). As purely scientific as well as public health problems, the polynuclear aromatic hydrocarbons offer a formidable challenge.

Mutagens In Cooked Food

A recent flurry of public and scientific interest in the chemistry of food cooking came about because of reports that mutagens were produced during the broiling (charring?) of hamburger (109) and were also present in the charred portions of cooked beef and fish (110). Although the mutagenic substances actually produced in the foods have not been identified, some important leads have been developed. It has been shown that some pyrolysates of amino acids and proteins contain mutagenic substances. Sugimura discusses some of the substances (Figure 5.24) in a short review (111). He also cautions that the data developed so far do not permit an estimate of human health risk from these and other products similarly arising during food preparation. It is clear, however, that a potential risk exists and needs to be vigorously explored.

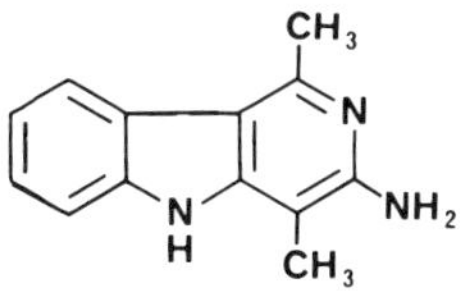

FIGURE 5.24 Trp-P-1, a pyrolysis produst of tryptophan possessing mutagenic properties.

REGULATIONS AND OTHER MEANS FOR CONTROLLING FOOD HAZARDS OF NATURAL ORIGIN

Superficial examination of the problem of food hazards of natural origin usually yields only two alternatives to dealing with foods bearing such substances: (*a*) completely avoid using the food or (*b*) do nothing and simply disregard the risks. Study of the history of attempts to deal with such problems reveals that these two approaches have been and continue to be commonly employed.

Certainly many early human inhabitants of this planet experienced severe distress or even lost their lives in their attempts to identify plants and animals that could be used for nourishment. Long before human beings were food producers, they were food gatherers, and it is during this gathering stage, probably mostly before about 8000 B.C., that many nonpoisonous plants and animals (at least as judged by their inability to kill or otherwise produce illness within a few hours or days of consumption) were identified and cultured.

Of course, the search for new sources of nourishment always has been a human occupation and continues to be to this day. In its present stage, it lies largely in the hands of the plant geneticists, those scientists who are able to manipulate genetic material to produce plants having almost any desired characteristic. More discussion of the role of plant geneticists in food safety appears later in this work, but for the present the major point is that because new foods have been introduced into the human diet throughout human history, it is not reasonable to argue that the human organism necessarily must have somehow "adapted" to all of the huge numbers of chemicals received through food.

Although many of the grains and other important foods have been in use for thousands of years, there are a number of common foods that have been widely used for much shorter periods of time. The white potato, for example, although widely used by Andean Indians for at least 2000 years, was not introduced into Europe until the late sixteenth century. The tomato, used only as an ornamental for a long time, was first introduced as a food item by the Italians in the late sixteenth century. Widespread cultivation of strawberries began as late as the midsixteenth century after wild cultivars were brought from the eastern United States. The consumption of grapefruit was limited to the West Indies until the nineteenth century, and only during the twentieth century has it gained worldwide popularity. Tea became an important beverage, on a worldwide scale, only in the seventeenth century.

This information, taken together with our knowledge of how chemical changes can result from genetic manipulation of plants, from cooking, and so on, should provide evidence that the natural components of food have not been with us since we first appeared on earth, and that many are relatively recent additions.

Satisfied that no special exemption need be given the natural components of food simply because of their "naturalness," the discussion can proceed to an examination of control capabilities, which surely extend beyond the two alternatives mentioned at the opening of this section. But first an excursion into the subject of safety assessment is appropriate for gaining an understanding of how hazards that need to be controlled can first be identified.

Food Additives

Throughout most of human history, the safety criteria for acceptance of a new food into the diet did not extend beyond assurance that no immediately observable harm would come to the consumer. During the past century, however, concern about the safety of substances intentionally added to food (so-called food additives) resulted in the passage of laws setting forth the conditions under which such substances could be used. In particular, the food additive laws require a showing that any such substance be "safe" under its proposed conditions of use before it is introduced into the food supply. Furthermore, new information that casts doubt on a substance's safety after it is approved for use may result in a withdrawal of its approval. In the United States the FDA has the authority to enforce this legal requirement.

A system for the safety evaluation of food additives had to be established to ensure that the legal requirements were met. In principle, the extent and the depth of toxicological testing required to ascertain the safety of a food additive are functions of the expected level of human exposure to the additive. In short, the higher the level of human exposure, the more extensive are the toxicity tests necessary to determine safety. With great oversimplification, but suitable for the present purpose, the safety assessment of food additives proceeds as follows (see Chapter 6 for more detail).

1. Knowledge of the expected level of human exposure to the additive is developed. Special population groups that might be at unusually high risk are identified.

2. Toxicity studies in experimental animals are designed to assess health risks, including those that might occur in exceptionally susceptible groups.

3. Levels of the additive capable of producing various forms of toxicity (acute poisoning, subacute effects, effects on reproduction or the developing fetus, cancer or other chronic disease, etc.) are identified.

4. Levels of the additive below which the observed health effects do not appear in the groups of animals studied (when compared to control animals) are also identified. These are termed the "no observed effect levels" (NOELs).

5. The lowest NOEL is divided by a "safety factor" to derive a level of human exposure that can be considered acceptable. The latter is termed the "acceptable daily intake" (ADI). The magnitude of the "safety factor" used depends on the depth and the quality of the toxicity studies and the severity of the effect against which protection is sought. A "safety factor" is used to compensate for the uncertainties involved in extrapolating from animals to humans. More recently the term "uncertainty factor" has been suggested by some as more descriptive than the term "safety factor." (In building or bridge design the term "safety factor" is applied to the additional safeguards built into a design beyond what are necessary to ensure safety. In the case of toxicity data, it cannot be ensured that a test animal NOEL is "safe" in the same sense. More discussion of this point appears later.)

6. If the known level of human exposure for a food additive is at or below the derived ADI, it is considered "safe" for use. The food additive laws exclude carcinogens and permit no ADI to be established for any such substance.

7. The actual quantitative relationship between the level of human exposure and the animal NOEL is called the "margin of safety." For food additives, the safety margin must be at least equal to the required "safety factor."

For several reasons there is no assurance that the procedure just described results in food additive exposures (ADIs) that are absolutely without risk to humans. Conversely, they may also permit the conclusion that no risk to humans exists even if the required margin of safety is not met (i.e., the ADI is exceeded). Our uncertainties in these matters are summarized as follows (2):

1. We cannot be certain that those adverse effects identified in experimental animals are also to be expected in exposed human beings. Conversely, we cannot be certain that human beings will not suffer adverse effects that are not manifested in all the animal toxicity tests we can think of conducting.

2. We cannot know with certainty the very important relationship between the level of exposure to a chemical and the resulting incidence of disease, except at the levels of exposure actually used in animal experimentation. More often than not, we need to know this relationship at a level of exposure well below those actually used in experimentation, because the lower level is that to which the human population will be exposed. The above-mentioned animal NOEL may appear to give us this information because no disease incidence is to be expected below the NOEL. The NOEL may, however, be simply an artifact of the animal test, dependent only on the number of animals used (i.e., the bioassay's inherent statistical limitations that prevent detection of disease incidences below about the 5 to 8% level).

3. We can never know with certainty the relative susceptibilities of experimental animals and humans to the toxic effects of a substance. Thus how much, if any, health risk remains at an established ADI is usually unknown. Epidemiology studies may sometimes provide useful verifications of the suitability of given margins of safety for food additives, but their relative insensitivity renders them somewhat impractical for all but the most visible kinds of health effect that may be associated with food. Perhaps the best that can be said is that the wider the margin of safety, the greater the degree of public health protection (i.e., the lower the risk). In many instances the risk to the human population may actually be zero, but we have no way of knowing this.

(Risk and the regulatory approach to its assessment are discussed in Chapter 7.)

Food Hazards of Natural Origin

The superficial discussion of safety assessment of food additives was offered to pave the way for a discussion of the criteria used to evaluate the safety of the substances that are the subject of this chapter. As a first point, it is worth emphasizing that the experimental methods used to detect and measure health risks of food additives (animal toxicity tests and, in very special circumstances, epidemiology studies) are as suited to the study of naturally occurring compounds as they are to synthetic ones. The same uncertainties appear and are dealt with in the same way. It would surely be an absurd safety assessment scheme that rested on the notion that an additive that was carcinogenic in experimental animals was a potential cancer threat to humans, but that a natural component of food that was similarly active in animals could not be similarly indicted.

However, beyond acceptance of the same methodologies for detecting and measuring these health risks, differences arise in the treatment of these two classes of substances. These differences are not only legal, but scientific as well. The procedure described in the preceding section for deriving an ADI for a food additive was designed and established to provide an operational definition of safety for this class of chemical. Without such a defined set of safety criteria, it could never be established whether a particular additive could be accepted into (or rejected from) the food supply.

The claim of this operational scheme is that at or below the ADI, the risks from a food additive are zero or, at most, vanishingly small. But the scheme was not designed, nor should it be used, for purposes of identifying hazards. That is, it should not be assumed that because a substance is present in food at a level that exceeds the level that would be considered suitable for a food additive (i.e., its margin of safety is not equal to or greater than the "safety factor" used for food additives), the substance must necessarily be considered a human health risk. Reiterating, all that can be said is that the margin of safety for such a substance is narrower than the margin that would be acceptable for a food additive. This last point is relevant to some of the known natural constituents or contaminants of food, inadequately studied as they are. For example, an approximately 4- to 10-fold margin of safety exists for solanine in white potatoes, whereas a 1000- or 2000-fold uncertainty factor would be applied to the NOEL for a substance producing similar short-term effects if the latter were destined for intentional addition to food. Solanine, if it were a food additive, could not be added to food at the level of its actual occurrence in potatoes. Carotatoxin and oxalate would probably fail the test of food additive safety. Certainly aflatoxin, a more complex problem to be discussed further later, would also fail. Although NOEL data (or, for that matter, any toxicity data) do not exist for most of the natural constituents or contaminants of food, it is highly likely that many could never satisfy the safety criteria long in use for food additives.

Although the failure of a compound to meet the safety standard of a food additive cannot be considered evidence that it is a risk to human health, the relatively narrow margins of safety exhibited by some of the natural constituents and contaminants of food are certainly cause for concern. This concern is compounded by our abysmal lack of chemical and toxicological information regarding this, the largest number of compounds to which we are exposed through food.

At present there are no established criteria, similar to those widely used for food additives and pesticides, that can be turned to in order to ascertain whether food hazards of natural origin are "safe" or "unsafe." In fact, except for substances exhibiting very high acute toxicity (e.g., PSP, botulinum toxin, and cyanide), the subject has received little review and discussion. Whether a country decides it must take steps to widen the margins of safety (sometimes even to infinity!) for naturally occurring food constituents or contaminants is a function of numerous factors, most of which can be thought of as deriving from a kind of benefit–risk assessment (2). It is highly unlikely that any of these can be reduced to the kind of simple operational criteria that have been applied to food additives, substances for which, in the United States at least, the law permits no benefit assessment to be entered into the decision on whether to allow their usage in food.

Following is a survey of some of the options available to widen margins of safety for natural constituents and contaminants of food, whenever it is considered necessary to do so. As stated in the first section, the five categories of food hazards of natural origin were organized according to the manner in which they became part of the food supply. Our ability to limit or eliminate (or otherwise control) human exposure to these substances is dependent on how they enter food, and the following survey is organized along the same lines as that of the second section.

Some Options for Control of Hazardous Substances That Are Intrinsic Components of Foods of Plant Origin

1. Completely reject the use of the plant as an item of food.

2. Search for varieties having the lowest concentration and use these exclusively (see cyanogenetic glycosides in lima beans). This will result in wider margins of safety.

3. Develop new varieties expressly for the purpose of reducing the concentration of or eliminating the offending constituent.

4. Identify food processing techniques to extract or otherwise remove the hazardous constituent.

5. Destroy the hazardous compound during food cooking (e.g., hemagglutinins).

6. Attempt to ensure the use of good food preparation practices in industry and the home. The sweet potato phytoalexins usually form in the part of the tuber subject to mold attack and can be significantly reduced in concentration by mechanical removal of the blemished portion. High solanine concentrations in potatoes can be avoided by ensuring that potatoes are stored in the dark.

Some Options for Control of Hazardous Natural Constituents of Soil and Water that Accumulate in Foods

1. Prohibit or limit the amounts of certain fertilizers (e.g., sewage sludge containing excessive amounts of cadmium).

2. Prohibit or limit the harvesting of marine life from waters containing excessive concentrations of any such hazardous substances.

3. Set limits on the concentration of any such hazardous substance in food. Reject food containing amounts in excess of such limits.

Some Options for Control of Hazardous Metabolites of Microorganisms that Grow on Food

1. Identify conditions of food production and storage needed to prevent growth of microorganisms and ensure their application.

2. Develop pesticides (e.g., fungicides) that prevent growth of the microorganism or its ability to produce toxins.

3. Develop plant varieties that are resistant to invasion by the hazardous microbes (e.g., peanuts or maize varieties resistant to *Aspergillus flavus* and hence to aflatoxin contamination have been a subject of considerable investigation).

4. Prohibit the production of foods in areas known to favor microbial growth and toxin production (e.g., some tropical areas may be unsuited for the production of maize if high aflatoxin exposure is to be avoided).

5. Set limits on the concentration of the hazardous metabolite that may be permitted in food and reject all food containing amounts above this level. This approach is now used in the United States for aflatoxins [limit, 20 ppb in all foods and feeds, with the FDA proposal to reduce the limit in peanut products to 15 ppb still pending (112)].

6. Develop methods to destroy or detoxify the hazardous metabolite. Depending on the treatment used, any treated food might become a food additive and thus require safety testing. Such is the case with treatment of aflatoxin-contaminated cottonseed meal and corn with ammonia, a process that appears to be effective but for which safety evaluation is not yet complete.

Some Options for Control of Compounds of Natural Origin Contaminating Edible Animal Products

1. For such substances that are of microbial origin and contaminate edible animal products because of their presence in animal feed, the options are the same as those described for hazardous metabolites.

2. Prevent livestock from grazing on poisonous range plants or, at least, ensure that any such animals are denied such opportunities long before they are used as food. Enforce limits on the amounts of such substances permitted in edible animal products.

Some Options for Control of Hazardous Compounds Produced During Food Storage, Processing, or Preparation

1. Identify conditions of storage, processing, or preparation required to minimize or eliminate the production of any such hazardous substances. Require the use of such methods in commercial operations. Educate homemakers in similar practices.

2. Some of the same options presented for foods of plant origin also pertain to this group.

Conclusions

The experimental and epidemiological methods that have been developed and refined over the last few decades for purposes of identifying health risks from synthetic chemicals should now be used for a systematic examination of the natural background of risk that may be hidden in food. Such an effort should not be limited to the nutrients, but should also include the extremely large number of chemicals of natural origin that are present in or that can contaminate the plants and animals we use as foods, spices, and beverages.

The methods used to identify risks from synthetic chemicals are probably appropriate for the examination of naturally occurring chemicals. The latter are ordinarily more difficult to assess because they usually occur as components of very complex mixtures. Nevertheless, it is difficult to find reasons to continue avoiding a deep examination of naturally occurring substances for their possible effects on human health and well-being. It is hoped that the information in this chapter promotes increased understanding of this aspect of food safety that has, for the most part, been treated with far less seriousness than it deserves.

REFERENCES

1. J. V. Rodricks, *Fed. Proc.*, 37, 2587 (1978).

2. J. V. Rodricks, *J. Assoc. Food Drug Off. U.S.*, 43, 3 (1979).

3. P. M. Zambreski and A. Hodgkinson, *Br. J. Nutr.*, 16, 627 (1962).

4. O. L. Oke, *World Review of Nutrition and Dietetics*, S. Karger, Basel, 1969.

5. R. F. Crampton and F. A. Charlesworth, *Br. Med. Bull.*, 31, 209 (1975).

6. D. W. Fassett, "Oxalates," in National Academy of Sciences, *Toxicants Occurring Naturally in Foods*, 2nd ed., Washington, D.C., 1973, pp. 346–362.

7. E. F. Kohmann, *J. Nutr.*, 18, 233 (1939).

8. O. G. Fitzhugh and A. A. Nelson, *J. Am. Pharm. Assoc.*, 36, 217 (1947).

9. D. K. Salunkhe and M. T. Wu, *Crit. Rev. Food Sci. Nutr.*, 9, 265 (1977).

10. N. Sapieka, *Food Pharmacology*, Thomas, Springfield, Ill., 1969.

11. S. G. Willimot, *Analyst (Lond.)*, 58, 431 (1933).

12. R. Ruhl, *Arch. Pharmacol.*, 284, 67 (1951).

13. J. M. Kingsbury, *Poisonous Plants of the United States and Canada*, Prentice-Hall, Englewood Cliffs, N. J., 1964.

14. B. O. Osumtokar, G. L. Monekosso, and J. Wilson, *Br. Med. J.*, **1**, 547 (1969).

15. R. D. Montgomery, *Am. J. Clin. Nutr.*, 17, 103 (1965).

16. E. E. Conn, "Cyanogenetic Glycosides," in National Academy of Sciences, *Toxicants Occurring Naturally in Foods*, 2nd ed., Washington, D.C., 1973, pp. 229–308.

17. N. Sharon and H. Lis, *Science*, 177, 949 (1972).

18. W. G. Jaffe, "Toxic Proteins and Peptides," in National Academy of Sciences, *Toxicants Occurring Naturally in Foods*, 2nd ed., Washington, D.C., 1973, pp. 106–129.

19. M. Nagao, T. Sugimura, and T. Matsishima, *Annu. Rev. Genet.*, 12, 117 (1978).

20. Y. Hashimoto, *Mutation Res.*, 66, 191 (1979).

21. V. L. Singleton and F. H. Kratzer, "Plant Phenolics," in National Academy of Sciences, *Toxicants Occurring Naturally in Foods*, 2nd ed., Washington, D.C., 1973, pp. 309–345.

22. D. Crosby and N. Ahranson, *Tetrahedrin*, 23, 465 (1967).

23. K. Muller, *Phytopathol. Acta* 27, 237 (1956).

24. L. T. Burka and B. J. Wilson, "Toxic Furanosesquiterpenoids in Mold-Damaged Sweet Potatoes," in J. V. Rodricks, Ed., *Mycotoxins and Other Fungal Related Food Problems*, Adv. Chem. Series No. 149, American Chemical Society, Washington, D.C., 1976, pp. 387–399.

25. G. Lincoff and D. H. Mitchell, *Toxic and Hallucinogenic Mushroom Poisoning*, van Nostrand Rheinhold, New York, 1977.

26. M. Randaskoski and H. Pysolo, *Z. Naturforach.*, 33, 472 (1978).

27. C. H. Van Etten and I. A. Wolff, "Natural Sulfur Compounds," in National Academy of Sciences, *Toxicants Occurring Naturally in Foods*, 2nd ed., Washington, D.C., 1973, pp. 210–234.

28. C. H. Van Etten, "Goitrogens," in I. E. Liener, Ed., *Toxic Constituents of Plant Foodstuffs*, Academic, New York, 1969, pp. 103–142.

29. P. M. Holscher and J. Natzschka, *Deutsch. Med. Wochenschr.*, 89, 1751 (1964).

30. D. W. Fassett, "Nitrates and Nitrites," in National Academy of Sciences, *Toxicants Occurring Naturally in Foods*, 2nd ed., Washington, D.C., 1973, pp. 7–25.

31. S. J. Van Rensburg and B. Altenkirk, "Claviceps Purpurea—Ergotism," in I. F. H. Purchase, Ed., *Mycotoxins,* Elsevier Scientific, Amsterdam, 1974, pp. 69–96.
32. A. Z. Joffe, "Toxicity of Fusarium Poae and Its Relation to Alimentary Toxic Aleukia," in I. F. H. Purchase, Ed., *Mycotoxins,* Elsevier Scientific, Amsterdam, 1974, pp. 229–262.
33. M. Emomoto and I. Ueno, "Penicillium Islandicum (Toxic Yellowed Rice)-Luteoskyrin, Islanditoxin, Cyclochlorotine," in I. F. H. Purchase, Ed., *Mycotoxins,* Elsevier Scientific, Amsterdam, 1974, pp. 303–326.
34. J. V. Rodricks and R. M. Eppley, "Stachybotrys and Stachybotryotoxicosis," in I. F. H. Purchase, Ed., *Mycotoxins,* Elsevier Scientific, Amsterdam, 1974, pp. 181–198.
35. L. Stoloff, "Aflatoxins—An Overview," in J. V. Rodricks, C. W. Hesseltine, and M. Mehlman, Eds., *Mycotoxins in Human and Animal Health,* Pathotox Publishers, Park Forest South, Ill., 1978, pp. 7–28.
36. J. V. Rodricks and L. Stoloff, "Aflatoxin Residues from Contaminated Feed in Edible Tissues of Food Producing Animals," *Mycotoxins in Human and Animal Health,* Pathotox Publishers, Park Forest South, Ill., 1978, pp. 67–80.
37. Natural Poisons, in W. Horwitz, Ed., *Official Methods of Analysis of the Association of Official Analytical Chemists,* Association of Official Analytical Chemists, Washington, D.C., 1975, Chapter 26.
38. Food and Agriculture Organization of the United Nations, *Mycotoxin Surveillance—A Guideline,* FAO Control Series No. 4, Rome, 1977.
39. J. V. Rodricks, *Food Nutr.,* 2, 9 (1976).
40. K. A. V. R. Krishnamachri, *Lancet,* 1, 1061 (1975).
41. T. C. Campbell and L. Stoloff, *J. Agr. Food Chem.,* 22, 1006 (1974).
42. M. Enomoto, "Carcinogenicity of Mycotoxins," in K. Uraguchi and M. Yomazaki, Eds., *Toxicology, Biochemistry and Pathology of Mycotoxins,* Wiley, New York, 1978, pp. 246–262.
43. P. M. Newberne, *Environ. Health Perspec.,* 9, 1 (1974).
44. G. N. Wogan, "Aflatoxin Carcinogenesis," in H. Busch, Ed., *Methods in Cancer Research,* Vol. VIII, Academic, New York, 1973, pp. 309–344.
45. S. J. Van Resburg, "Role of Epidemiology in Elucidation of Mycotoxin Health Risks," in J. V. Rodricks, C. W. Hesseltine, and M. Mehlman, Eds., *Mycotoxins in Human and Animal Health,* Pathotox Publishers, Park Forest South, Ill., 1977, pp. 669–712.
46. L. Tomatis, C. Asthe, and H. Bartsch, *Cancer Res.,* 38, 877 (1978).
47. International Agency for Research on Cancer, *Some Naturally Occurring Substances,* IARC Monograph on Evaluation of Carcinogenic Risks of Chemicals to Man, Vol. 10, International Agency for Research in Cancer, Lyon, France, 1976.
48. *Mycotoxins, Food and Nutrition Paper 2,* Food and Agricultural Organization of the United Nations, Rome, 1977.
49. *Global Perspective on Mycotoxins,* Food and Agriculture Organization of the United Nations, Rome, 1977.

50. E. Chain, H. W. Florey, and M. A. Jennings, *Br. J. Exp. Pathol.*, 29, 202 (1942).
51. R. B. Woodward and G. Singh, *J. Am. Chem. Soc.*, 71, 758 (1949).
52. C. D. De Rosnay, C. Martin-Dupont, and R. Jensen, *J. Med. Bordeaux*, 129, 189 (1952).
53. F. Dickens and H. E. H. Jones, *Br. J. Cancer*, 15, 85 (1961).
54. G. M. Ware, *J. Assoc. Off. Anal. Chem.*, 85, 754 (1975).
55. L. S. Stoloff, *N. Y. State Agr. Exp. Sta. Spec. Rep.*, 19, 51, 1975.
56. A. Ciegler, "Patulin," in J. V. Rodricks, C. W. Hesseltine, and M. Mehlman, Eds., *Mycotoxins in Human and Animal Health*, Pathotox Publishers, Park Forest South, Ill., 1977, pp. 608–624.
57. E. A. Buxton and S. Legenhausen, *Vet. Med.*, 22, 451 (1927).
58. C. J. Mirocha, S. V. Pathre, and C. M. Christensen, "Chemistry of Stachybotrysin Mycotoxins," in T. D. Wyllie and L. G. Morehouse, Eds., *Mycotoxic Fungi, Mycotoxins and Mycotoxicoses*, Vol. 1, Marcel Dekker, New York, 1977, pp. 365–420.
59. W. H. Urry, J. L. Wehrmeister, E. B. Hodge, and P. H. Hidy, *Tetrahedron Lett.*, 27, 3109 (1966).
60. C. J. Mirocha, S. V. Pathre, and C. M. Christensen, "Zearalenone," in J. V. Rodricks, C. W. Hesseltine, and M. Mehlman, Eds., *Mycotoxins in Human and Animal Health*, Pathotox Publishers, Park Forest South, Ill., 1977, pp. 345–364.
61. G. M. Ware and C. W. Thorpe, *J. Assoc. Off. Anal. Chem.*, 61, 1058 (1978).
62. A. W. Hayes, *Mycopathologia*, 65, 29 (1979).
63. Y. Ueno, "Trichothecenes Overview Address," in J. V. Rodricks, C. W. Hesseltine, and M. Mehlman, Eds., *Mycotoxins in Human and Animal Health*, Pathotox Publishers, Park Forest South, Ill., 1977, pp. 189–207.
64. A. N. Leonov, "Current View of the Chemical Nature of Factors Responsible for Alimentary Toxic Aleukia," in J. V. Rodricks, C. W. Hesseltine, and M. Mehlman, Eds., *Mycotoxins in Human and Animal Health*, Pathotox Publishers, Park Forest South, Ill., 1977, pp. 323–336.
65. R. Schoental, A. Z. Joffe, and B. Yagen, *Cancer Res.*, 39, 2179 (1979).
66. S. V. Pathre and C. J. Mirocha, "Assay Methods for Trichothecenes," in J. V. Rodricks, C. W. Hesseltine, and M. Mehlman, Eds., *Mycotoxins in Human and Animal Health*, Pathotox Publishers, Park Forest South, Ill., 1977, pp. 229–253.
67. G. M. Ware and C. W. Thorpe, paper 177 presented at the 93rd Annual Meeting of the Association of Official Analytical Chemists, Washington, D.C., October 15–18, 1979.
68. F. S. Chu, S. Grossman, R. Wei, and C. J. Mirocha, *Appl. Environ. Microbiol.*, 37, 104 (1979).
69. I. Hsu, E. B. Smalley, F. M. Strong, and W. E. Rebelin, *Appl. Microbiol.*, 24, 684 (1972).

70. A. Ciegler, *J. Food Prot.*, 4, 399 (1978).

71. P. M. Scott, "Penicillum Mycotoxins," in T. D. Wyllie and L. G. Morehouse, Eds., *Mycotoxic Fungi, Mycotoxins and Mycotoxicoses*, Marcel Dekker, New York, 1977, pp. 283–291.

72. M. H. Brown, G. M. Szczeck, and B. P. Purnealis, *Toxicol. Appl. Pharmacol.*, 37, 331 (1976).

73. M. Kanisawa, and S. Suzuki, *Gann*, 10, 599 (1978).

74. R. C. Doster, R. D. Sinnkuber, J. H. Wales, and D. J. Lee, *Fed. Proc.*, 30, 578 (1971).

75. P. Krogh, "Ochratoxins," in J. V. Rodricks, C. W. Hesseltine, and M. Mehlman, Eds., *Mycotoxins in Human and Animal Health*, Pathotox Publishers, Park Forest South, Ill., 1977, pp. 489–498.

76. L. Stoloff, "Occurrence of Mycotoxins in Foods and Feeds," in J. V. Rodricks, Ed., *Mycotoxins and Other Fungal Related Problems*, Adv. Chem. Series No. 149, American Chemical Society, Washington, D.C., 1976, pp. 23–50.

77. T. Hamasaki and Y. Hatsuda, "Sterigmatocystin and Related Compounds," in J. V. Rodricks, C. W. Hesseltine, and M. Mehlman, Eds., *Mycotoxins in Human and Animal Health*, Pathotox Publishers, Park Forest South, Ill., 1977, pp. 579–607.

78. G. M. Shannon and O. L. Shotwell, *J. Assoc. Off. Anal. Chem.*, 59, 963 (1976).

79. R. L. Hall, *Proceedings of Marabou Symposium on Food and Cancer*, Caslon Press, Stockholm, 1978.

80. R. Schoental, "Carcinogens in Plants and Microorganisms," in C. E. Searle, Ed., *Chemical Carcinogens*, Monograph 173, American Chemical Society, Washington, D.C., 1976, pp. 626–689.

81. O. Mohabbat, M. S. Younos, and A. A. Merzod, *Lancet*, 2, 271 (1976).

82. A. E. Stillman, R. J. Huxtable, D. Fox, M. Hart, and P. Bergeson, *Ariz. Med.*, 34, 545 (1977).

83. C. A. Linsell, *J. Toxicol. Environ. Health*, 5, 173 (1979).

84. L. Stoloff, U.S. Food and Drug Administration, Washington, D.C., private communication, 1979.

85. M. L. Deinzer, P. A. Thomson, D. M. Burgett, and D. Issacson, *Science*, 195, 497 (1977).

86. Y. Hashimoto, *Marine Toxins and Other Bioactive Marine Metabolites*, Japan Scientific Societies Press, Tokyo, 1979.

87. B. Dale and C. M. Yentsch, *Oceanus*, Woods Hole, Mass., 1978.

88. C. M. Yentsch and F. C. Mague, "Motile Cells and Cysts: Two Probable Mechanisms of Intoxication of Shellfish in New England Water," in D. L. Taylor and H. H. Seliger, Eds., *Toxic Dinoflagellate Blooms*, Elsevier/North-Holland, New York, 1957, pp. 127–130.

89. E. J. Schantz, V. E. Ghazarossina, J. B. Mold, D. W. Stanger, J. Shavel, J. P. Bowden, J. M. Lynch, R. S. Wyler, B. Riegel, and H. Sommer, *J. Am. Chem. Soc.*, 79, 5230 (1957).

90. E. J. Schantz, V. E. Ghazarossina, H. K. Schnoes, F. M. Strong, J. P. Springer, J. O. Pessanita, and J. Clardy, *J. Am. Chem. Soc.*, 97, 1238 (1975).

91. R. M. Gershey, R. A. Neve, D. L. Musgrave, and P. B. Reichardt, *Am. J. Med.*, 34, 559 (1977).

92. L. J. Buckley, Y. Oshima, and Y. Shimizu, *Anal. Biochem.*, 85, 157 (1978).

93. J. A. Berger and L. R. Berger, *Rev. Int. Oceanogr. Med.*, 53 (1979).

94. J. R. Sylvester, A. E. Dammann, and R. A. Dewey, *Marine Fish. Rev.*, 39, 14 (1979).

95. Y. Hirata, *Pure Appl. Chem.*, 50, 979 (1978).

96. K. Tsuda, *Naturwissenschaften*, 53, 171 (1966).

97. Y. Kishi, T. Fukyuoma, M. Aratani, F. Nakatfubo, T. Goto, S. Inoue, H. Tanino, S. Sugiura, and H. Kakoi, *J. Am. Chem. Soc.*, 94, 9219 (1972).

98. K. Suenaga, *Jap. Forensic J.*, 32, 97 (1978).

99. F. A. Fuhrman, *Toxic Constituents of Animal Foodstuffs*, Academic, New York, 1974.

100. K. J. Motel and N. S. Scrimshaw, *Toxicol. Lett.*, 3, 219 (1979).

101. W. Lovenbert, "Some Vaso- and Psychoactive Substances in Food: Amines, Stimulants, Depressants, and Hallucinogens," in National Academy of Sciences, *Toxicants Occurring Naturally in Foods*, 2nd ed., Washington, D.C., 1973, pp. 170–188.

102. P. N. Magee, R. Montesano, and R. Preussman, "*N*-Nitroso Compounds and Related Carcinogens," in C. E. Searle, Ed., *Chemical Carcinogens*, Monograph 173, American Chemical Society, Washington, D.C., 1976, pp. 245–315.

103. B. Spiegelhalder, G. Eisenbranda, and R. Preussman, *Cosmet. Toxicol.*, 17, 29 (1979).

104. T. H. Matton, "Potential Toxicity of Food Lipids," in National Academy of Sciences, *Toxicants Occurring Naturally in Foods*, 2nd ed., Washington, D.C., 1973, pp. 189–209.

105. M. Lo and E. Sandi, "Polynuclear Aromatic Hydrocarbons," in F. A. Gunther and J. D. Gunther, Eds., *Residue Reviews*, Vol. 69, Springer-Verlag, New York, 1978.

106. A. Dipple, "Polynuclear Aromatic Carcinogens," in C. E. Searle, Ed., *Chemical Carcinogens*, ACS Monograph 173, American Chemical Society, Washington, D.C., 1976, pp. 245–315.

107. W. Lijinsky and A. E. Ross, *Food Cosmet. Toxicol.*, 5, 343 (1967).

108. J. W. Howard and T. Fazio, *J. Agr. Food Chem.*, 17, 527 (1969).

109. B. A. Commoner, A. J. Vithayathil, P. Dolara, S. Nair, P. Modyostha, and G. C. Cura, *Science*, 201, 913 (1978).

110. M. Nagao, M. Honda, Y. Seino, T. Yahagi, T. Kawachi, and T. Sugimura, *Cancer Lett.*, 2, 22 (1977).

111. T. Sugimura, *Mutat. Res.*, 55, 149 (1978).

112. U.S. Food and Drug Administration, *Fed. Reg.*, 39, 42748 (1974).

CHAPTER 6

Food Additives

H. R. ROBERTS

Much of the current concern and confusion associated with food additives stems from a lack of definition and, hence, understanding of the term. As Hall (1) has pointed out, discussion of food additives requires not only differentiation among the terms food—and ingredients, components, and constituents of foods—but also recognition of the legal and practical definitions of these terms. As we see in this chapter, quite different legal standards of "safety" are applied to various classes of substances associated with our food supply. These legal differences are especially misunderstood.

The material in this chapter is based on a broad connotation of the term "food additive," namely, a (normally) minor part (ingredient, constituent, or component, as the reader may prefer) of a food item, intentionally added to produce a functional or technical effect or unintentionally added as a consequence of the production, distribution and/or processing of a food item. It should be noted, however, that microorganisms and their toxins, nutrients naturally present in foods, chemical contaminants and mycotoxins, and natural contaminants have been covered in previous chapters and are excluded from the substances discussed herein as food additives.

One other definition needed for a discussion of food additives is that of food itself. For this purpose the legal definition in Section 201(f) of the FFD&C Act (2) will suffice: "The term 'food' means (1) articles used for food or drink for man or other animals, (2) chewing gum, and (3) articles used for components of any such article." The primary reason for including this legalistic definition is to point out that food includes food and drink as well as animal food. As a consequence, food additives by our definition involve the intentionally and unintentionally added substances in both foods and beverages for humans and for food-producing animals as well.

This chapter includes a brief look at the history of food additives, a description of the technical use of the term, human exposure to food additives, specifications, and legal definitions. Safety considerations for food additives are discussed in terms of the legal subsets: generally recognized as safe (GRAS) substances, direct food additives, indirect additives, color additives, and drugs in food-producing animals.

DESCRIPTION AND USE OF FOOD ADDITIVES

Food additives, in the general sense of the term, have been used by humans for centuries and, in some cases, for thousands of years. The first food additive may well have been smoke, accidentally discovered by neolithic man (along with drying and freezing), to be a useful technique for preserving excess meat and fish. It is likely that as agriculture developed toward the end of the Stone Age, food processing also had its beginnings—and with it the use of other additives. Fermentation products certainly were among the first of these. The advent of simple unleavened breads was probably followed by experimentation that lead to the first beer, and as the ancient civilizations in Egypt and Sumer developed, so did a variety of wines (3).

Among the first major food additives was salt, used thousands of years ago for the preservation of meats and fish. One account (4) points out the use of desert salts (likely containing nitrates as impurities) in the ancient Jewish Kingdom as early as 1600 B.C. Salt was used extensively by the Romans to cure pork and fish products, and in later Roman times, Cato (234 to 149 B.C.) and others recorded these curing procedures. In medieval times salt, in combination with saltpeter (potassium nitrate), was commonly used in meat curing. The process ultimately developed into the modern meat-curing process with its controversy surrounding nitrosamines as well as nitrite itself.

Spices also have a long history of food additive use, and the spice trade was a major political factor in Roman times as well as later in the Middle Ages. Great importance was attached to pepper, cloves, nutmeg, cinnamon, and ginger for masking the flavor of spoiled foods and enhancing the flavor of other foods. The long journey of these spices from the Indian Ocean Islands to Tyre and Sidon and then throughout the Roman Empire is ample evidence of this. It is interesting to note that after the Visigoth seige of Rome in 408 A.D. a major part of the extracted tribute was 300 pounds of pepper (5).

Many other additives also have a long history of use. The ancient Chinese burned kerosene (thus producing, without realizing it, the modern additives ethylene and propylene) to ripen bananas and peas. Honey for sweetening can be traced back to Ancient Egypt, as can the first color additives, fruit and vegetable juices. The vanilla bean as a flavoring agent counted among its devotees, Cortes, the sixteenth century explorer (6).

Toward the end of the eighteenth century the British Navy instituted the practice of issuing a daily ration of lemon juice to sailors to combat the widespread incidence and high loss of life from scurvy. (Lemons were then called "limes" and the British sailors became known as "limeys"). The vitamin C in the lemon juice, representing one of the first examples of the addition of nutrients to the diet, was not recognized as the curative factor until 130 years later. Since then, nutrient fortification of foods has played a significant role in the United States in the eradication of nutrient deficiency diseases—vitamin D in milk to combat rickets, iodized salt for goiter, and niacin in bread and flour for pellagra.

The modern use of and, perhaps more so, the current concern about food additives began with the industrial revolution. With the rapid growth of urban areas and improvement in road systems, an organized food industry came into being, and with it the first faint stirring of modern food technology. A major event in this era was the initial development of canning. In 1809 a Frenchman, Nicholas Appert, published directions for the preservation of a variety of foods in cork-stoppered glass bottles by heating in boiling water. Although the role of microorganisms in food spoilage was not then known, Appert derived canning methods successfully used commercially and in the home for years thereafter (7).

The nineteenth century is also noted for the recognition in the laboratory that much of the food supply was adulterated. Some adulterants were fairly harmless from other than the economic point of view, for example, floor sweepings in pepper and ash leaves in tea. Others, such as copper and lead salts used to color candy and cheese, were not. Common adulterants in this period also included alum in bread, acorns in coffee, brick dust in cocoa, copper salts in pickles, and prussic acid in wine (3).

At the end of the nineteenth century food was considered either pure or adulterated. The objective of identifying and eliminating the undesirable "additives" (i.e., deliberate adulterants) and of establishing food safety control on a federal basis found a champion in Dr. Harvey Wiley, who approached the problem of food safety assessment directly, with his human volunteer "poison squad." His efforts in large part resulted in the Pure Food and Drug Act of 1906, the basis for today's control of food safety (8). Much has changed since that time however including the law itself, lifestyles and food technology and, in recent years interest in food additives has greatly increased.

Current concern about food additives is in many ways associated with the adoration of mother nature—the idea that "natural" food is good and "artificial" food (i.e., processed foods containing additives) is bad. This back-to-nature movement is not necessarily a recent development. The first great proponent of natural foods was the Reverend Sylvester Graham who throughout the late 1800s extolled the virtues of

Garden of Eden foods—fruits, nuts, farinaceous seeds, and roots and warned against "artificial preparation." The particular flour he advocated still bears his name today, and apparently his doctrine has also survived (9, 10).

Other factors contributing to food additive concerns include the expanding ingredient information appearing on food labels; a growing public interest, fueled by the media, in health and nutrition; increasing scientific scrutiny of food; governmental actions relative to food ingredients; and the campaigns of consumer activist groups. Whatever the individual contribution of these factors, the net result is concern about the risks presented by food additives.

Individual consumers, consumer activist groups, and many professionals in other than food fields consistently rank food additives at or near the highest in terms of risks in the food supply (11). Yet, in terms of the food safety criteria of severity, incidence, and onset, food additives must be ranked in the lowest risk category (12–15). In its annual technical program planning the Bureau of Foods of the FDA has consistently ranked the food additive program lowest in risk, but other components of the FDA have often placed food additives in a higher risk category among all agency programs. Part of the discrepancy between these perceptions of risk derives from differences in understanding of the term "food additive." To place the safety of food additives in perspective, it is necessary to take into account the technical connotation of the term, the legal definitions, uses and consumption of additives, the major classes of additives, and what is known about food additive safety.

Food Additives—Technical Use of the Term

Corresponding to the broad definition of food additive used in this chapter is the technical definition of the Food Protection Committee of the Food and Nutrition Board, National Academy of Sciences: "a substance or mixture of substances other than a basic foodstuff which is present in a food as a result of any aspect of production, processing, storage or packaging." Note that, technically, a food additive may be a substance intentionally incorporated into a product or a substance that becomes a component of food as a consequence of its journey from the field to the table. The former (intentional additives) are called *direct additives* and are present to serve some functional purpose in the food such as adding flavor, sweetness, or color or to prevent spoilage. The latter are called *indirect additives.* Indirect additives normally are present only in trace amounts and result from contact of the food with agricultural chemicals, with processing equipment or processing aids, and with the food container.

Another group of food additives meeting the technical definition is that of drugs and feed additives for food-producing animals. In one respect these substances are somewhere between direct and indirect additives. They are intentionally added to animal feed or administered to the animal, but they may or may not be (detectably) present in the food products (meat, milk, and eggs). Further, the parent compound may be metabolized by the food animal, thus giving rise, indirectly, to other substances in the edible by-products.

Although popular discussion of food additives usually focuses on a few items—the food colors such as Red No. 2, sweeteners such as cyclamate and saccharin, preservatives such as sodium nitrite, and container materials such as polyvinyl chloride, there are literally thousands of food additives. The exact number depends on whether only officially recognized additives are included and on how detailed a breakdown is employed. The National Science Foundation (16) refers to 1850 officially recognized substances added to foods for specific technical effects. According to the FDA, there are about 2800 direct additives (17). By either count the vast majority of the direct additives are spices and flavors, and such familiar items as salt, pepper, sugar, mustard, and yeast, although probably not considered as additives by the average consumer, are also included.

In the case of indirect additives, no one really knows how many there are. The National Science Foundation (16) has noted the existence of nearly 3000 additives permitted for use in food packaging materials. Some of these with multiple uses are counted more than once, whereas others are themselves composed of many different constituents. The FDA has estimated that there may be in excess of 10,000 (17). Whatever the number of these primary constituents may be, only a small fraction become additives in the sense that they actually end up in food.

Substances that are naturally present in food can also be food additives when their natural levels are increased by humans. For example, as shown in Chapter 4, lead is a natural component of soil and thus is present in fruits, vegetables, and the food of food-producing animals. Lead can also be present in food as a result of the lead solder used to seal the seams of tin cans. In the latter case, lead would technically be an indirect food additive (18).

To discuss the "risks" presented by food additives, it is necessary to look at all the substances in this very large category, rather than just those few of popular interest.

Human Exposure to Food Additives

Any potential hazard to humans posed by a food additive, of course, depends not only on its inherent toxicity, but also on the level of that additive ingested. As the FDA has pointed out (19), four quite familiar food ingredients account for about 93% by weight of all the direct food additives used in the United States—sugar (sucrose), salt, corn syrup, and dextrose. On a per capita basis, annual use in 1970 of sucrose was about 100 pounds; salt, 15 pounds; corn syrup, 8 pounds; and dextrose, 4 pounds. All the remaining direct additives together are used at the rate of somewhat less than 10 pounds per person annually. If the following familiar items are also included, which collectively account for about 3 pounds per person annually, 95% of direct additive consumption is covered:

Modified starch	Caramel
Yellow mustard	Citric acid
Sodium bicarbonate	Carbon dioxide
Yeasts	Black pepper

It is interesting to look at the principal additives other than the big four (sugar, salt, corn syrup, and dextrose) from the point of view of their uses. About 90% by weight (9 out of 10 pounds annual per capita consumption) of these additives are accounted for by the following (1, 20):

Flavors/Flavor Enhancers. Monosodium glutamate, mustard, black pepper, and hydrolyzed vegetable protein.

Stabilizers/Thickeners. Sodium caseinate, gum arabic, and modified starch.

Leavening Agents. Yeasts, monocalcium phosphate, sodium aluminum phosphate, and sodium acid phosphate.

Leavening/Acidity Control. Sodium carbonate, calcium carbonate, dicalcium phosphate, and disodium phosphate.

Acidity Control. Sodium bicarbonate, hydrogen chloride, citric acid, sulfuric acid, sodium citrate, sodium hydroxide, acetic acid, phosphoric acid, and calcium oxide.

Emulsifiers. Lecithin, monoglycerides, and diglycerides.

Miscellaneous. Sulfur dioxide (preservative), calcium chloride (firming agent), calcium sulfate (processing aid), carbon dioxide (effervescent), sodium tripolyphosphate (curing humectant), and caramel (color).

Employing the convenient figures of 2000 direct additives, altogether consumed at the annual rate of 10 pounds per person gives an average use of 0.08 of an ounce (ca. 2.25 g) for each of the additives other than sugar, salt, corn syrup, and dextrose. But the average is misleading. As shown previously, 9 out of the 10 pounds used are accounted for by only 33 additives. Thus the vast majority of additives are used at much smaller levels than the average. More meaningfully, the median per capita usage is about 0.5 mg (<0.0002 of an ounce); that is, half of the additives are used at the annual per capita rate of 0.5 mg or less (about the weight of a grain of salt). It should also be emphasized that these are usage estimates; actual consumption figures are lower because of losses in food production and distribution. The National Science Foundation (16) has estimated that about 0.5% of our food supply consists of intentional additives. This includes nutrient supplements added at the level of a few percent and ranges through preservatives used at fractions of a percent down to flavor constituents used at a few parts per hundred million.

Human exposure to indirect additives is more difficult to estimate. Commercial usage of an indirect additive, a packaging component, for example, would have to take into account the amount actually present in the packaging material and would have to be factored by the migration into foods estimated from laboratory tests (21, 22). By way of illustration, a substance migrating at a level of 50 ppb into a food consumed at a rate of 50 g/day would result in 2.5 μg ingested per day or less than 1 mg annually. Similarly, migration at 10 ppm would result in less than 200 mg of annual ingestion.

Another way of putting human exposure to food additives in perspective is by examining the levels of some of the natural toxicants in our food supply. The potato is an excellent example. Based on an average annual consumption of 119 pounds of potatoes, it has been estimated that we ingest annually 9700 mg of the toxic alkaloid, solanine (20). (Recall that the median annual ingestion of direct food additives is about 0.5 mg.) Similarly, consumption of 0.85 pounds of lima beans results in ingestion of about 40 mg of hydrogen cyanide. (Bamboo shoots contain even more.) Ingestion of arsenic from seafood amounts to about 14 mg annually. Then, too, there is nitrite. Celery, radishes, beets, and leafy vegetables are rich sources of nitrates (15,000 ppm have been found in radishes), which are converted to nitrites by oral and intestinal bacteria. Nitrites from ingested foods other than cured meats may amount to 8 to 10 mg/day as compared to 1 to 2 mg/day from cured meats (23, 24). As noted in Chapter 3, even some nutrients that are essential to human life at low levels display overt toxicity at higher levels. Obviously, humans can tolerate without readily discernible harm a variety of substances with known or suspected toxic effects, at least when they are ingested in small amounts over a period of time as they are in food. Although this is illustrated by these few examples, it becomes much more apparent from a comprehensive review such as that of the National Academy of Sciences (25).

The preceding data clearly show that the vast preponderance of food additives are ingested at very low levels. Usage data, for example, indicate that for many additives, it would take several years' worth of total annual production to obtain enough of the additive even to conduct an animal feeding study. Risk, of course, depends on toxicity as well as exposure, but the low exposure levels and what we know about toxicity, together with human experience with the natural toxicants, have to be at least somewhat reassuring.

Specifications for Food Additives

Another aspect to the safety of food additives is that of specifications. Even though an additive has been thoroughly tested, its safety in use cannot be assured if it contains impurities not present in the standard substance. Therefore, specifications are needed to assure the purity as well as the quality and the uniformity of additives. Food and Drug Administration regulations require as part of any food additive petition pertinent information relative to the chemical identity and the composition of the substance; its physical, chemical, and biological properties; content of desirable components; identification and limits of impurities; and related data such as methods of manufacturing or processing. In some cases specifications have been incorporated into regulations directly and in other cases by reference. A significant source of such reference specifications, defining minimum standards for food grade substances, are those generated under the aegis of the Committee on Food Protection of the National Academy of Sciences and published as the *Food Chemicals Codex* (26). A second edition of the *Food Chemicals Codex* was published in 1972 (27). It contains specifications for 639 direct and indirect additives (processing aids). For each listed substance, the data include a description, specifications, associated tests, packaging and storage instructions, and functional use in food. To assure the purity of food additives, the FDA requires that they be of food grade quality as defined by *Food Chemicals Codex* and/or by FDA regulations. A third edition of *Food Chemicals Codex* is planned for publication in 1980.

More and more emphasis is being placed on specifications for food additives by the FDA and other world regulatory bodies. This is the case for not only synthetic additives, but also those of natural derivation. Perhaps as more is learned about the composition of natural additives, all additives can be evaluated from a more rational perspective.

Food Additives—Legal Use of the Term

Regulatory control of a given food additive depends not only on its toxicity and the level of human exposure, but also on its legal classification. Therefore, it is critical to the understanding of food additive safety to consider the pertinent legal definitions. The FFD&C Act (2) contains the following definition in Section 201(s):

> The term "food additive" means any substance the intended use of which results or may reasonably be expected to result, directly or indirectly, in its becoming a component of or otherwise affecting the charac-

teristics of any food (including any substance intended for use in producing, manufacturing, packing, processing, preparing, treating, packaging, transporting, or holding food; and including any source of radiation intended for any such use), if such substance is not generally recognized, among experts qualified by scientific training and experience to evaluate its safety, as having been adequately shown through scientific procedures (or, in the case of a substance used in food prior to January 1, 1958, through either scientific procedures or experience based on common use in food) to be safe under the conditions of its intended use; except that such term does not include—

1. a pesticide chemical in or on a raw agricultural commodity; or

2. a pesticide chemical to the extent that it is intended for use or is used in the production, storage, or transportation of any raw agricultural commodity; or

3. a color additive; or

4. any substance used in accordance with a sanction or approval granted prior to the enactment of this paragraph pursuant to this Act, the Poultry Products Inspection Act (21 U.S.C. 451 and the following) or the Meat Inspection Act of March 4, 1907 (34 Stat. 1260), as amended and extended (21 U.S.C. 71 and the following); or

5. a new animal drug.

The legal definition, in the beginning, corresponds to the technical definition. Both direct and indirect additives are included. Further, any radiation treatment of food is formally defined as a food additive. But then the exemptions to the definition are given. Excluded from the legal category of food additives are the so-called GRAS substances, pesticide chemicals for use on raw agricultural produce and residues thereof, color additives, substances formally approved by the FDA or the USDA prior to the enactment of the food additives law in 1958 (these are termed *prior-sanctioned* substances), and new animal drugs.

Those substances legally exempt from the food additives definition are by no means exempt from federal scrutiny as to safety. The GRAS substances are evaluated in accordance with regulations issued by the FDA (see next section). Pesticide chemicals are subject to control by both the FDA and the EPA under Section 508 of the FFD&C Act. Color additives are separately regulated by the FDA under Section 706 of the act and new animal drugs under Section 512 of the act.

From the legal point of view, then, the definition of a food additive is much more limited than the technical definition or even the popular connotation of the term. In general, food additives, legally, are only those substances that were approved as direct or indirect additives (with the exceptions noted) after the food additives amendments became part of the FFD&C Act in 1958. (The color additives amendments became part of the FFD&C Act in 1960, and the animal drug amendments were incorporated in the act in 1968). These differences understandably contribute to the public confusion and hence concern about food additives.

In this chapter the term "food additive" is used in the generic technical sense. The major legal classifications of GRAS, direct additives, indirect additives, color additives, and drugs used in food-producing animals are employed as a convenience in describing the approach to food safety and regulatory control and status in each case.

GRAS SUBSTANCES

The 1958 food additive amendments to the FFD&C Act established the requirement for premarket safety approval for food additives but exempted ingredients in common use at that time. These latter ingredients have become known as GRAS substances. As noted in the legal definition of a food additive, ingredients in use before January 1, 1958 could be considered GRAS either because of experience based on common use in food or through scientific evaluation procedures. In addition, any food ingredient can be classified as GRAS if it is "generally recognized, among scientific experts qualified by scientific training and experience to evaluate its safety . . . to be safe under the conditions of its intended use."

Starting in 1958, then, legally defined food additives had to have their safety demonstrated by industry prior to approval; that is, substantial evidence of a lack of hazard had to be provided. On the other hand, GRAS substances were collectively approved by law and could have their approval limited or terminated only if the FDA demonstrated that they might render food injurious to health at the levels and for the uses involved. (That regulatory distinction between approved and new additives is not well understood and again leads to popular confusion about food additives.) Since the GRAS category is so extensive and includes the bulk of direct food additives, it is important to consider this category in some detail.

The GRAS List

With the establishment of the GRAS classification, the FDA developed several lists of substances that might be considered GRAS. After public review these lists were modified and published in the Code of Federal Regulations. The published GRAS lists are not exhaustive. The FDA has noted in Part 182 of the Code of Federal Regulations (28) that:

> It is impracticable to list all substances that are generally recognized as safe for their intended use. However, by way of illustration, the Commissioner regards such common food ingredients as salt, pepper, sugar, vine-

gar, baking powder and monosodium glutamate as safe for their intended use.

The published GRAS list "includes additional substances that when used for the purposes indicated, in accordance with good manufacturing practice are regarded by the Commissioner as generally recognized as safe for such uses."

There are several sections of FDA regulations pertaining to GRAS substances. Part 182 of the Code of Federal Regulations (28) lists those substances that are considered GRAS when used in accord with good manufacturing practices. (In general, this is considered by the FDA to mean that no more of the substance is used than is required to accomplish the intended functional or technical effect in the case of a directly added substance, or that residual levels in foods are as low as reasonably possible in the case of indirect additives.)

By way of illustration, Table 6.1 shows the listed GRAS flavors and spices and Table 6.2 shows multiple purpose GRAS substances. Part 182 of the Code of Federal Regulations also includes:

- Essential oils, solvent free oleoresins and natural extractives including distillates.
- Natural extractives and other natural substances used in conjunction with spices, seasonings and flavorings.
- Certain other spices, seasonings, essential oils, oleoresins and natural extracts.
- Synthetic flavoring substances and adjuvants.
- Substances migrating from cotton and cotton fabrics used in dry food packaging.
- Substances migrating to food from paper and paperboard products.
- Adjuvants for pesticide chemicals.

The GRAS lists in Part 182 include slightly over 600 substances, many of which are quite familiar and have no significant questions of safety associated with their use. As noted by the FDA, however, these substances are being reevaluated to determine whether they either can be affirmed as GRAS (and if so, under what conditions of use), must be regulated as food additives or interim food additives (requiring additional safety information), or should be prohibited for use in food.

As a result of the GRAS review, a wide variety of substances have had their status reevaluated. The FDA has developed regulations for some of these in Part 184 (direct additives) and Part 186 (indirect additives) of the Code of Federal Regulations. In each case general food use under good manufacturing practices or specified food uses and levels of use are prescribed for the direct additives. Substances ap-

TABLE 6.1 FDA GRAS List—Spices and Other Natural Seasonings and Flavorings

Alfalfa, herb and seed	Glycyrrhiza
All spice	Grains of paradise
Ambrette seed	Horehound
Angelica, root and seed	Horseradish
Angostura	Hyssop
Anise, anise star	Lavender
Balm	Licorice
Basil, bush and sweet	Linden flowers
Bay	Mace
Calendula	Marigold, pot
Camomile (English and German)	Marjoram, pot and sweet
Capers	Mustard, black, brown, white or yellow
Capsicum	Nutmeg
Caraway and black caraway	Oregano
Cardamon	Paprika
Cassia (Chinese, Padang, Saigon)	Parsley
Cayenne pepper	Pepper, black, red and white
Celery seed	Peppermint
Chervil	Poppy seed
Chives	Rosemary
Cinnamon (Ceylon, Chinese, Saigon)	Saffron
Clary	Sage
Cloves	Sage, Greek
Coriander	Savory, summer and winter
Cumin	Sesame
Cumin, black	Spearmint
Elder flowers	Tarragon
Fennel, common and sweet	Thyme
Fenugreek	Thyme, wild or creeping
Galanga	Turmeric
Geranium	Vanilla
Ginger	Zedoary

Source. Code of Federal Regulations, Title 21, U. S. Government Printing Office, Washington, D.C., 1979.

TABLE 6.2 FDA GRAS List—Multiple Purpose GRAS Food Substances

Acetic acid	Glyceryl monostearate
Adipic acid	Helium
Citric acid	Hydrogen peroxide
Glutamic acid	Lecithin
Glutamic acid hydrochloride	Magnesium carbonate
Hydrochloric acid	Magnesium hydroxide
Lactic acid	Magnesium oxide
Malic acid	Magnesium stearate
Phosphoric acid	Methylcellulose
Potassium acid	Monoammonium glutamate
Sodium acid pyrophosphate	Monopotassium glutamate
Succinic acid	Nitrogen
Sulfuric acid	Nitrous oxide
Tartaric acid	Papain
Aluminum sulfate	Potassium bicarbonate
Aluminum ammonium sulfate	Potassium carbonate
Aluminum potassium sulfate	Potassium citrate
Aluminum sodium sulfate	Potassium hydroxide
Ammonium bicarbonate	Potassium sulfate
Ammonium carbonate	Propane
Ammonium hydroxide	Propylene glycol
Ammonium phosphate	Rennet
Ammonium sulfate	Silica aerogel
Bentonite	Sodium acetate
Butane	Sodium bicarbonate
Caffeine	Sodium carbonate
Calcium carbonate	Sodium carboxymethylcellulose
Calcium chloride	Sodium caseinate
Calcium citrate	Sodium citrate
Calcium gluconate	Sodium hydroxide
Calcium hydroxide	Sodium pectinate
Calcium lactate	Sodium phosphate
Calcium oxide	Sodium aluminum phosphate
Calcium phosphate	Sodium sesquicarbonate
Caramel	Sodium potassium tartrate
Carbon dioxide	Sodium tripolyphosphate
Ethyl formate	Triacetin
Glycerin	Triethyl citrate
	Carnauba wax

Source. Code of Federal Regulations, Title 21, U. S. Government Printing Office, Washington, D.C., 1979.

TABLE 6.3 Additives Affirmed as GRAS by FDA Regulations

Direct Additives		Indirect Additives
Aconitic acid	Locust (carob) bean gum	Caprylic acid
Benzoic acid	Karaya gum (sterculia gum)	Dextrans
Caprylic acid	Gum tragacanth	Acacia (gum arabic)
Calcium iodate	Methylparaben	Guar gum (technical grade)
Clove and its derivative	Rapeseed oil	Locust (carob) bean gum
Cocoa butter substitute from palm oil	Potassium iodide	Pulp
L-Cysteine	Potassium iodate	Sodium thiosulfate
L-Cysteine monohydrochloride	Propyl gallate	Sorbose
Dill and its derivatives	Propylparaben	
Ethyl alcohol	Rue	
Garlic and its derivatives	Oil of rue	
Acacia (gum arabic)	Sodium benzoate	
Gum ghatti	Sodium thiosulfate	
Guar gum	Sorbitol	
	Beeswax (yellow and white)	
	Baker's yeast extract	
	Ox bile extract	

Source. Code of Federal Regulations, Title 21, U. S. Government Printing Office, Washington, D.C., 1979.

proved as GRAS for direct addition to food may also be used in food contact articles, subject to any separate indirect additive regulations. The regulations for indirect GRAS substances address both the uses in food contact articles as well as the levels that can migrate to foods. The substances currently affirmed as GRAS are shown in Table 6.3.

Review of GRAS status may, of course, indicate that a previously used substance is not considered GRAS for human consumption, or that existing toxicity information does not justify a GRAS classification. Additionally, independent of any review, new information may become available that leads to regulatory restriction. These situations are illustrated by the previously used substances, now prohibited, shown in Table 6.4. This, of course, is by no means a complete list of prohibited substances. Such a list, like a complete GRAS list, would be impractical to compile. In this regard the interested reader may wish to pursue the thinking at the time the 1958 amendments were being debated in the U.S. Congress. Then Commissioner Larrick, in his testimony on the proposals, presented several examples of both GRAS substances and substances that should not be used in food (29).

Other GRAS Lists

It is important to note that the 1958 amendments did not reserve the decision about GRAS status only to the FDA. Any group of "experts qualified by training and experience" could make a GRAS classification. One such group that did so was the Flavor and Extract Manufacturers' Association (FEMA). The FEMA expert committee has approved over 1000 flavoring ingredients that subsequently were listed by the FDA and now appear in the Code of Federal Regulations in Title 21, Section 172, Subpart F. (Part 172 lists "food additives permitted for direct addition to food for human consumption.")

A series of reports have been published in the scientific literature dealing with the findings of the FEMA expert committee starting in 1960. In the 1979 report, Oser and Ford (30) note the continuing progress in the isolation and characterization of components of natural flavors and attendant progress in the effective simulation of those components. These new flavoring substances as well as others inspired by continued development of new foods such as meat and cheese substitutes necessitate continuing safety review. The FEMA expert panel has followed the practice of listing those substances, and their maximum use levels in various food categories, that have been determined to be GRAS. New food uses or high use levels in previously listed categories would likely still be considered GRAS, but such changes are evaluated by the panel to ensure that this is the case.

The FDA has formally recognized FEMA's flavor GRAS lists numbers 3 through 11 and is expected to do likewise with list number 12

TABLE 6.4 Substances Prohibited from Use in Human Food

Direct Additives	Indirect Additives
Calamus and its derivatives	Flectol H
Cobaltous salts and its derivatives	Mercaptoimidazoline and 2-mercaptoimidazoline
Coumarin	4,4'-Methylenebis (2-chloroanaline)
Cyclamate and its derivatives	
Diethylpyrocarbonate (DEPC)	
Dulcin	
Monochloroacetic acid	
Nordihydroguaiaretic acid (NDGA)	
P-4000	
Safrole	
Thiourea	
Chlorofluorocarbon propellants	

Source. Code of Federal Regulations, Title 21, U. S. Government Printing Office, Washington, D.C., 1979.

published in 1979. The FDA plans to review flavor ingredients as part of their overall food additive cyclic review program.

The GRAS Review

In 1969 the U.S. President directed FDA to update the safety assessment of GRAS substances according to current scientific knowledge and the available data. The FDA initiated the GRAS review, with a contract compilation of the world's scientific literature from 1920 to 1970 on each of the GRAS substances. The National Academy of Sciences, National Research Council was contracted by the FDA to survey the food industry to obtain data on usage of GRAS substances and to estimate their daily consumption. Evaluation of safety was provided for by a contract with the Life Sciences Research Office of the Federation of American Societies for Experimental Biology (FASEB). The FASEB established a select committee on GRAS substances (SCOGS), a group of scientific experts, to make the safety evaluations with FASEB staff support.

After some initial work by the SCOGS, a GRAS review procedure was developed that is in use today. This consists of the development of a tentative evaluation report, on a given substance or group of related substances, which is then made public by the FDA. An invitation is issued in the Federal Register providing an opportunity for the public to submit oral and/or written views and/or data to the SCOGS. If requested, a public hearing before the committee can be held to receive data or air views on the substance involved. A final report is then prepared based on the tentative report and additional data and submitted to the FDA. Again, the final report is also made public.

Although the SCOGS reports are the result of scientific evaluations of safety, they are, of course, intended for regulatory use by the FDA. Thus conclusions and recommendations are made within the context of the FFD&C Act, including the fact that to officially change GRAS status, the legal burden of demonstrating any question of GRAS status falls on the FDA. Safety conclusions are therefore stated by SCOGS in one of five ways (31):

1. **Reaffirmation of GRAS Status.** "There is no evidence in the available information on ____________ that demonstrates or suggests reasonable grounds to suspect a hazard to the public when it is used at levels that are now current or that might reasonably be expected in the future." (FDA interpretation—continue in GRAS status with only good manufacturing practices limitations.)

2. **Reaffirmation of GRAS Status at Current Levels of Use.** "There is no evidence in the available information on ____________ that demonstrates or suggests reasonable grounds to suspect a hazard to the public when it is used at levels that are now current and in the manner now practiced. However, it is not possible to determine without additional data, whether a significant increase in consumption would constitute a dietary hazard." (FDA interpretation—continue in GRAS status with limitations on amounts used in foods.)

3. **Reaffirmation of Safety but Additional Research Required.** "While no evidence in the available information on ____________ demonstrates a hazard to the public when it is used at levels that are now current and in the manner now practiced, uncertainties exist requiring that additional studies should be conducted." (FDA interpretation—require additional testing within a stated period under terms of an interim food additive regulation, and continue in GRAS status until tests completed and evaluated.)

4. **Insufficient Information to Reaffirm Safety.** "The evidence on ____________ is insufficient to determime that the adverse effects reported are not deleterious to the public health when it is used at levels that are now current and in the manner now practiced." (FDA interpretation—establish safe usage conditions or rescind GRAS status. Petitions may be submitted proposing safe usage conditions.)

5. **Insufficient Information to Judge Safety.** "In view of the almost complete lack of biological studies, the Select Committee has insufficient data upon which to evaluate the safety of ____________ as a food ingredient." (FDA interpretation—provide opportunity to submit relevant data for evaluation or rescind GRAS status.)

The FASEB–SCOGS has completed their safety review and issued preliminary or final reports on a wide variety of GRAS substances. The reports issued by the FASEB as of mid-1979 are presented in Table 6.5.

It is interesting to note that, for those GRAS ingredients evaluated through mid-1978, approximately 71 percent fell in category 1, that is, were reaffirmed as GRAS subject only to good manufacturing practices limitations. Those classified in category 2, reaffirmed as GRAS at current levels of use, amounted to about 15%. Category 3, reaffirmation of safety but additional research required, contained 6%. In category 4, information insufficient to reaffirm safety, there were about 4%. Finally, the remaining 4 percent were classified in category 5, information insufficient to judge safety (31).

Those substances classified with conclusions 2 through 5 (on the basis of the final or tentative report, as noted) as of mid-1979, with their SCOGS report number, are as follows (32):

Conclusion 2: Reaffirmation of GRAS Status at Current Levels of Use

Final reports

1. Gum arabic.

3. Carob bean gum.

4. Gum tragacanth.

5. Sterculia gum.

12. Gum ghatti.

13. Guar gum.

14. Oil of rue.

15. *Sulfiting agents.* Potassium bisulfite, potassium metabisulfite, sodium bisulfite, sodium metabisulfite, sodium sulfite, and sulfur dioxide.

21. *Zinc salts.* Zinc sulfate, zinc oxide, zinc acetate, zinc carbonate, and zinc chloride.

23. Agar agar.

24. *Alginates.* Ammonium, calcium, potassium, sodium, and propylene glycol alginate.

25. *Cellulose.* Methyl cellulose and hydroxy propylmethyl cellulose.

28. *Glycyrrhiza.* Glycyrrizzia, licorice, and ammoniated glycyrrhizin.

37a. *Glutamates.* Monoammonium glutamate, monopotassium glutamate, glutamic acid, and glutamic acid-HCl (supplemental report in process).

TABLE 6.5 Life Sciences Reasearch Office GRAS Reports*

"Evaluation of the Health Aspects of ______________ as (a) Food Ingredient(s)"

Acetic acid, sodium acetate, and sodium diacetate, 1977
Aconitic acid, 1974
Adipic acid, 1977
Agar-agar, 1974
Algae, certain red and brown, 1974
Alginates, 1974
Aluminum compounds, 1976
Ammonium salts, 1974
Beeswax (yellow or white), 1976
Bentonite and clay (kaolin), 1977
Benzoic acid and sodium benzoate, 1973
Bile salts and ox bile extract, 1975
Biotin, 1978
Butylated hydroxyanisole, 1978
Butylated hydroxytoluene, 1973
Caffeine, 1978
Calcium oxide and calcium hydroxide, 1975
Calcium salts, 1975
Caprylic acid, 1975
Caramel, 1973
Carbonates and bicarbonates, 1975
Carnauba wax, 1976
Carob bean gum, 1972
Carrageenan, 1973
Cellulose and certain cellulose derivatives, 1974
Choline chloride and choline bitartrate, 1976
Citric acid, sodium citrate, potassium citrate, calcium citrate, ammonium citrate, triethyl citrate, isopropyl citrate, and stearyl citrate, 1977
Coconut oil, peanut oil, oleic acid and linoleic acid, 1977
Corn silk, 1977
Corn sugar (dextrose), corn syrup and invert sugar, 1976
Dextrans, 1975
Dextrin and corn dextrin, 1975
Dill, 1973
Formic acid, sodium formate and ethyl formate, 1977
Garlic and oil of garlic, 1973
Gelatin, 1975
Gluconates, 1978
Glutamates, 1978
Glycerine and glycerides, 1975

TABLE 6.5 Life Sciences Reasearch Office GRAS Reports (Continued)

"Evaluation of the Health Aspects of ______________ as (a) Food Ingredient(s)" - (continued)

Glycerophosphates, 1977
Guar gum, 1973
Gum arabic, 1973
Gum ghatti, 1973
Gum guaiac, 1976
Gum tragacanth, 1973
Hydrogenated fish oil, 1976
Hydrogenated soybean oil, 1977
Hypophosphites, 1977
Inositol, 1976
Lactic acid and calcium lactate, 1978
Licorice, glycyrrhiza, and ammoniated glycyrrhizin, 1975
Magnesium salts, 1977
Malic acid, 1976
Mannitol, 1973
Methyl paraben and propyl paraben, 1973
Mustard and oil of mustard, 1975
Nutmeg, mace, and their essential oils, 1973
Oil of cloves, 1973
Oil of rue, 1973
Pantothenates, 1978
Papain, 1977
Pectin and pectinates, 1977
Phosphates, 1976
Potassium iodide, potassium iodate, and calcium iodate, 1975
Propyl gallate, 1973
Propylene glycol and propylene glycol monostearate, 1974
Protein hydrolyzates, 1978
Pyridoxine, 1977
Rennet, 1977
Sodium hydroxide and potassium hydroxide, 1977
Sodium thiosulfate, 1975
Sorbic acid and its salts, 1976
Sorbitol, 1973
Sorbose, 1975
Stannous chloride, 1974
Sterculia gum, 1973
Succinic acid, 1975
Sucrose, 1976
Sulfuric acid and sulfates, 1975
Sulfiting agents, 1977
Tallow, hydrogenated tallow, stearic acid, and calcium stearate, 1976
Tannic acid, 1977
Thiamin hydrochloride and thiamin mononitrate, 1978
Tocopherols and α-tocopheryl acetate, 1976
Urea, 1978
Vitamin B_{12}, 1978
Vitamin D_2 and D_3, 1978
Zinc salts, 1974

TABLE 6.5 Life Sciences Reasearch Office GRAS Reports (Continued)

"Evaluation of the Health Aspects of ____ as it (They) May Migrate to Foods from Packaging Materials"
Coconut oil, peanut oil, oleic acid and linoleic acid, 1977
Hydrosulfites, 1976
Japan wax, 1976
Lard and lard oil, 1977
Monomeric and polymeric ethyl acrylate and methyl acrylate, 1977
Pulps, 1974
Sodium oleate and sodium palmitate, 1977
Sulfamic acid, 1976
Tall oil, 1976

"Review of the Recent Literature on the Health Aspects of ____ as (a) Food Ingredient(s)"
Copper salts
Gases
Lecithin
Niacin and niacinamide
Riboflavin and riboflavin-5'-phosphate
Vitamin A, vitamin A acetate and vitamin A palmitate
Vitamin B_{12}

* Final reports issued by the Life Sciences Research Office, Federation of American Societies for Experimental Biology, under contract with the Food and Drug Administration to evaluate the safety of substances on the generally recognized as safe (GRAS) list.

37b. *Protein hydrolyzates.* Soy sauces (supplemental report in process).

48. *Tannic acid.* Hydrolyzable gallotannins.

50. *Corn sugar.* Corn sugar, invert sugar, corn syrup.

69. Sucrose.

83. Dextran.

95. *Vitamin D.* Vitamins D_2 and D_3.

Tenatative reports

35. *Iron salts.* Reduced iron, electrolytic iron, carbonyl iron, ferrous sulfate, ferric phosphate, ferric pyrophosphate, ferrous fumerate, ferrous carbonate, and ferric citrate

59. *Ascorbates.* Erythorbic acid and sodium erythorbate

108. *Niacin.* Nicotinic acid and nicotinamide

115. *Starches.* Acetyl distarch adipate, acetylated distarch phosphate, acetylated diamylopectin phosphate, monostarch phosphate, distarch phosphate, phosphated distarch phosphate, and starch acetate

Conclusion 3: Reaffirmation of Safety but Additional Research Required

Final reports

2. BHT

6. Carrageenan

18. *Oil of nutmeg.* Nutmeg, mace, nutmeg oil, mace oil, and mace oleoresin.

30. *Glycerides.* Oxystearin only.

37a. *Glutamates.* Monosodium glutamate (supplemental report in process).

37b. *Protein hydrolyzates.* Acid hydrolyzates, enzyme hydrolyzates, and yeast autolyzates (supplemental report in process).

55. BHA.

89. Caffeine (includes a minority report with conclusion 4).

Tentative reports

94. Starter distillate.

101. Soy protein isolate.

115. *Starches.* Distarchoxypropanol, acetylated distarchoxypropapanol, hydroxypropyl distarch phosphate, starch sodium succinate, starch sodium ocetenyl succinate, starch aluminum octenyl succinate, and sodium hypochlorite oxidized starch.

Conclusion 4: Information Insufficient to Reaffirm Safety

Final reports

37a. *Glutamates.* Glutamic acid, glutamic acid HCl, monoammonium glutamate, monopotassium glutamate, and monosodium glutamate (revised report in process).

37b. *Protein hydrolyzates.* Soy sauces, enzyme hydrolyzates, acid hydrolyzates, yeast autolyzates (supplemental report in process).

102. Sodium chloride.

115 *Starches.* Distarch glycerol, hydroxypropyl distarch glycerol, acetylated distarch glycerol, and succinyl distarch glycerol.

116. *Lactates.* D and DL lactic acid and calcium D and DL lactate (infants).

Conclusion 5: Information Insufficient to Judge Safety

Final reports

30. *Glycerides.* Mono- and diglycerides of sulfoacetate, mono- and diglycerides of monosodium phosphate, and monoglyceride citrate.

46b. Japan wax (packaging).

47. Carnauba wax.

87. Corn silk.

88. *Methyl and ethyl acrylates.* Polymeric methyl acrylate (packaging).

Tentative reports

35. *Iron salts.* Ferrous ascorbate, ferrous citrate, ferrous gluconate, ferrous lactate, ferric ammonium citrate, ferric oxide, iron peptonate, iron polyvinylpyrrolidone, sodium ferric EDTA, sodium ferricitropyrophosphate, or ferric sodium pyrophosphate.

61. *Silicates.* Sodium metasilicate, sodium zinc metasilicate, and methyl polysilicones.

67. *Manganese salts.* Manganous oxide.

115. *Starches.* Starch gelatinized with NaOH.

DIRECT FOOD ADDITIVES

Those directly added food ingredients, other than GRAS substances and color additives that have been regulated by the FDA are listed in Part 172 of the Code of Federal Regulations and are entitled "Food Additives Permitted for Direct Addition to Food for Human Consumption." The regulated direct food additives numbering over 100 have resulted from food additive petitions submitted by industrial sponsors or from the FDA's own evaluation of the substance.

Under current law, direct food additives, in addition to safety requirements, must be shown to accomplish their intended physical or other technical effect and are subject to limitations on the amounts that can be used in certain foods. The FDA has defined by regulation (Code of Federal Regulations, Section 170.3) a listing of 32 "physical or technical functional effects for which direct human food ingredients may be added to foods":

Anticaking and free-flow agents
Antimicrobial agents
Antioxidants
Colors and color adjuncts
Curing and pickling agents
Dough strengtheners
Drying agents
Emulsifiers and emulsifier salts
Enzymes
Firming agents
Flavor enhancers
Flavoring agents and adjuvants
Flour treating agents
Formulation aids
Fumigants
Humecants
Leavening agents
Lubricants and release agents
Non-nutritive sweeteners
Nutrient supplements
Nutritive sweeteners
Oxidizing and reducing agents
pH control agents
Processing aids
Propellants, aerating agents, and gases
Sequestrants
Solvents and vehicles
Stabilizers and thickeners
Surface-active agents
Surface-finishing agents
Synergists
Texturizers

The technically defined individual physical and functional effects result practically in the smaller number of categories in which the FDA has grouped the regulated direct food additives:

Food preservatives
Coating, films, and related substances
Special dietary and nutritional additives
Anticaking agents
Multipurpose additives
Flavoring agents and related substances
Gums, chewing gum bases, and related substances
Other specific usage additives

The direct food additives listed in Part 172 include a wide variety of substances on the basis of their chemical nature as well as their notoriety. It is impractical to describe all of these several hundred substances but examples along with GRAS substance examples are discussed in the following section.

EXAMPLES OF GRAS AND DIRECT ADDITIVES

BHA and BHT

In the food preservative category are the well-known and frequently criticized antioxidants BHA (butylated hydroxyanisole) and BHT (butylated hydroxytoluene). Both are GRAS substances, limited by a total antioxidant content of not more than 0.02% of the fat or oil content of foods. In addition, they have regulated food additive uses in dry cereals, shortenings and potato shreds, granules and flakes ranging from 10 to 200 ppm, in combination (BHA is also a permitted food additive in active dry yeasts, in dry beverage and dessert mixes, and in beverages and desserts made from dry mixes). Both BHA and BHT are also prior sanctioned as antioxidants in food packaging materials subject to a migration limit of 0.005% in foods. Evaluation in the GRAS review resulted in conclusion 3 (reaffirmation of safety but additional research required) for both substances (BHT in 1973 and BHA in 1978; see Table 6.5). The FDA has also proposed tolerances for the uses of BHT and will do likewise for BHA, based on the results of the GRAS review.

Butylated hydroxytoluene provides a typical example of the close scrutiny currently devoted to widely used food additives. Initially approved by FDA in 1954 and listed as GRAS in 1959, BHT prevents degenerative oxidation of fats that can lead to undesirable flavor and the destruction of fat-soluble vitamins and essential fatty acids. The oxidation may also produce toxic by-products in foods (33). In 1976 about 9 million pounds of BHT were produced for food uses. In addition to the GRAS review and FDA evaluation, BHT was also tested in the Carcinogenesis Testing Program of the National Cancer Institute. Observation of rats and mice fed BHT at 3000 and 6000 ppm produced the conclusion that BHT was not carcinogenic for rats or mice (33). Scrutiny of BHT continues, however, and it will be periodically reviewed under the FDA's cyclic review program.

Nitrites

The nitrate and nitrite salts also provide an interesting direct food additive example. Food and Drug Administration food additive regulations for these compounds cover their uses in fish products (e.g., cod roe and in smoked and cured sablefish, salmon, shad, tunafish, and chub) and in preparations for the home curing of meats and meat products. The much larger volume uses of these compounds in meat and poultry products are directly regulated by the USDA. The latter uses were for many years considered to be prior sanctioned, but in 1978 the USDA, although affirming the prior sanction of uses in meat products, denied the prior sanction of uses in poultry products. Also, in 1978 concern about nitrates and nitrites was extended from their potential role in nitrosamine formation to the possibility of direct induction of cancer based on a study conducted at the Massachusetts Institute of Technology (MIT) for the FDA (34).

On the one hand the uses of nitrites and nitrates may pose some human risk—how much risk has not been determined. The FDA has estimated an annual cancer risk of about 1 in 1 million from average consumption of cured meat products, based on the uncorroborated MIT study (35). On the other hand, nitrites in cured meats serve the function of preventing the growth of *Clostridium botulinum* spores and hence the production of the highly lethal botulinum toxin. Then too, nitrates are widely present in drinking water and vegetables, and nitrites are produced endogenously in the human digestive tract; these two natural sources produce perhaps as much as 5 to 70 times more nitrite than does the average ingestion of cured meats. Thus nitrites provide an excellent illustration of the many factors involved in food safety risk assessment (see also Chapter 5).

Sweeteners

At least a brief mention of some of the sweeteners other than sucrose (a GRAS substance) is also warranted (sucrose is discussed in Chapter 3). Additives such as cyclamate and, more especially, saccharin have received widespread attention from government regulators and scientists, industry and university scientists, and hence the media. Somewhat lesser attention has been devoted to aspartame, xylitol and other polyols, and fructose.

Cyclamate

As pointed out earlier, cyclamate is specifically listed as a prohibited substance, as it was banned in 1970 after it was associated with tumors in rats fed a 10:1 cyclamate–saccharin mixture in a study sponsored by Abbott (36). Cyclamate (cyclohexylsulfamic acid and its calcium and sodium salts) was discovered in 1937 and was first marketed in 1950. In the 1950s cyclamate began to be used in combination with saccharin in a variety of dietary products and was included on the FDA's GRAS list in 1959. (Cyclamate's indictment led to the presidential directive for the GRAS review.) In 1973 a petition for the reapproval of cyclamate was filed, but after detailed evaluation by FDA and a National Cancer Institute panel, the petition was denied in 1976; the FDA took the position that evidence of safety required for approval had not been demonstrated. Objection to the petition denial, after a prolonged administrative proceeding, was finally resolved in 1980 and cyclamate remains prohibited. As with several other controversial additives, opinion varies concerning the carcinogenicity of cyclamate. In addition, cyclohexylamine, a cyclamate metabolite, has been associated with testicular atrophy and other effects in test animals (37). Thus if cyclamate were to be reapproved, it would be subject to at least some limitation based on its rate of conversion to cyclohexylamine and the no-effect level for cyclohexylamine.

In 1980 the Joint Expert Committee on Food Additives (JECFA) continued its temporary ADI of 4 mg/kg for cyclamate pending completion of studies on the conversion rate for cyclohexylamine and studies on reproduction.

Saccharin

The safety of saccharin, discovered in 1879 and in commercial use for about 80 years, has probably been investigated more and certainly has been reviewed and debated more than that of any food additive (36–45). Like cyclamate, saccharin has been a GRAS substance and is currently officially regulated as an "interim" food additive. (Although the FFD&C Act does not specifically provide for interim status, the FDA has continued approval of several additives contingent on the conduct of certain research to resolve safety questions. Regulations covering the interim status for food additives are given in Part 180 of the Code of Federal Regulations (28). This classification is analagous to the provisional listing of color additives that is provided for in the FFD&C Act.) Current attention on saccharin derives from the 1977 FDA proposal (46) to terminate its approval based on a Canadian two-generation (*in utero*) rat test that showed an increased tumor incidence in second-generation males (47). A moratorium on regulatory action against saccharin was mandated by Congress in the Saccharin Study and Labeling Act, passed in 1977 when a public outcry arose at the possibility of losing the only noncaloric sweetener. That act also directed a review of the safety of saccharin and a study of food safety policy by the National Academy of Sciences.

The National Academy of Sciences (NAS) saccharin safety study (42), which utilized the legal approach to the definition of a carcinogen contained in the FFD&C Act (i.e., "if found to induce cancer when ingested by man or animal"), concluded that "In rats saccharin is a carcinogen of low potency relative to other carcinogens." The food safety policy study (43) concluded that food additive regulation should be modified to incorporate "risk–benefit" considerations. Although a minority of the study group (seven members) preferred a phaseout of saccharin, the majority (30 members) felt that saccharin should be regulated under the provisions of the recommended new food safety policy.

The extensive and continuing research on saccharin going back to at least 1886 (45) has not yet resolved all the questions posed by today's state of the art in safety testing. For example, Sweatman and Renwick (48) confirmed earlier studies indicating that saccharin is not metabolized and concluded that such results show that saccharin does not act as a classic electrophilic carcinogen. The absence of metabolism, together with evidence of tumorigenicity only in F1 males resulting from two-generation, high-dose exposure, poses many questions about saccharin. This is reflected in the 1980 JECFA decision to extend the temporary ADI for saccharin of 0 to 2.5 mg/kg pending further investigation of carcinogenicity in rodents, including the mechanism of high-dose tumor production.

Current scientific questions about saccharin have led to probably the largest animal study of its kind ever conducted—a study of dose response and *in utero* exposure of saccharin in the rat. This study (49, 50), involving 2500 F1 male rats and six saccharin feeding levels from 1 to 7.5%, has as its objectives to determine the dose response for urinary bladder and other toxic effects, to provide a data base for risk assessment, to evaluate the role of *in utero* exposure, and to evaluate the role of excess sodium and the specificity of saccharin to produce bladder tumors.

In addition to this animal study, the saccharin controversy has also led to probably the largest epidemiology study of its kind ever conducted. Major reviews of the substantial body of epidemiologic data bearing on saccharin consumption by the Office of Technology Assessment (37), the FDA (51), and the American Council on Science and Health (45) have demonstrated an overall lack of association between saccharin consumption and human bladder cancer but have also revealed shortcomings in existing studies. As a consequence, the Interagency Saccharin Working Group (National Cancer Institute and FDA scientists) developed a large-scale epidemiology study involving approximately 3000 cases and 6000 controls. The study included five states—New Jersey, Connecticut, Iowa, New Mexico, and Utah and four metropolitan areas—Detroit, San Francisco/Oakland, New Orleans, and Atlanta.

The preliminary report of the National Cancer Institute study (52) showed no difference in the risk of bladder cancer between users and nonusers of artificial sweeteners (saccharin, cyclamate, or their combination). That lack of association also held separately for males and for females in the study population. Analysis of subgroups failed to show any statistically significant difference in relative risk for long-term use of artificial sweeteners or for the highest frequency of daily use. Two subsequent studies (53, 54) also failed to show any increased risk from consumption of artificial sweeteners. Despite the lack of association shown in these studies, no study, of course, can "prove" that saccharin (or any other substance) is free of all risk. Until that basic truth is realized about food safety, the debate about saccharin and other substances receiving popular attention will continue often at the expense of more important problems.

Aspartame

The first commercially available noncarbohydrate nutritive sweetener, aspartame, was approved for table use and for uses in dry mixes by the FDA in 1974 (55). However, an objection to the approval on safety grounds and questions about the authenticity of the test data led to a stay of that approval. Subsequent audit of the test data resolved the authenticity question, but the safety objection (the possibility that aspartame alone or in combination with glutamate poses a risk of contribution to mental retardation, brain damage, or undesirable effects on neuroendocrine systems) remains to be administratively resolved. The appeal of aspartame lies in its sweetness (approximately 200 times sweeter than sucrose). Although it provides about 4 calories/g, its sweetness results in insignificant caloric contribution in most uses. Aspartame is composed of two amino acids, L-aspartic acid and L-phenylalanine, both of which occur naturally in foods.

High cooking temperatures (e.g., in baking and frying) result in the breakdown of aspartame to its amino acid constituents and a degradation product, diketopiperazine. Such breakdown, which results in a loss of sweetness, also occurs in liquids and acidic foods and thus somewhat limits the utility of aspartame.

The Canadian Health Protection Branch (HPB) has indicated its intent to recommend to the Minister of National Health and Welfare the approval of aspartame for uses in tabletop sweeteners, beverage mixes, breakfast cereals, desserts and topping mixes, chewing gum and soft drinks (56). Canadian HPB scientists in considering the possible risk of brain damage have taken into account the hypothalamic lesions in neonatal rats but have noted the absence of such lesions in infant monkeys and have concluded that "Aspartame is considered to be of limited toxicological significance." Because persons with phenylketonuria, an inborn error of metabolism, must restrict their intake of phenylalanine from all sources, the presence of aspartame in foods would be required by the HPB to be prominently noted. (Aspartame was approved in France in 1979 and elsewhere in Europe in 1980 and received an ADI of 40 mg/kg body weight from JECFA in 1980.)

Polyols

Also of interest in the sweetener category are the sugar alcohols (polyols) xylitol, mannitol, and sorbitol. Xylitol occurs widely in the plant kingdom and has found commercial application in several countries. Limited clinical trials suggest the possible utility of xylitol in the dietary management of diabetes and other studies indicate a reduction in the incidence and severity of human dental caries when xylitol replaces sucrose in the diet (57). Xylitol is formally regulated in Section 172.395 of the Code of Federal Regulations (28) as a sweetener in foods for special dietary uses but has found significant use in the United States only in chewing gum. Contemplated changes in the regulation of xylitol as well as continuing human research in the United States were sidetracked when the report of a Huntingdon Research Center study became available (58). This study showed increased hyperplasia, metaplasia, and neoplasia of the transitional epithelium of the urinary bladder in male mice fed 20 and 10% xylitol in the diet. Results at the 2% feeding level were negative. The effects at the 20 and 10% feeding levels in each case were associated with macroscopically observed calculi. The results illustrate one of the many regulatory and scientific quandries relative to the definition of a carcinogen—what meaning should be attributed to effects occurring only at very high feeding levels, especially when an intermediate effect (i.e., macroscopic calculi) is involved?

Mannitol, a six-carbon alcohol, is commercially produced by the reduction or the hydrogenation of solutions of glucose or fructose. It occurs naturally in plants as well as in animal tissues. Used as a formulation aid in both drugs and foods, mannitol was estimated to have a daily per capita consumption of 20 to 30 mg in 1975 (59). Food uses that include candies, chewing gum, confections, frostings, and nonstandard jams and jellies are limited by its laxative effects at levels of 20 g or more. It has about half the sweetness of sucrose.

Initially a GRAS substance, mannitol was favorably reviewed by the FASEB–SCOGS in 1972, and the FDA proposed affirmation of GRAS status in 1973. Subsequently, an association of mannitol with benign thymic tumors in female Wistar rats led to its regulation as an interim food additive pending additional research (Code of Federal Regulations, Section 180.25). Further lifetime feeding studies have shown an absence of thymic effects in female rats of the Wistar, Fischer, and Sprague–Dawley strains (60). The 1979 FASEB–SCOGS review concluded that mannitol is neither mutagenic, teratogenic, nor carcinogenic (59).

As with other sweeteners, the association of mannitol with dental caries has been investigated in several studies using the rat model. Recognizing the difficulties of comparing these caries studies because of differences in design, Allison (59) has concluded that mannitol appears to be less cariogenic in the rat than glucose, dextrin, or sucrose.

It has been demonstrated that mannitol is absorbed and metabolized in humans (61). Mannitol is apparently metabolized in the liver and contributes to blood glucose, but the process for and the extent of metabolism is not completely understood.

Sorbitol, also a six-carbon alcohol differing from mannitol primarily in its optical rotation, is commercially produced by the catalytic reduction of glucose. It occurs naturally in a variety of fruits and berries and also is found with other polyols in animal tissues (62).

Sorbitol is used in a wide variety of foods, including candies, chewing gum, desserts, and beverages. Consumption estimates vary widely; an average per capita daily intake of 79 mg was calculated for 1970 (63). Like mannitol, sorbitol has about half the sweetness of sucrose (64) and is associated with laxation in adults at about the 50-gr level.

The GRAS status of sorbitol was reviewed by the FASEB–SCOGS in 1972 and reaffirmed at current and expected levels of use. It is listed in the Code of Federal Regulations (Section 184.1835) among the substances affirmed as GRAS.

Although long-term feeding studies of sorbitol are fairly limited, no adverse effects have been reported. Teratology and mutagenic studies have also been largely negative (63). The metabolism of sorbitol has been extensively studied, by Förster (65) and Thomas et al. (66), for example, and it is known that sorbitol is absorbed and rapidly metabolized through normal glycolitic pathways in humans. Sorbitol is absorbed from the alimentary canal more slowly and to a lesser extent than is sucrose and leads to less postprandial hyperglycemic peaking. However, its use as a sucrose substitute by diabetics is limited by its low degree of sweetness and laxation effects (63).

Fructose

The sweetener fructose, a monosaccharide, has been receiving increased attention in recent years. Fructose, also referred to as *levulose* or *fruit sugar*, occurs widely in nature in sweet fruits and berries and in honey. It occurs in combination with glucose as half of the disaccharide sucrose molecule (67, 68). The basis for recent attention on fructose is twofold: (*a*) the popularity of fructose among health food advocates and (*b*) the increasing availability of high-fructose corn syrup.

Commercial production of fructose became possible in 1970 with the development of a process for hydrolyzing sucrose into glucose and fructose and separating the components. Fructose can also be crystallized from aqueous solutions of hydrolyzed sucrose (68). Since fructose is produced from sucrose, it is much more expensive than table sugar; nonetheless, it has become increasingly popular because it has been identified as the "natural" sweetener, because it is relatively sweeter than sucrose, and because of claims of reduced dental caries and other special dietary benefits. Part of the popularity of fructose derives from the "fabulous 14-day fructose diet" (69).

According to Moskowitz (64), fructose is approximately 8 to 70% sweeter than sucrose, but perceived sweetness is dependent on pH, temperature, and concentration and varies among different foods. Thus claims of fewer calories when fructose is substituted for sucrose are often exaggerated, and such substitution likely produces only a minor advantage (68).

Fructose is absorbed largely unchanged from the gastrointestinal tract (70). Because it is absorbed more slowly than glucose and results in a less precipitous increase in blood sugar levels than does sucrose or glucose, nutritional advantages have been claimed for both diabetics and nondiabetics. However, such claims are still the subject of controversy, and Kimura and Carr (68) have cautioned against the notion that fructose could be used by diabetics without exchange for other caloric sources. In their review of fructose Kimura and Carr (68) concluded that fructose is well tolerated by normal subjects, but that substitution of fructose for sucrose would not significantly change the incidence of dental caries, nor would the substitution of fructose for glucose produce clinical advantages in any disease state.

In recent years the so-called high-fructose corn syrups (HFCSs) produced by enzymatic isomerization of corn starch hydrolyzates have become increasingly available. Their costs and sweetness have made HFCS competitive with sugar in food applications where liquid sweeteners can be used. High-fructose corn syrups have seen a dramatic increase in use, from less than 1 pound per capita (ca. 0.5% of total caloric sweeteners) in 1970 to 15 pounds in 1979 (ca. 12% of total caloric sweeteners) (71). These corn-derived sweeteners are normally composed of 42 or 55% fructose (the remainder is primarily glucose). High-fructose corn syrup with 90% fructose is also being produced. (This is appearing in the health food market as "natural liquid sweetener.") If present trends continue, corn sweeteners (predominantly HFCS) could account for nearly half of the caloric sweeteners in our food supply by the end of the next decade (71). Although there are no obvious indications of food safety problems associated with the use of this GRAS sweetener (nor are there obvious benefits), should fructose consumption significantly increase in the years ahead, the safety of fructose would have to be reassessed in light of that new human exposure.

Caffeine

One of the most interesting GRAS substances, and again one that is the subject of much attention, is caffeine. Caffeine (1,3,7-trimethylxanthine) is one of the xanthine derivatives occurring naturally in coffee beans, tea leaves, kola nuts, cocoa beans, and mate and hence in beverages made from these plant sources. It is listed in the Code of Federal Regulations (28) as a multipurpose GRAS food substance (Section 182.1180) and is a mandatory ingredient in cola-type beverages as provided in the standard of identity for soda water (Section 165.175).

Caffeine has been under scientific study since its isolation in the laboratory in 1820 (72). As part of the GRAS review by the SCOGS, its GRAS status as an ingredient of cola-type beverages was evaluated and reported in 1978 (73). The SCOGS concluded:

> While no evidence in the available information on caffeine demonstrates a hazard to the public when it is used in cola-type beverages at levels that are now current and in the manner now practiced, uncertainties exist requiring that additional studies be conducted.

Popular attention on caffeine has centered largely on the question of birth defects. Activist groups have requested that the FDA restrict the GRAS status and uses of caffeine and require label statements on coffee and tea, warning against its consumption by pregnant women. The SCOGS considered this question and concluded in summary:

> Many animal tests showed that teratogenic effects are generally absent at caffeine doses up to 50 mg per kg body weight. At doses up to 75 mg per kg of body weight, teratogenic effects are neither striking nor consistently demonstrated.

Such animal effects have not been of great concern since even direct extrapolation to humans would yield a considerable margin of safety. For example, 75 mg/kg would translate to consumption of over 200 cola-type drinks for an average-weight woman. Perhaps more reassuring is the fact that the absence of any association between caffeine intakes and abnormalities in offspring in retrospective studies of more than 14,000 human mothers was also noted by the SCOGS (73).

The animal teratology on caffeine is an excellent example of the difficulties in food safety assessment. For example, there is scientific controversy over the interpretation of animal effects associated with large bolus doses administered by intubation. In addition, there are questions concerning how appropriate the rat, or any other animal, is as a model for assessing effects in humans. [Garattini (74) has pointed out a number of major differences between caffeine metabolism in rats and humans, including the fact that paraxanthine, the major metabolite in humans, is not found in rats whereas theophylline and theobromine, the major metabolites in rats, represent only a small percentage of metabolites in humans.]

The extensive research being conducted on caffeine (72) undoubtedly will gradually clarify the scientific picture on this important GRAS substance and hopefully will also contribute to the general problem of food safety assessment.

INDIRECT ADDITIVES

The very large class of indirect food additives technically includes not only those that may originate through migration from food packaging, but also those arising from food contact surfaces and those used as processing aids. Legally, as with direct additives, there are indirect additives classified as food additives, as GRAS, and as prior-sanctioned substances. Definition, safety assessment, and hence regulation are much more straightforward in the case of most processing aids than for many of the packaging components. This is the case since the processing aids are normally discrete substances, used at fixed levels and in rather specific applications. In contrast, migration characteristics of a given plastic packaging material, for example, may depend on the basic combination of monomers used; the type and the degree of polymerization, the resulting molecular weight, and the spatial orientation; the plasticizers used; components such as extenders and colors; the method of forming the container; the size of the container or the amount of packaging material; and finally the food product involved and its processing, transportation, and storage conditions.

Processing Aids

Included among the food processing aids are those defined by the FDA as "secondary direct food additives permitted in food for human consumption." These are divided into four major categories:

- Polymer substances for food treatment.
- Enzyme preparations and microorganisms.
- Solvents, lubricants, release agents, and related substances.
- Specific usage additives.

The processing aids include a few well-known additives, but for the most part they are familiar only to the food technologist. For example, there are special purpose resins used as flocculents to clarify cane and beet sugar juice and ion-exchange membranes to adjust the citric acid: total solids ratio in grapefruit juice. More widely used are the ion-exchange resins employed in the purification of water and a variety of foods. A variety of enzymes and microorganisms have been approved for use in clam and shrimp processing and in the production of citric acid, vinegar, distilled spirits, dextrose, sucrose, and cheeses.

Among the permitted solvents and related substances are more familiar compounds such as acetone, isopropyl alcohol, and methyl alcohol. This class of additives has, among its major uses, extraction of spices and solvent vehicles for flavors. One of the compounds, methylene chloride, is the major substance used in extracting caffeine from green coffee beans in the preparation of decaffeinated coffee. Methylene chloride replaced trichloroethylene (TCE) as the major caffeine extractive when TCE became a suspected animal carcinogen. Although it has been argued that the decaffeinated coffee itself is without adverse effects, traces of TCE can remain in the decaffeinated coffee (the tolerance had been 10 ppm for instant decaffeinated coffee). Therefore, industry terminated its use and the FDA proposed the banning of TCE. Trichloroethylene provides yet another example of the problem of realistically controlling trace residues in foods.

The safe use of the substances in this category is prescribed by specifying the food processing uses permitted and the amounts of the additives that can be employed in each case and/or by setting limits on the residues that can occur in or on foods. Of course, the Delaney Clause applies to the regulated indirect additives in this category. Therefore, according to FDA policy, no additive of this type that has been associated with cancer in animals can be permitted if it might become a component of food. (Such a residue does not necessarily have to be detectable for this to apply.)

Packaging Components

By far the largest and most complex set of additives is that associated with food containers and packaging. Indeed, food packaging has become so complex that a separate industry has evolved and with it a specialized technical field—packaging engineering. The complexity can be seen by considering some of the total requirements that can be imposed on packaging, such as (*a*) nontoxic, (*b*) compatible with the specific food, (*c*) sanitary protection, (*d*) moisture and/or fat protection, (*e*) gas and odor protection, (*f*)light protection, (*g*) impact or handling resistance, (*h*) transparency, (*i*) tamperproofness, (*j*) ease of opening, (*k*) accessibility of product, (*l*) reseal feature, (*m*) size, weight, and shape limitations, (*n*) appearance and printability, (*o*) cost, (*p*) environmental considerations, such as ease of disposal and biodegradability, and (*q*) special features (67).

An indication of the scope and the complexity of food safety questions associated with packaging components can be seen by examining Part 175 of the Code of Federal Regulations. This part lists several hundred substances that may be used as components of adhesives and coatings in food packaging applications. Unless such substances are separated from contact with foods by a functional barrier, they are subject to restrictions based on their toxicological and migration characteristics. Again, except for GRAS and prior-sanctioned items, these indirect additives are subject to the Delaney Clause.

It is impractical to discuss individual substances in this very large category, but some general comments on adhesive components are in order. Adhesive components have posed significant hurdles for the implementation of a major packaging innovation, the retortable flexible pouch. Weight, labor, and convenience advantages derived from flexible foil and laminated plastic packaging have led to widespread use of flexible containers in such applications as condiments and other single-service foods. The development of flexible pouches for foods requiring retort processing pointed to even greater expansion of flexible packaging. However, the potential for adhesive components to come in contact with the packaged foods and safety concern about those components have imposed delays. For example, the National Cancer Institute has concluded that 2,4-diamino toluene is carcinogenic in rats and mice. This substance could arise from toluene diisocyanate, which is a component of adhesives used in certain retortable pouches that are under evaluation (75).

The three other, also very large, classes of indirect additives are paper and paperboard components; polymers; and adjuvants, production aids, and sanitizers. These substances and their conditions of use are enumerated in Parts 176, 177, and 178, respectively, in the Code of Federal Regulations. Although these additives are widely divergent in their chemical and technical characteristics, they share the food safety questions common to all indirect additives: Can they become a component of food? If so, under what conditions and at what levels? What is their toxicological significance?

The vast preponderance of the indirect additives, familiar only to the packaging specialists and food technologists, are likely of no significant toxicological concern. However, polyvinyl chloride (PVC) and acrylonitrile polymers in food contact articles have received a great deal of professional and public attention. It was discovered in 1973 that residual vinyl chloride monomer in PVC packaging material could migrate to foods (76). Previously, residual monomer levels had been assumed to be reduced to insignificant levels during the processing of the PVC resin and fabrication of the packaging material. Again, advances in analytical chemistry had led to identification of a new problem. The acute toxic effects of vinyl chloride have been known since the 1930s, but in 1974 hemangiosarcomas were associated with vinyl chloride at 250 ppm in animal inhalation studies by Maltoni and Lefemine (77). Subsequent studies produced hemangiosarcomas at 50 ppm. [An extensive bibliography on vinyl chloride toxicity is given in Frank and Carlin (78).] These results led to the elimination of the PVC container for alcoholic beverages and proposed restrictions on the use of rigid and semirigid PVC packaging materials. (The uses of vinyl chloride as a propellant in hair sprays and other products were also eliminated and occupational exposure limits established.)

Increased analytical detection capability also came into play relative to the uses of acrylonitrile and its copolymers. In 1976 the FDA established an interim regulation limiting acrylonitrile extraction in food simulating solvents to 0.3 ppm and conditioning the continued use of acrylonitrile copolymers in food applications on additional toxicological testing (79). The following year FDA stayed the interim regulations permitting use of acrylonitrile copolymers in beverage containers and proposed reduction of the extraction limit in other food applications from 0.3 ppm to 0.05 ppm (80, 81). The 1977 action was based in part on proliferative brain lesions observed at 300 ppm and 100 ppm of acrylonitrile in the drinking water of rats after 13 months of a two-year study (82).

These two major indirect additive problems have not been completely resolved. The FDA took the position in the acrylonitrile case that theoretical demonstration of the presence in food of a migrant, shown to have adverse effects, is sufficient for disapproval of the packaging material—even though extraction might not be analytically demonstrated. In the judicial review of the FDA's position, Judge Leventhal, for the U.S. Court of Appeals (District of Columbia Circuit), first noted:

> This case brings into court the second law of thermodynamics, which C. P. Snow used as a paradigm of technical information well understood by all scientists and practically no persons of the culture of humanism and letters.

The court remanded the acrylonitrile decision to the FDA for further consideration, and Judge Leventhal (83) made the significant statement:

> Congress has granted to the Commissioner a limited, but important area of discretion. Although as a matter of theory the statutory net might sweep within the term "food additive" a single molecule of any substance that finds its way into food, the Commissioner is not required to determine that the component element of the definition has been satisfied by such an exiguous showing. The Commissioner has latitude under particular circumstances to find migration "insignificant" even giving full weight to the public health and welfare concerns that must inform his discretion.

Judge Leventhal's decision provides formal, legal justification for not insisting on absolute zero in the case of indirect additives, thus implicitly recognizing the impracticality of attempting to provide the impossible, absolute safety. This represents a significant first step toward the ultimate realization of a food safety policy based on regulation of additives and other food ingredients, according to the relative risk involved.

COLOR ADDITIVES

The addition of colors to foods has probably been criticized more than that of any other subset of additives. Critics argue that color additives serve only a cosmetic purpose, that consumers have been conditioned to expect an attractive appearance, and hence that the addition of colors to foods should be stopped. Others argue that "we taste with our eyes" and point to nature's colors such as the green of chlorophyll in lettuce and peas, the orange of carotene in carrots and corn, and the purple of anthocyanins in grapes and blueberries.

The use of colors in foods is receiving worldwide attention and, for example, has been carefully examined recently by the United Kingdom Food Additives and Contaminants Committee. Their report in part states:

> There is evidence to show that the use of colouring matter—or rather the omission of it—can have a significant effect on consumer demand. . . . Of course, consumer opinions and tastes can and do change. We think that such changes are, in general, best catered for by the law of supply and demand operating in a free market. In such a market it should be possible for both coloured and uncoloured foods to be produced and sold. We have no reason to doubt that if there is a big enough demand for both types of foods, then food manufacturers will seek to satisfy that demand (84).

Regulatory Requirements for Colors

Following the concepts embodied in the Food Additives Amendments, Congress in 1960 passed the Color Additives Amendments to the FFD&C Act. The 1960 amendments replaced the necessity of the FDA to prove a color unsafe in order to deny its use, with the requirement that the manufacturer demonstrate the safety of the color for its intended use. Also provided in the new law was the requirement that safe conditions of use be established by regulation and that all color additives be batch certified unless exempted by the secretary of HEW. (Certification entails testing by the FDA of a sample from each production batch against chemical specifications. Any batch not conforming to specifications is denied permission for use.) Colors on the market when the new law became effective were "provisionally listed"; that is, their continued use was permitted on an interim basis pending completion of analytical and/or toxicological tests required for their final regulation.

Color additives find uses not only in foods, but also in drugs, cosmetics, and medical devices. The 1960 amendments provided for separate listings of the uses for which colors could be safely employed. Thus, for example, a given color might be approved for external drug and cosmetic use, but not approved for food use or ingested drug use.

Color additives are officially designated as either "subject to certification" (synthetic colors that are required to be batch certified) or "exempt from certification" (mostly "natural" colors that may be used without batch certification).

Like food additives, color additives also have an elaborate definition in the FFD&C Act:

> 1. The term "color additive" means a material which—
>
> A. is a dye, pigment, or other substance made by a process of synthesis or similar artifice, or extracted, isolated, or otherwise derived, with or without intermediate or final change of identity, from a vegetable, animal, mineral, or other source, and
>
> B. when added or applied to a food, drug, or cosmetic, or to the human body or any part thereof, is capable (alone or through reaction with other substances) of imparting color thereto:
>
> except that such term does not include any material which the Secretary, by regulation, determines is used (or intended to be used) solely for a purpose or purposes other than coloring.
>
> 2. The term "color" includes black, white, and intermediate grays.
>
> 3. Nothing in subparagraph (1) of this paragraph shall be construed to apply to any pesticide chemical, soil or plant nutrient, or other agricultural chemical solely because of its effect in aiding, retarding, or otherwise affecting, directly or indirectly, the growth or other natural physiological processes of produce of the soil and thereby affecting its color, whether before or after harvest.

Certified Colors

Perhaps surprisingly, there are relatively few synthetic color additives approved for food use in the United States. There are only six synthetic food colors that are "permanently" approved for food use (see Table 6.6).

In the Federal Register of September 23, 1976 (85), the FDA addressed all the colors that were then provisionally listed for food, drug, or cosmetic uses. It was pointed out that three of the colors, FD&C Blue No. 1, FD&C Red No. 3, and FD&C Yellow No. 5—each approved for food and ingested drug use—were only provisionally listed for ingested cosmetic use. Chronic toxicity feeding studies were required for these three colors as a condition of continuing the provisional listing. Noting that the safety data supporting these and some of the other colors did not meet current scientific standards, the FDA advised that new chronic feeding studies, if not performed at that time, would likely be required in the future to support continued approval of food and drug uses. It was emphasized that existing data indicated no concern for the safety of these colors. However, as in other additive situations, the need to provide toxicology information meeting current standards was stated.

It is illustrative of color additive safety to briefly examine each of the six permanently listed synthetic food colors. Although FD&C Blue No. 1, as noted, is approved for general food use, it is only provisionally approved for cosmetic use until January 31, 1981. The FDA has re-

TABLE 6.6 Approved Synthetic Food Colors

Official and (Common) Name	Permitted Uses
FD&C Blue No. 1 (Brilliant Blue FCF)	General, including dietary supplements
Orange B	Casing of surfaces of frankfurters and sausage, up to 150 ppm
Citrus Red No. 2	Skins of oranges not intended for processing, up to 2 ppm in whole orange
FD&C Red No. 3 (Erythrosine)	General, including dietary supplements
FD&C Red No. 40 (Allura Red AC)	General, including dietary supplements
FD&C Yellow No. 5 (Tartrazine)	General, including dietary supplements

Source. Code of Federal Regulations, Title 21, U. S. Government Printing Office, Washington, D.C., 1979.

quired that chronic feeding studies (two generations in the rat and mouse) be satisfactorily completed prior to that date. The outcome of the chronic feeding studies will obviously provide an up-to-date assessment of the safety of FD&C Blue No. 1 in food as well as cosmetic uses.

Orange B provides an interesting example of a current food safety regulatory dilemma. In 1978 the FDA proposed to terminate the approval of Orange B because of trace amounts of a contaminant (β-naphthylamine) detected in some batches of the intermediate and final color (86). Again, dramatic advances in analytical chemistry had revealed the presence of a previously unsuspected trace contaminant. Although Orange B could likely be produced with less than 1 ppb of β-naphthylamine, it is not possible to guarantee absolute absence of the contaminant, and the only U.S. manufacturer has discontinued production. Orange B thus provides a precedent setting possibility for a wide variety of other trace contaminant situations.

Citrus Red No. 2 is approved only for use in coloring the skins of oranges not intended for processing. Therefore, it is not subject to the testing required for colors that have some provisionally approved use. However, along with the other colors, Citrus Red No. 2 is subject to safety evaluation in the FDA's cyclic review of food additives.

The colorant FD&C Red No. 3 is also provisionally listed for cosmetics, and chronic feeding studies in the mouse and rat as well as a multigeneration rat reproduction study have been required for final approval. Also under investigation is the bioavailability of the iodine fraction of the FD&C Red No. 3 molecule. Human clinical studies are being conducted to assess the contribution of FD&C Red No. 3 to the generally high intake of iodine, the bioavailability of which has not yet been determined.

As a result of a chronic mouse feeding study conducted to gain Canadian approval of FD&C Red No. 40, questions arose concerning the possible acceleration of time for lymphoma development. (Canada permits the use of FD&C Red. No. 2 but not FD&C Red No. 40. The converse is true in the United States.) A second and larger mouse study has been conducted and evaluated by an interagency government working group that apparently concluded that FD&C Red No. 40 produced neither an increased incidence of lymphomas nor an acceleration in the time to lymphoma development (87).

Likewise, FD&C Yellow No. 5 has provisional approval in cosmetic use, and chronic feeding studies in the mouse and rat have been required by the FDA. Although there have been no adverse chronic effects as yet associated with FD&C Yellow No. 5, this color has been connected with allergic responses in some individuals. The FDA has, therefore, issued regulations requiring that foods and drugs containing FD&C Yellow No. 5 be labeled to indicate its presence (88).

There are three other synthetic colors, shown in Table 6.7, which are only provisionally approved for food use until January 31, 1981 subject to the requirement of chronic feeding studies.

Noncertified Colors

In contrast to the certifiable, synthetic colors, little safety concern has heretofore been expressed about the colors exempt from certification. This is undoubtedly due in part to the popular belief that "natural is good." However, current regulatory scrutiny in both the United States and Europe is beginning to be focused on food and color additives derived from natural as well as synthetic sources.

TABLE 6.7 Provisionally Approved Synthetic Food Colors

Official and (Common) Name	Permitted Uses
FD&C Blue No. 2 (Indigotine)	General Food Use
FD&C Green No. 3 (Fast Green FCF)	General Food Use
FD&C Yellow No. 6 (Sunset Yellow FCF)	General Food Use

Source. Code of Federal Regulations, Title 21, U. S. Government Printing Office, Washington, D.C., 1979.

The FDA has recognized by regulation, in Part 73 of the Code of Federal Regulations, some 25 color additives exempt from certification:

Dried algae meal
Annatto extract
Dehydrated beets (beet powder)
β-Apo-8'-Carotenal
β-Carotene
Toasted, partially defatted, cooked cottonseed flour
Ferrous gluconate
Grape skin extract
Synthetic iron oxide
Fruit juice
Vegetable juice
Paprika
Canthaxanthin
Caramel
Carrot oil
Cochineal extract, carmine
Corn endosperm oil
Paprika oleoresin
Riboflavin
Saffron
Tagetes (Aztec marigold) meal and extract
Titanium dioxide
Turmeric
Turmeric oleoresin

These colors are exempt from certification in part because of the impracticality of attempting their precise chemical characterization. Both natural and synthetic sources are encompassed in the list. The actual colorants are obtained by chemical extraction, heat treatment, steeping, drying of natural materials, or, in some cases, synthetically by chemical reaction.

Cochineal extract is an interesting example. It is defined by FDA regulations as "the concentrated solution obtained after removing the alcohol from an aqueous–alcoholic extract of cochineal [*Dactylopius coccus costa* (*Coccus cactil*)]." *Dactylopius coccus* is a scale insect, found mainly in the Canary Islands, South America, and Mexico, which feeds on the opuntia cactus. These insects, especially the females, have a brilliantly red body fluid that chemically is mainly carminic acid.

Some of the colors that are exempt from certification in the United States have recently had their safety reviewed by the United Kingdom Food Additives and Contaminants Committee. Attention is also being focused on these and the synthetic colors by the EEC Scientific Committee for Food and by the FAO/WHO Joint Expert Committee on Food Additives. All colors currently approved in the United States will have their safety reviewed in the forthcoming FDA cyclic review program.

The British review of food colors, in a manner analagous to the GRAS review, classified several colors as acceptable but called for further information and for their subsequent review. Among the colors placed in this category were the caramels, not because of any safety concern, but rather because of their wide use and the apparently large number of varieties being produced.

Caramel does warrant some special mention since it is by far the most widely used food color. (It is also a GRAS flavoring substance listed in Section 182.1235 of the Code of Federal Regulations.) The FASEB/SCOGS affirmed caramel as GRAS in its 1973 evaluation report (89). Food and Drug Administration regulations identify caramel color (Section 73.85 of the Code of Federal Regulations) as "the dark-brown liquid or solid material resulting from the carefully controlled heat treatment of the following food grade carbohydrates: Dextrose, Invert sugar, Lactose, Malt syrup, Molasses, Starch hydrolysates and fractions thereof, Sucrose." These regulations also specify the food grade acids, alkalis, and salts that may be used to assist caramelization.

Caramel colors have been used in a variety of foods for at least the past 100 years with the different types arising to meet the particular color and compatibility requirements of the foods involved. They are known by common names such as *burnt sugar caramel* and *brewer's, baker's, beverage,* or *confectioner's caramel.* The International Technical Caramel Association (ITCA) formally characterizes commercial caramels in four classes: Caramel Color (Plain), also known as *spirit caramel;* Caramel Color (Caustic Sulfite); Caramel Color (Ammonia Process), also known as *beer caramel;* and Caramel Color (Sulfite Ammonia Process), also known as *soft drink caramel.* These classes differ primarily in the reactants (caramelization aids) and time-temperature profiles used.

Toxicity studies on the caramels conducted over the past 20 years in rodent and nonrodent species indicate the safety of these colors. Concern was expressed in 1977 by the Joint Expert Committee on Food Additives about the small but significant reduction in leukocytes observed in some studies of Ammonia Process Caramel (but not observed with other caramel types). These studies employed a type of commercial rat food (Spratt's) that, as subsequently shown by analysis, contained 2.3 ppm of pyridoxine compared to the generally recommended level of 7.5 ppm. Supplementation of the Spratt diet with 10 ppm of pyridoxine produced no hematological changes in rats fed 8% Ammonia Process Caramel (90).

In addition to research concerning the antipyridoxine factor in ammonia process caramel, extensive short-term toxicity testing is under way, and chronic feeding studies are contemplated. Research is also under way to develop chemical specifications for the individual classes and types of caramel color (91). Current FDA regulations permit the use of caramel colors generally in foods subject only to good manufacturing practices.

The concentrated attention that has been focused on color additives is illustrated not only by the foregoing, but also by the colors that have been specifically prohibited in recent years. Prohibited colors include:

Carbon Black	Orange No. 1
FD&C Red No. 1	FD&C Violet No. 1
FD&C Red No. 2	FD&C Yellow No. 2
FD&C Red No. 4	FD&C Yellow No. 3
FD&C Green No. 2	

Close scrutiny of food colors is continuing and, as noted, is becoming a worldwide phonomenon. In the United States, the extensive testing devoted to the provisional colors as well as the safety evaluation of all colors in the FDA cyclic review program will for the first time permit an assessment by current standards of all food colors.

DRUGS IN FOOD-PRODUCING ANIMALS

One of the most unique and complex problem areas with respect to food additives is that presented by animal drugs (92). This is the case since, in effect, there is a "biological filter," that is, the animal, between a substance added to animal feed or directly administered to the animal, and the residues of that substance in the food products (meat, milk, and eggs). Depending on the compound, the animal species involved, the route, and levels and times of administration of the compound, a variety of residues may occur in these foods. The residues can include not only the parent compound, but also several of its metabolites produced by the "biological filter."

In evaluating an animal drug, the basic criteria are that it be effective for its intended use in the animal, that it be safe with respect to the animal, and that the residues in the food end-products be safe for human consumption (93). Food safety assessment for animal drugs requires answers to the following questions:

1. What residues are there in which edible portions of the animal? At what levels do they occur?

2. What is the significance of these residues to human health (i.e., toxicity and extent of human exposure)?

3. What conditions of use must be imposed to assure the safety of residues (i.e., in which species, what dosage, and what time periods)?

As a condition of approval for a new animal drug, it may be necessary to require a withdrawal period, that is, a prescribed period of time prior to slaughter during which the drug may not be used, so that residues of the parent drug may deplete to safe levels. In the case of dairy cattle it may be necessary to specify a period during which milk may not be marketed. It may also be necessary to prescribe an enforcement program to assure that permitted conditions of use are followed in practice. Such a situation in turn requires that a practicable method of analysis be developed to monitor residues in food.

Nature and Uses of Animal Drugs

A wide spectrum of products find existing and proposed uses as animal drugs. Some compounds have quite limited uses, such as an anesthetic for surgery or a drug used to treat a specific disease condition in a single species. Certain drugs may be used only early in an animal's life to treat or prevent a condition unique to the young. Other compounds have very broad usage, especially with today's concentration and mechanization of beef, pork, and poultry production. These latter compounds include growth promotants and broad-spectrum antibiotics used for prophylaxis as well as therapy (94). Those antibiotics used in food-producing animals that are also used by humans (e.g., penicillin and tetracycline) present a special problem. Controversy has existed for several years over the possibility that the pool of bacteria (e.g., salmonella) resistant to antibiotics is being increased by the subtherapeutic use of antibiotics in food-producing animals, thus diminishing the effectiveness of antibiotic treatment of human diseases.

Because of the variety of compounds and uses involved and the fact that animal drugs, unlike other food additives, are used because they are pharmacologically active, food safety assessment is complex indeed. The residues to which humans are exposed may include not only those of the parent compound, but also those of a variety of metabolites. Thus practical consideration must be given to the chemical identification of the metabolites and the levels at which they may be present and their toxicological significance. The toxicity of the parent compound and metabolite residues can cover the gamut from no significant effects to carcinogenicity.

Depletion of the residues of some drugs may be rapid with little or no residue storage in edible tissue. Others may result in persistent residues requiring a long withdrawal period or disapproval, depending not only on toxicity, but also on whether a long withdrawal period is "reasonably certain to be followed in practice." For example, for a drug used with broiler chickens, the maximum practical withdrawal time is about 5 days, but with laying hens, no egg discard time is considered practical.

Finally, it should also be noted that residue (parent compound and metabolites) sites can also vary, with muscle tissue the predominant site in some cases and less frequently consumed tissue, such as liver and kidney, the predominant site in other cases. Any tolerance must be adequate to assure the safety of all residues resulting from the drug use and thus must be based on the individual residues and their rates of depletion, the residue sites, and practical methods for enforcing the permitted residues and/or withdrawal times.

Drugs that are chemically identical to substances occurring naturally in animal tissue, that metabolize to residues identical to endogenous substances or that otherwise alter the naturally occurring levels of endogenous substances, present special problems of safety assessment. Certain endogenous compounds, such as hormones, although occurring naturally, may be of toxicological concern when their use substantially increases the levels normally found in food-producing animals. In such cases, background endogenous levels must first be assessed taking into account such variables as breed, sex, age, state of estrus, and geographic location.

As is the case with other food additives, a conservative position is normally taken with respect to safety assessment of animal drugs. These compounds are tested in animal bioassays at levels much higher than expected human exposure levels, and various safety factors are applied with their magnitude depending on the toxicity and the extent of testing involved. Also, like other food additives, there is the problem of reevaluating the safety of animal drugs that were in use when new requirements were added to the FDD&C Act.

Legal Provisions for Animal Drugs

The FFD&C Act defines animal drugs in the following way:

> (w) The term "new animal drug" means any drug intended for use for animals other than man, including any drug intended for use in animal feed but not including such animal feed—
>
> 1. the composition of which is such that such drug is not generally recognized, among experts qualified by scientific training and experience to evaluate the safety and effectiveness of animal drugs, as safe and effective for use under the conditions prescribed, recommended, or suggested in the labeling thereof; except that such a drug not so recognized shall not be deemed to be a "new animal drug" if at any time prior to June 25, 1938, it was subject to the Food and Drug Act of June 30, 1906, as amended, and if at such time its labeling contained the same representations concerning the conditions of its use; or
>
> 2. the composition of which is such that such drug, as a result of investigations to determine its safety and effectiveness for use under such conditions, has become so recognized but which has not, otherwise than in such investigations, been used to a material extent for a material time under such conditions; or
>
> 3. which drug is composed wholly or partly of any kind of penicillin, streptomycin, chlortetracycline, chloramphenicol, or bacitracin, or any

> derivative thereof; except when there is in effect a published order of the Secretary declaring such drug not to be a new animal drug on the grounds that (A) the requirement of certification of batches of such drug, as provided for in section 512 (n), is not necessary to insure that the objectives specified in paragraph (3) thereof are achieved and (B) that neither subparagraph (1) nor (2) of this paragraph (w) applies to such drug.

It should be noted that the legal definition of animal drugs includes the same exemption for GRAS substances as does the legal definition of food additives. Not surprisingly, substances GRAS for human food are generally also GRAS for animals. Specific listings of GRAS substances for animals are shown in Section 582 of the Code of Federal Regulations (28). Food additives for animal uses are listed in Section 573. (Of the 47 substances listed, 16 are unique animal uses.)

There is one important regulatory distinction between animal drugs and other food additives (i.e., those regulated under Section 409). A food additive regulation allows the manufacture and use of the additive by anyone if the conditions of the regulations are met. For animal drugs, however, there is the equivalent of a license granted only to the holder of the approval. Only the approved sponsor(s) may manufacture and distribute the drugs and the information supporting effectiveness and safety is proprietary.

The FFD&C Act also defines animal feed:

> (x) The term "animal feed," as used in paragraph (w) of this section, in section 512, and in provisions of this Act referring to such paragraph or section, means an article which is intended for use for food for animals other than man and which is intended for use as a substantial source of nutrients in the diet of the animal, and is not limited to a mixture intended to be the sole ration of the animals.

The current law with respect to animal drugs is contained in Section 512 of the Act and is based on amendments passed in 1968. These amendments consolidated the prior requirements for approval of animal drugs according to both the human drug and food additive sections into a single set of requirements. Section 512 describes the procedures entailed for gaining approval of an animal drug through the submission of a new animal drug application (NADA).

One of the most interesting aspects of Section 512 is that, like the food additive provisions (Section 409), it contains a "Delaney Clause." This clause prohibits approval of a new animal drug if "such drug induces cancer when ingested by man or animal." (Food and color additives for animal use are also covered by this clause.) The interesting part of Section 512 is the added proviso: "[unless] no residue of such drug will be found (by methods of examination prescribed or approved by the Secretary by regulations . . .), in any edible portion of such animals after slaughter or in any food yielded by or derived from the living animals." That exemption is often referred to as the *DES proviso*. It was incorporated into the Act as part of the 1962 drug amendments and was retained in the 1968 animal drug amendments. The addition of the no residue exemption was prompted in large part by the existence of prior sanctions which permitted production of DES by some manufacturers (those obtaining approval before 1958) but denying it to others (94). Diethylstilbestrol, a synthetic hormone that has been associated with cancer in test animals, is a widely used growth promotant in beef cattle. It has also been used as a human drug for protection against abortion and premature delivery and as a postcoital contraceptive (78). Although the FDA supported the DES proviso in 1962, it subsequently moved to ban DES when residues (in parts per billion) were detected in edible meat.

Sensitivity Requirements for Carcinogenic Animal Drug Residues

The DES proviso represents the logical concept that it is not the original animal drug, but rather the residues of that drug and its metabolites in food that are of human safety significance. (The indirect food additive situation is somewhat analagous.) Despite the logic involved, the regulatory interpretation of this clause has entailed several years of difficulty. In 1973, the FDA issued a proposed regulation for the operational definition of the no-residue requirement (95). That proposal was subsequently withdrawn, and a new proposal for a "sensitivity of method" regulation was issued in 1979 (96). The basic concept of this regulation is that the analytical method must be at least sensitive enough to detect residue levels corresponding to a specified degree of lifetime risk of cancer to humans consuming the affected food. (The FDA has chosen one in one million as the specified risk.) Not only is this concept of importance in assuring the safety of food derived from animals, but it also explicitly recognizes the fact that since absolute food safety is impossible the regulation of food safety should be based on the objective of maintaining the risks involved at or below negligible levels. Hence, this concept is of fundamental importance relative to practical regulatory control of all food additives.

Future Regulation of Animal Drugs

The unique food safety problems involved in the use of animal drugs have focused considerable attention on their regulation. This attention is illustrated, for example, by the 1979 report (97) of the congressional Office of Technology Assessment. The benefits of animal drugs in terms of disease prevention and treatment, increased weight gain, and improved food efficiency have led to their widespread use. The Office of Technology Assessment reported that nearly 100% of poultry, 90%

of swine and veal calves, and 60% of cattle receive antibacterial feed supplementation and that 70% of our beef comes from cattle that have received weight-promoting feed supplementation (97). Whereas the benefits of animal drugs are fairly clear, there is considerable controversy over the risks to humans posed by their use.

Concern about risk involves primarily the contribution of antibacterials used in animals to what is believed to be a growing pool of drug-resistant bacteria and the residues of drugs and metabolites that may pose a carcinogenic risk to consumers. In this regard the FDA has taken the position that low levels of penicillin in animal feeds should be banned, similar uses of tetracyclines should be restricted to situations where replacements are not available, and all uses of the nitrofurans and DES should be banned. The FDA continues to support the "sensitivity of method" approach with its associated risk limit of one in a million but apparently believes that neither DES nor the nitrofurans can meet the proposed criteria.

The Office of Technology Assessment identified five options for Congress: (*a*) allow the FDA to regulate drug use, subject to congressional oversight, (*b*) require the FDA to make economic as well as scientific assessments of benefits and risk, (*c*) modify the special approach required for cancer-causing drugs, (*d*) require the FDA to decrease the use of antibacterials in humans and livestock feed, and/or (*e*) require that in the future only those drugs that have proven more effective than those now in use be approved. What course Congress will follow is not clear at this point, but in any event consideration will have to be given to the fact that, like all other aspects of food safety, the use of animal drugs poses some risks. Future advances in science will help elucidate those risks but can by no means completely eliminate them. In the absence of congressional direction to the contrary, it appears that the concept embodied in the "sensitivity of method" approach—that of minimizing risk—will be pursued. That concept, although difficult to implement, should permit the eventual regulation of animal drugs on a rational basis and perhaps point the way toward more rational regulation of other components of our food supply.

CYCLIC REVIEW OF FOOD ADDITIVES

Rapidly advancing science and changing consumption patterns as well as consumer concern about food additives point to the advisability of periodic review of this large class of substances. (The same sort of attention is, of course, warranted for other aspects of food safety.) The FDA committed to such periodic review beginning in 1977 (98). The cyclic review program covers the original GRAS review activity and a review of safety information for direct additives, indirect additives, and color additives. All of the more than 2000 direct food and color additives (about 1700 are flavors, many of which are chemically closely related) are included, as are those color additives approved only for drug and/or cosmetic use and on the order of 10,000 indirect additives.

It should be noted that the cyclic review program was not instituted because of any suspicion of significant hazards from food additives. Rather, it reflects primarily the changing state of the art in chemistry and toxicology. Therefore, as in the GRAS review, the cyclic review involves evaluation of existing safety information on each substance. Those compounds that meet minimum requirements for test information by the standards of the time can then be evaluated in depth, and in priority order, relative to their uses and the resulting consumer exposure. For the additives whose safety information does not meet current testing standards, continued approval is contingent on the provision by the sponsor of required data.

The cyclic review of indirect additives is similar to that of the direct additives except for one important difference. In this case the migration characteristics must also be assessed (21, 22, 99). Of primary importance here are the components of food packaging materials. Safety evaluation for these components entails the complex investigation of migration characteristics as a function of package composition, intended food uses, and toxicological significance.

A companion effort to the food additives cyclic review is the cyclic review of drugs for food-producing animals (100). Currently marketed animal drugs are reviewed generically as a group, and the various uses are separately evaluated within each group. There are approximately 200 generic animal drugs and over 700 separate uses. As with the other safety review programs, existing safety information is evaluated relative to current standards. Individual drugs and uses are assessed in priority order depending on toxicity, residues, and the magnitude of use. An integral part of the animal drug cyclic review is evaluation of the sensitivity and practicability of the analytical detection methodology used for residue monitoring. Thus the cyclic review of an animal drug can result in the requirement for detection method refinement as well as for additional toxicity testing.

The animal drug cyclic review has another unique feature, the aspect of metabolism. Most of the currently approved drugs were originally evaluated without complete knowledge of the total drug residue exposure (parent compound and metabolites). Experience has shown that frequently the parent compound is only a small fraction of the total residue. For example, the total residue for some of the nitrofurans is several orders of magnitude greater than the residue of parent compounds. Consideration of the total residue pool is thus a critical part of the animal drug cyclic review.

In addition to the cyclic review programs, the FDA in 1977 initiated inspection of the more than 100 laboratories involved in the testing of food additives (101). Subsequently, good laboratory practices regulations (102) were issued specifying operating procedures for the conduct of acceptable animal tests.

The food additive cyclic review programs entail a systematic review of additives in priority order. Further new safety testing must be conducted in accordance with the good laboratory practices regulations. In this way safety information will continually be updated to meet the changing standards of chemistry and toxicology and provide the best possible assurance to consumers of food additive safety.

REFERENCES

1. R. L. Hall, "Food Ingredients and Additives," in F. Clydesdale, Ed., *Food Science and Nutrition: Current Issues and Answers*, Prentice Hall, Englewood Cliffs, N.J., 1979, pp. 116–150.
2. *Federal Food, Drug, and Cosmetic Act, As Amended*, U.S. Government Printing Office, Washington, D.C., 1979.
3. R. Tannahill, *Food In History*, Stein and Day, New York, 1973.
4. E. F. Binkerd and O. E. Kolari, *Food Cosmet. Toxicol.*, 13, 655 (1975).
5. S. A. Goldblith, "Food and History: An Anecdotal Development of Man's Food and the Role of Gastronomy," *Nestle Research News 1976/77*, 5 (1977).
6. E. M. Whelan and F. J. Stare, *Panic in the Pantry*, Atheneum, New York, 1973.
7. W. C. Frazier, *Food Microbiology*, 2nd ed., McGraw-Hill, New York, 1967.
8. O. Anderson, *The Health of a Nation: Harvey Wiley and the Fight for Pure Food*, Univ. Chicago Press, 1958.
9. J. H. Young, "The Agile Role of Food: Some Historical Reflections," in *Nutrition and Drug Interrelations*, Academic, New York, 1978, pp. 1–18.
10. R. L. Hall, *Chem. Technol.*, 3 (7), 412 (1973).
11. Food and Drug Administration, *Discussion of Priorities—Fiscal Year 1981 Planning Process*, Washington, D.C., May, 1979.
12. "Wodicka Rates Food Additive Hazard as Low," *Food Chem. News*, 12 (11), 49 (1971).
13. V. O. Wodicka, "Food Safety in 1973," *Proceedings of the Flavor and Extract Manufacturers Association*, Washington, D.C., 1973.
14. H. R. Roberts, "America's Food Supply—The Safest in the World," *The Food Protection Paradox Proceedings*, University of Minnesota, St. Paul, 1976.
15. R. L. Hall, *J. Inst. Can. Sci. Technol.*, 6 (1), 17 (1973).
16. National Science Foundation, *Chemicals & Health*, U.S. Government Printing Office, Washington, D.C., 1966.
17. P. Lehmann, *FDA Consumer*, 13 (3), 10 (1979).

18. Food and Drug Administration, *Fed. Reg.*, 44, 51233 (1979).
19. T. Larkin, *FDA Consumer*, 10 (5), 4 (1976).
20. R. L. Hall, *Nutr. Today*, 8 (4), 20 (1973).
21. T. Fazio, *Food Technol.*, 33 (4), 61 (1979).
22. S. G. Gilbert, *Food Technol.*, 33 (4), 63 (1979).
23. J. W. White, Jr., *J. Agr. Food Chem.*, 23 (5), 886 (1975).
24. J. W. White, Jr., *J. Agr. Food Chem.*, 24 (2), 202 (1976).
25. National Academy of Sciences, *Toxicants Occurring Naturally in Foods*, 2nd ed., Washington, D.C., 1973.
26. National Academy of Sciences, *Food Chemicals Codex*, Washington, D.C., 1966.
27. National Academy of Sciences, *Food Chemicals Codex*, 2nd ed., Washington, D.C., 1972.
28. *Code of Federal Regulations, Title 21: Food and Drugs*, U.S. Government Printing Office, Washington, D.C., 1979.
29. Committee on Interstate and Foreign Commerce, *Food Additives* (hearings), U.S. Government Printing Office, Washington, D.C., 1958.
30. B. L. Oser and R. A. Ford, *Food Technol.*, 33 (7), 65 (1979)
31. G. W. Irving, *Nutr. Rev.*, 36 (12), 351 (1978).
32. K. D. Fisher, *Federation of American Societies for Experimental Biology*, personal communication to H. R. Roberts, July 26, 1979.
33. National Cancer Institute, *Bioassay of Butylated Hydroxytoluene (BHT) for Possible Carcinogenicity*, Carcinogenesis Technical Report Series, No. 150, Washington, D.C., 1979.
34. Food and Drug Administration, "Study of Dietary Nitrite in the Rat," MIT Report on Contract No. 223-74-2181, Washington, D.C., 1978.
35. H. R. Roberts, *Food Drug Cosmet. Law J.*, 34 (3), 153 (1979).
36. National Academy of Sciences, *Sweeteners: Issues and Uncertainties*, Washington, D.C., 1975.
37. Office of Technology Assessment, *Cancer Testing Technology and Saccharin*, U.S. Congress, Washington, D.C., 1977.
38. National Academy of Sciences, *The Safety of Artificial Sweeteners for Use in Foods*, Washington, D.C., 1955.
39. National Academy of Sciences, *Non-Nutritive Sweeteners*, Interim Report to the FDA, Washington, D.C., 1968.
40. National Academy of Sciences, *Safety of Saccharin for Use in Foods*, Washington, D.C., 1970.
41. National Academy of Sciences, *Safety of Saccharin and Sodium Saccharin in the Human Diet*, Washington, D.C., 1974.
42. National Academy of Sciences, *Saccharin: Technical Assessment of Risks and Benefits*, Washington, D.C., 1978.
43. National Academy of Sciences, *Food Safety Policy: Scientific and Societal Considerations*, Washington, D.C., 1979.

44. M. F. Cranmer, *Final Report on Saccharin,* Food and Drug Administration, Washington, D.C., 1978.

45. American Council on Science and Health, *Saccharin,* New York, 1979.

46. Food and Drug Administration, *Fed. Reg.*, 42, 19996 (1977).

47. D. L. Arnold, C. A. Moodie, B. Stavric, D. R. Stoltz, H. C. Grice, and I. C. Munro, *Science,* 197, 320 (1977).

48. T. W. Sweatman and A. G. Renwick, *Science,* 205, 1019 (1979).

49. G. P. Schoenig, *Procs. Saccharin Working Group of the Toxicology Forum,* Washington, D.C., 1979.

50. F. W. Carlborg, *Proceedings of the Saccharin Working Group of the Toxicology Forum,* Washington, D.C., 1979.

51. Food and Drug Administration, *Preliminary Findings and Recommendations of the Interagency Saccharin Working Group,* Washington, D.C., 1977.

52. *Progress Report to the Food and Drug Administration from the National Cancer Institute Concerning the National Bladder Cancer Study,* Bethesda, Maryland, 1979.

53. A. S. Morrison and J. E. Buring, *New Engl. J. Med.*, 302, 537 (1980).

54. E. L. Wynder and S. D. Stellman, *Science,* 207, 1214 (1980).

55. Food and Drug Administration, *Fed. Reg.*, 39, 27317 (1974).

56. Health and Welfare Canada, *Proposal on Aspartame,* Information Letter No. 564, Ottawa, 1979.

57. Federation of American Societies for Experimental Biology, *Dietary Sugars in Health and Disease—II. Xylitol,* Bethesda, Maryland, 1978.

58. B. Hunter, C. Graham, R. Heywood, D. E. Prentice, F. J. C. Roe, and D. N. Noakes, *Tumorigenicity and Carcinogenicity Study with Xylitol in Long Term Dietary Administration to Mice,* Huntingdon Research Centre, Huntingdon, England, 1978.

59. R. G. Allison, *Dietary Sugars in Health and Disease—IV. Mannitol,* Federation of American Societies for Experimental Biology, Bethesda, Maryland, 1979.

60. L. E. Gongwer, *Mannitol II: Lifetime Feeding Study of A-132-01320 in Three Strains of Female Rats,* ICI Americas Inc., Wilmington, Del., 1978.

61. H. E. Ginn, "The Renal Metabolism and Uses of Mannitol," in H. L. Sipple and K. W. McNutt, Eds., *Sugars in Nutrition,* Academic, New York, 1974, pp. 607–612.

62. O. Touster, "The Metabolism of Polyols," in H. L. Sipple and K. W. McNutt, Eds., *Sugars in Nutrition,* Academic, New York, 1974, pp. 229–239.

63. Federation of American Societies for Experimental Biology, *Evaluation of the Health Aspects of Sorbitol as a Food Ingredient,* Bethesda, Maryland, 1972.

64. H. R. Moskowitz, "The Psychology of Sweetness," in H. L. Sipple and K. W. McNutt, Eds., *Sugars in Nutrition*, Academic, New York, 1974, pp. 38–64.

65. H. Förster, "Comparative Metabolism of Xylitol, Sorbitol, and Fructose," in H. L. Sipple and K. W. McNutt, Eds., *Sugars in Nutrition*, Academic, New York, 1974, pp. 259–280.

66. D. W. Thomas, J. B. Edwards and R. G. Edwards, "Toxicity of Parenteral Xylitol," in H. L. Sipple and K. W. McNutt, Eds., *Sugars in Nutrition*, Academic, New York, 1974, pp. 567–590.

67. N. N. Potter, *Food Science*, 3rd ed., AVI, Westport, Conn., 1978.

68. K. K. Kimura and C. J. Carr, *Dietary Sugars in Health and Disease, I. Fructose*, Federation of American Societies for Experimental Biology, Bethesda, Maryland, 1976.

69. Anonymous, "The Fabulous 14-Day Fructose Diet," *Family Circle*, February 20, 1979.

70. R. H. Herman, "Hydrolysis and Absorption of Carbohydrates, and Adaptive Responses of the Jejunum," in H. L. Sipple and K. W. McNutt, Eds., *Sugars in Nutrition*, Academic, New York, 1974, pp. 146–172.

71. H. R. Roberts, "Sweeteners for the Food Industry," presented at 6th Annual ABC Technical Seminar, Gainesville, Florida, 1980.

72. Anonymous, *Nutr. Rev.*, 37 (4), 124 (1979).

73. Federation of American Societies for Experimental Biology, *Evaluation of the Health Aspects of Caffeine as a Food Ingredient*, Bethesda, Maryland, 1978.

74. S. Garattini, "Selected Comparative Data on Metabolic and Kinetic Differences Between Human and Other Species," unpublished report to the International Life Sciences Institute, Washington, D.C., 1980.

75. "Retortable Pouch Petitions Held Up by New NCI Report," *Food Chem. News*, 21 (23), 37 (1979).

76. W. L. Pines, *FDA Consumer*, 9 (10), 4 (1976).

77. C. Maltoni and G. Lefemine, *Environ. Res.*, 7, 387 (1974).

78. A. L. Frank and M. D. Carlin, *Cancer*, Matthew Bender, New York, 1978.

79. Food and Drug Administration, *Fed. Reg.*, 41, 23940 (1976).

80. Food and Drug Administration, *Fed. Reg.*, 42, 13546 (1977).

81. Food and Drug Administration, *Fed. Reg.*, 42, 13562 (1977).

82. Manufacturing Chemists Association, *Progress Report to FDA on Two Year Study of Acrylonitrile in Drinking Water of Rats*, Washington, D.C., 1977.

83. Judge Leventhal, Opinion for the U.S. Court of Appeals for the District of Columbia Circuit, No. 77-2023 and 77-2024, November 6, 1979.

84. Ministry of Agriculture, Fisheries and Food, *Food Additives and Contaminants Committee Interim Report on the Review of the Colouring Matter in Food Regulations*, Her Majesty's Stationery Office, London, 1979.

85. Food and Drug Administration, *Fed. Reg.*, 41, 41852 (1976).
86. Food and Drug Administration, *Fed. Reg.*, 43, 45611 (1978).
87. "Statisticians Discount Acceleration Effects of Red 40," *Food Chem. News*, 21 (14), 3 (1979).
88. Food and Drug Administration, *Fed. Reg.*, 44, 37212 (1979).
89. Federation of American Societies for Experimental Biology, *Evaluation of the Health Aspects of Caramel as a Food Ingredient*, Bethesda, Maryland, 1973.
90. E. J. Sinkeldam, *Influence of Pyridoxine Supplementation on Leucocyte Counts of Rats Fed Plain Ammonia Caramel in Civo Diet*, report to International Technical Caramel Association, Washington, D.C., 1978.
91. International Technical Caramel Association, *Caramel Color Update*, Washington, D.C., 1979.
92. M. K. Cordle, "When is a Residue a Residue?," presented at the Relay Toxicity and Residue Bioavailability Symposium, Paris, 1978.
93. M. K. Perez, *J. Toxicol. Environ. Health*, 3, 837 (1977).
94. Committee on Interstate and Foreign Commerce, *A Brief Legislative History of the Food, Drug, and Cosmetic Act*, U.S. Government Printing Office, Washington, D.C., 1974.
95. Food and Drug Administration, *Fed. Reg.*, 38, 19226 (1973).
96. Food and Drug Administration, *Fed. Reg.*, 44, 17070 (1979).
97. Office of Technology Assessment, *Drugs in Livestock Feed*, Volume I: Technical Report, U.S. Government Printing Office, 1979.
98. H. Hopkins, *FDA Consumer*, 11 (5), 8 (1977).
99. R. B. Davis, *Food Technol.*, 33 (4), 55 (1979).
100. M. K. Perez, "Drug Residue Safety," paper presented at the First Symposium on Veterinary Pharmacology and Therapeutics, Baton Rouge, La., 1978.
101. J. M. Taylor, R. K. Biskup, and C. O. Schulz, *FDA By-Lines*, No. 4 (1979).
102. Food and Drug Administration, *Fed. Reg.*, 43, 59986 (1978).

CHAPTER 7

Food Safety and Toxicology

J. DOULL

Previous chapters in this book have examined five major areas that impact on food safety. These areas were assigned a relative ranking for hazards presented to humans on the basis of what is known and unknown (mostly the latter) about the risks involved in each case. As was noted, there is relatively little information derived from actual evidence of hazards to humans. Existing food safety data consist in part of isolated observations of acute effects arising from accidental exposures, some clinical information (e.g., therapeutic treatment with vitamins), and a limited number of epidemiology studies. For the most part, however, the inventory of toxicity data consists of results from animal studies. (Recently an increased amount of data from genetics toxicology testing is also beginning to become available.)

Because of the heavy dependence of our food safety knowledge on the results of animal toxicology, it is of value to examine this area to further assist in putting all aspects of food safety in proper perspective.

In the last 20 or even in the last 10 years, toxicology has advanced tremendously in importance and in application. Nonetheless, the conduct of animal toxicity tests and their evaluation is in a state of flux. Toxicology as a scientific discipline is under stress not only from the demands it is receiving for manpower and facilities, but also from external and internal criticism which is impacting on its credibility.

Growing concern and confusion about the safety of food additives and chemicals in our environment, together with what has been described as a national cancer phobia, have led to numerous reviews of safety for existing substances, testing of old and new substances, and the formulation of new controls on testing. Examples of the former include the FDA review of GRAS substances, the FDA cyclic review of direct and indirect food additives (see Chapter 6), the National Cancer

Institute's Carcinogenesis Testing Program and the National Toxicology Program. Examples of the latter include the many attempts at guidelines prescribing how to conduct and evaluate animal toxicity tests such as the efforts of the Interagency Regulatory Liaison Group (The Consumer Product Safety Commission, the EPA, the FDA, and the Occupational Safety and Health Administration). The Interagency Regulatory Liaison Group (IRLG) has published draft guidelines for several acute toxicity tests, including eye irritation; oral toxicity in rodents; dermal toxicity; inhalation in rats; and teratogenicity testing in rats, mice, and rabbits (1). Guidelines for chronic testing are also being developed by the IRLG. In a similar fashion, the United Kingdom has developed draft guidelines for the testing of chemicals for mutagenicity as well as carcinogenicity (2, 3), showing the international nature of such trends. The FDA has also promulgated Good Laboratory Practices (GLP) regulations (4) prescribing the conduct, monitoring, and reporting of animal studies.

The movement toward standardized testing is viewed by some with reservation, however. For example, the Select Committee on GRAS Substances (SCOGS) has cautioned against the development of rigidly standardized protocols for safety evaluation, noting that such an approach fails to recognize that all chemicals do not act in the same fashion. What is needed according to the SCOGS is a set of principles and guidelines that permit tailoring the tests to the particular nature and intended uses of the substance in question (5).

The current emphasis on toxicology has, in turn, led to a recognition of an insufficient number of toxicologists and testing facilities to do all the review, testing, and evaluation that might appear to be desirable. Efforts to define proper priority for attention such as that proposed by the Food Safety Council (6) have been a logical consequence of this recognition.

With respect to the training of new toxicologists and those in related disciplines, and with respect to increasing available testing facilities, the problem can be alleviated, at least in the long term, with proper attention and financial resources. However, a perhaps more worrisome concern is that of the credibility problem.

The public appears to have growing skepticism about the validity of animal tests that seem to produce a new carcinogen every week. This is especially the case when our diet foods and beverages; bacon, ham and sausage; and even our beer and scotch whiskey come under attack. The proposed banning of saccharin led to such a public rebellion that Congress interceded to delay any regulatory action. The anticipation of similar regulatory action in the case of nitrite brought pressure on the FDA to reevaluate the latest animal studies. At the same time there appears to be some concern that, if long-used substances are only now being associated with harm, there must be many other harmful substances that have not yet been identified. Finally, there is the lurking fear that if careful scientific scrutiny is not widely applied, there could be a thalidomide-type disaster.

In general, the response of the regulatory agencies has been toward stricter controls. As noted, for example, the FDA has put into effect its GLP regulations. Individual federal agencies and the IRLG are developing standardized testing guidelines and protocols for the whole range of toxicological tests. There are, however, also some signs of a tendency toward a more flexible regulatory policy. For example, the IRLG has issued a report addressing the question of risk assessment procedures (7). The natural consequence of stricter control has been the charge of bureaucratic red tape by some; and others in industry, the academia and Congress, are questioning what costs will be involved and what benefits will be derived from such trends.

SCIENCE AND PUBLIC POLICY

Although the scientific questions are difficult enough, the controversy relative to food safety also involves substantial questions of public policy. Even if the toxicologists could precisely determine the risk involved with a given food component, there would still remain the public policy question of how to regulate that component. Hutt (8), for example, has pointed out, in discussing the Delaney Clause, that the food safety controversy is in large part due to the failure to distinguish between the questions involving scientific determinations and those in which a public policy determination is required. By the same token, public policy and scientific determination each affect the other. The debate focusing on the Delaney Clause is a case in point. If a public policy determination is going to be made that the Delaney Clause should be retained as is, extended to effects other than cancer, revoked entirely, or modified in some way, that determination should be based on the best scientific information available.

The passage by Congress in 1958 of the Food Additives Amendment, including the Delaney Clause, to the FFD&C Act was an expression of both the public policy and the scientific knowledge of that time. Although there was debate about inclusion of the Delaney Clause, there was probably a consensus in its favor in 1958. Since then, however, its opponents have increased, as has the level of debate. In 1962 a modified Delaney Clause was incorporated into the FFD&C Act to permit the use of a carcinogenic drug or feed additive in food-producing animals, provided that no residue of such substance would be detectable in edible portions of the food animal. Perhaps this modification was a conscious reflection of a public policy changing to reflect scientific practicality, or perhaps it was only a somewhat clearer statement of the original intent. In any event, the regulatory implementation of a definition of detectability is still to be accomplished, and the Delaney Clause has remained unchanged.

The advances of science since 1958 have had a significant impact on the debate over public policy for food safety. This is especially true relative to analytical methodology. The IFT Expert Panel on Food Safety and Nutrition has noted that "At this point the ability of scientists to detect minute quantities of chemicals has outstripped their ability to interpret the findings" (9). Using DES as an example, the panel cites the current sensitivity of detection as 2 ppb and in some cases in parts per trillion—an increase in sensitivity about fiftyfold or more since 1958. A maximum of 2 ppb in liver compares to 4000 ppb DES equivalent in wheat germ, 1500 ppb in wheat bran, 2200 ppb in soybean oil, and 1800 ppb in peanut oil. On the order of 2500 to 25,000 times as much estrogenic activity occurs naturally in a 120-pound woman as would result from the average daily ingestion of liver containing 2 ppb of DES (9).

In addition to advances in analytical chemistry, toxicology testing has expanded with two-generation *in utero* exposure at maximum tolerated doses becoming the norm in chronic testing and more detailed histopathology becoming customary. Thus scientific knowledge of adverse effects is increasing, posing additional questions in the public policy arena.

It is certainly in order to separate, to the extent possible, scientific determination and public policy, but if food safety is to be better understood, more than that separation is required. We need to recognize that toxicity tests do not always provide answers to the public policy questions that are asked. That point can be explored, and at the same time the role of toxicology in food safety can be put in perspective by exploring three areas: the yes–no toxicology question, toxicity versus safety testing, and animal-to-human extrapolation.

YES–NO TOXICOLOGY QUESTIONS

There is a natural tendency on the part of the public and a growing tendency on the part of regulatory officials to demand yes or no answers to questions about the safety of food constituents—is it safe or is it not? As the reader has seen previously in this book, the answer to the question "Is substance X toxic?" must be "yes," since all substances have some toxic effects under some conditions of exposure (even the essential nutrients discussed in Chapter 3). As Paracelsus, a Swiss alchemist and physician (and recently the patron saint of toxicology), noted nearly 500 years ago, "All things are poisons, for there is nothing without poisonous qualities. It is only the dose which makes a thing a poison." (Paracelsus presumably spoke from first-hand experience since he is credited with making mercury, lead, and arsenic a part of the pharmacopoeia.) Thus no toxicological testing is necessary to answer such a question; the answer is "yes" to begin with.

The Delaney Clause (and indeed the basic safety requirements of the FFD&C Act with respect to food additives) asks the yes–no question with respect to carcinogenesis, "Is the substance a carcinogen (in humans or animals) or is it not?" No consideration is given to what is meant by the term "carcinogen," to the type of tumor involved, or to the relative potency of the compound. Neither is consideration given to the dose levels entailed, the frequency, and the route of exposure or other factors routinely involved in toxicological evaluation. All these excluded facets are of fundamental importance to toxicological evaluation. For example, Kolbye (10) has explained the wide variety of situations encompassed in the general term "carcinogen" and the need to make scientific and regulatory distinctions. Numerous other difficulties with the yes–no approach such as a compound metabolized to carcinogenic form in an animal species but not in humans can be determined without too much thought.

The yes–no philosophy, if extended, could lead to automatic disapproval of almost any substance considered. If the yes–no approach to carcinogenesis is logical, the next biological step might be mutagenesis (Is the compound mutagenic in any test at any test level?), then reproductive and teratological effects, then effects on any specific organ, and so forth.

There is a danger that the yes–no approach can adversely impact on rational assessment of food safety in other areas such as the trend toward standardized protocols and regulations for the conduct of animal studies. This potential is illustrated by the possibility that a negative study, departing in any regard from a rigid checklist, would be unacceptable for regulatory purposes; yet a study indicating positive effects, no matter how many departures from sound toxicology were involved, would classify the substance as unsafe. There are two major problems with these trends: they can lead to unnecessary studies, diverting scarce resources from higher priority problems, and they in no way assure that the right questions are asked on a case-by-case basis about the particular substance involved. We would do well to remember the admonition of the National Academy of Sciences:

> There is no substitute for the vigilance of an inquiring and skeptical mind which has assumed the full responsibility for planning, conducting and evaluating the results of toxicity tests in making safety assessments. If that responsibility is lessened by an exclusive dependence on a "checklist" approach, the major assurance has been lost that a responsible, perceptive and efficient investigation will be conducted (11).

The trend of the regulatory agencies to rely more and more on the "checklist" approach rather than what might be called *toxicological evaluation* is understandable. The FDA operates under a basic law that in many respects is inflexible. Then, too, the checklist approach makes the regulatory job much easier and removes the necessity to utilize and justify the scientific judgment that would otherwise be required. Finally, this approach appeals to the legal mind. Since many or even most of the difficult questions are ultimately decided by the courts, the yes–no answer is much more likely to be sustained on the basis of the record than is the scientific answer, which probably is "maybe." Although the black or white classification of answers is a regulatory convenience, toxicological problems like most other aspects of life come only in various shades of gray.

The public policy question involved in all of this is fundamentally whether, as a society, we want the regulatory agencies to rely on the yes–no approach. If that is the case, then we should embrace the Delaney Clause and the food additive safety provisions of the FFD&C Act and extend them to all toxic effects that might be associated with food and indeed beyond food to other environmental effects. It should be realized, however, that extension of the yes–no approach can ultimately lead only to disapproval of the entire food supply, for as we have seen in the previous six chapters, absolute food safety is impossible.

If, instead, the public policy question is resolved in favor of making scientific judgment an essential part of the safety decision process, there is hope. However, that path is not without difficulties. Quality toxicity testing will be required, but some flexibility must be built into the protocols and into the conduct, reporting, and evaluation rules. Further, there must be recognition of the fact that toxicology is a developing science. Therefore, differentiation must be allowed between accepted, standard techniques and those that are still research techniques. Information from both areas can and should be considered, but the flexibility to attach proper weighting to each must be permitted.

TOXICITY VERSUS SAFETY TESTING

The considerations involved in the safety assessment of food additives have been discussed in some detail in both Chapter 5 and Chapter 6 and are not repeated here. However, to understand food safety and the role of toxicology in its evaluation, a careful distinction must be made between toxicity testing and safety testing. To the toxicologists, toxicity is defined as the inherent ability of a substance to produce injury or adverse effects. Toxicity tests are carried out first to (*a*) identify what those adverse effects are and (*b*) determine the extent to which those effects occur as a function of dose. The basic purpose of the toxicity tests, then, is to find out what adverse effects the test substance produces regardless of the frequency, type, or level of exposure and to determine the dose-response characteristics of the substance relative to those effects. To elicit these answers, dose levels that are unrealistic in terms of human exposure will likely have to be utilized. Such dose levels, in fact, may be so high that the effects evidenced are due to some secondary mechanism rather than primary toxicity. A test substance cannot be characterized relative to toxicity without employing exposure conditions severe enough to produce the inherent toxic effects and to indicate the dose-response relationship involved.

Although the fact that the dose levels employed for toxicity evaluation may be unrealistically high in terms of human exposure is irrelevant to the toxicity question, it is highly relevant to the safety question. Recalling that absolute safety is impossible to demonstrate, safety can be defined as the probability that injury or other adverse effects will not occur from exposure to the test substance at a particular dose level and under specified conditions. It is, therefore, entirely possible to conduct an animal study that will answer the safety question for some particular exposure conditions but will give no information concerning the inherent toxicity of the test substance. Conversely, of course, a toxicity study can provide rather complete information about toxic effects but cannot answer the safety question unless there is also information about actual human exposure.

The distinction between toxicity testing and safety evaluation is not well understood, but the results of such toxicological thinking have been generally accepted in a public policy sense for the control of food safety. The exception to that statement is the case in which the toxic end-point is cancer. A separate treatment of the cancer situation, mandated by law (the Delaney Clause) has generated a wide investigation of and debate about the applicability of animal toxicity tests and how the results of such tests should be used for safety evaluation.

ANIMAL-TO-HUMAN EXTRAPOLATION

The information that we have on the toxicity of various components of our food supply has been derived mainly from animal tests, usually with rats and mice. This is especially the case relative to carcinogenesis testing, although in recent years an extensive body of information from a variety of genetics toxicology testing has also been generated. Public skepticism about the relevance to humans of animal tests conducted at dose levels much higher than human consumption levels has been widespread. The two basic questions in this regard relate to whether animal toxicity tests are appropriate for predicting hazards to man and, if so, how the extrapolation should be made.

The Animal Model for Toxicity

The question as to the appropriateness of animal toxicity tests must be answered with a qualified "yes." Toxicologists are generally committed to the belief that the effects observed in properly conducted animal tests, when suitably interpreted, are applicable to humans and that such tests represent the best currently available approach to the problem without actually doing the testing in humans. For example, the Office of Technology Assessment (12) strongly endorses the animal approach for carcinogenicity testing:

> Animal tests are the best current method for predicting the carcinogenic effect of substances in humans. All substances demonstrated to be carcinogenic in animals are regarded as potential human carcinogens; no clear distinctions exist between those that cause cancer in laboratory animals and those that cause it in humans. The empirical evidence overwhelmingly supports this hypothesis.

Some of the advantages of animal testing include the capability of studying a wide variety of effects, ranging from acute toxicity to long-term effects such as cancer, under controlled conditions. Effects on reproduction and offspring can also be studied. By sacrificing animals during the course of a study, the time dependence of an effect can be evaluated. In each case one or more routes of administration (e.g., in food, in drinking water, or by injection) can be used to mimic the manner in which human exposure occurs. A principal advantage of animal tests is time; normal life spans and gestation periods for rats and mice, for example, permit the assessment of reproductive effects, birth defects, and even chronic effects within a 2- to 3-year period.

The high doses utilized in animal studies are perhaps both advantageous and disadvantageous. Characteristically, we expect that the higher the dose, the greater the response. A dose that produces a response of, say, only 10% would not likely produce observed effects in an animal experiment, yet such a level of response occurring in the total U.S. population would be of serious consequence. The possibility of overlooking such a response is compensated for by testing at doses as high as possible, thus greatly increasing the likelihood of observing the effect. The disadvantage of that approach is that we gain no quantitative information about the dose-response behavior in the low-dose range (which probably corresponds to human exposure levels). In addition, care must be exercised in selecting the dose levels to be tested so that the results will be meaningful. This involves using, at most, the so-called maximum tolerated dose (MTD), that is, the highest dose that will not significantly affect survival or longevity or produce other toxicity in the test animals (13). (There is disagreement among toxicologists as to the MTD definition; arguments are advanced that in some animal testing the doses have been so high that they may have disrupted the normal physiology of the animal.)

There are several limitations to the use of animals for predicting toxicity in humans. Primary among these are the differences in kinetics between a given test species and humans, including differences in absorption, distribution, metabolism, and excretion. Further, the type of end-point and/or the target organ(s) may differ between animals and humans. Certain toxic effects such as those involving learning and behavior are at best quite difficult to assess, although this area is receiving increased emphasis, and the speciality of behavioral toxicology is coming into being (14, 15).

Generally speaking, mammals metabolize most of the compounds they ingest in similar fashion, but there can be pronounced differences between species. Such differences also occur with respect to other aspects of the kinetics involved. Therefore, the toxicologist attempts to select an animal model that handles the compound in similar fashion, if that is possible, and/or utilizes more than one species to increase the likelihood of detecting adverse effects. Of course, practical constraints of time, cost, and logistics have led to the popularity of rats and mice as the animal models.

Toxicity tests that are inappropriate for predicting the existence of a potential hazard in humans can produce either false negatives or false positives. Testing in more than one species helps to reduce the likelihood of both of these errors. However, regulatory toxicologists in particular must be more concerned about false negatives and, therefore, concentrate on that error, attempting to find the most sensitive species.

Specifically, with respect to cancer it is worth noting that virtually every form of human cancer has an experimental counterpart in animals and that every form of multicellular organism is subject to cancer, including plants, insects, and fish. There are, in fact, striking similarities between cancer in humans and in test animal species. Of course, there are differences, sometimes pronounced, in susceptibility between species, strains, and individuals for cancer, just as for other toxic effects. Therefore, especially with quantitative prediction, the same limitations and cautions apply to the cancer end-point as to any other toxic end-point.

In short, test animals can be useful models for predicting potential hazards in humans. The limitations associated with such models must be given careful consideration, however, in evaluating the potential human toxicity of a test substance. This requires maximum input by the toxicologist in the planning, execution, and evaluation of safety studies. Even the most detailed and comprehensive guidelines or legislative pronouncements are no substitute for the expertise provided by the experienced toxicologist. Then, too, it is critical that all existing information be utilized in making a safety evaluation. This should include the nature of the test compound (16) and its metabolism and kinetics in the different test species and in humans. Not only positive but also negative testing results should be considered, along with any auxiliary information, such as epidemiology studies, if they are available, and the results from genetics testing.

Risk Assessment Techniques

As noted previously, risk information relative to a given food constituent is usually available only in the form of animal or other laboratory test data. Usually, too, the response data are associated with dose levels several times in excess of normal human exposure. The problem is to infer what risk the food constituent presents to humans. Traditionally, the dose-response data obtained from the animal toxicity tests have been used to predict safe levels for humans. A threshold or a no-effect level (i.e., the dose that produces either no effect or for which the response is not different from that of the untreated control group) for the toxic effect under consideration is estimated from the animal data and then divided by a safety factor to obtain a safe level for humans.

Usually a safety factor of 100 is used to reflect both animal variability (by a factor of 10) and the potential difference in susceptibility between animals and humans (by a factor of 10). In some cases where human toxicity information is available a safety factor as low as 10 has been employed, and safety factors as high as 1000 or 2000 have also been used (e.g., for certain animal drugs) when there is uncertainty and concern about the toxic effect involved. Often further calculations are made to establish an ADI of the substance in question.

A safety factor of 100, or some other value, is admittedly somewhat arbitrary but can be in part justified by the noted animal variability in response and the potential for greater susceptibility in humans. Further, experience would indicate that the use of such a safety factor in the regulation of drugs, food additives, pesticides, and other chemicals to which humans are exposed has worked well over the years. However, when the toxic end-point of interest is cancer, the law (the Delaney Clause) proscribes the use of a safety factor in the case of a food or color additive or a drug used in a food-producing animal. (See Chapter 6 for a description of the legal aspects of these substances.)

The concept embodied in the Delaney Clause is that cancer-producing agents do not have a threshold or a no-effect level; however small the dose, there will be some response, and hence no carcinogen, regardless of its potency, is to be allowed. This aspect of the Delaney approach to safety evaluation is much criticized since the dose–response behavior of a substance is of fundamental importance in not only toxicology, but all biology including medicine. There is little, if any, argument in the case of carcinogens about whether they exhibit a dose-dependent response, since clearly they do. However, there is argument about whether some or all carcinogens have a threshold or a no-effect dose level at some point on the dose–response curve. Advocates of the no-threshold concept point out that the existence of a threshold has not been experimentally shown for any carcinogen. Their argument is obviously irrefutable since it is impossible to prove a negative. The absence of observed effects in even a very large number of animals proves only that there is at most a low degree of probability of effects in the total population.

The question of the existence of a threshold for carcinogens is an obviously important point with respect to food safety, since in safety evaluation it is generally the estimated no-effect level that determines the tolerance or other regulatory limit on the use of the test substance. As the argument about the existence of a threshold for carcinogens is incapable of absolute resolution, it will likely continue. In the meantime, it is clear that most toxicologists believe that even though there may not be a true threshold, there is, indeed, an effective or practical threshold for many carcinogens, with a response sufficiently low to be of no toxicological significance. Such practical thresholds can well be accepted as existing if one acknowledges the body's normal detoxification and repair mechanisms. Accepting the existence of effective thresholds then logically leads to the reasoning, advanced by many, that carcinogens should be regulated on a case-by-case basis, taking into account the observed dose–response evidence. In this way the same ground rules that are applied elsewhere in toxicology could be applied to carcinogens, thus permitting differentiation among carcinogens on the basis of potency rather than relying on the legislative approach of the Delaney Clause, which permits no differentiation.

The Delaney Clause has prompted a great amount of scientific and public-policy debate over its attributes and shortcomings. Along with the debate, a great deal of attention has been devoted to mathematical techniques for extrapolating observed animal dose–response data to provide estimates of the risk to humans, usually at dose levels much below those to which the animals have been exposed. The appeal of such risk estimation is that it avoids the classification of a substance as safe or unsafe and also avoids the necessity of entering the debate about thresholds. If some acceptably small level of risk is selected (that level must be a societal decision), the extrapolation techniques can be used to establish the corresponding dosage levels.

A detailed treatment of mathematical techniques for risk extrapolation is beyond the scope of this book, but a brief summary will assist the present discussion. It should first be noted that these extrapolation techniques have, for the most part, been applied only to carcinogenesis risk estimation. The Food Safety Council (6), however, has pointed out that an extrapolation model that encompasses a variety of forms for the theoretical dose-response function can be applied to any end-point whether it is from acute or chronic testing or from mutagenic or teratology studies. Roberts (17) has gone a step farther by advocating the replacement of safety factors with extrapolation techniques, arguing that "There is as little justification for the assumption that a non-carcinogenic response always has a threshold as there is for the assumption that a carcinogenic response never has a threshold."

There are various mathematical models that have been applied to the risk assessment problem. These have been reviewed by the Food Safety Council (6) and Roberts (17), among others. (The interested reader is referred to those reviews and to the references following for more detail.) The two most popular extrapolation models have been the Mantel–Bryan Model (18, 19) and the so-called one-hit model (20). Both have primarily been applied to carcinogenic risk estimation. Other major models include the Armitage–Doll multistage model (21), the Cornfield pharmacokinetic model (22), and the gamma multihit model (23); the last was recommended by the Food Safety Council. In addition, mathematical models (24, 25) have been proposed for risk extrapolation which are based on the time to response. Generally, however, response time data are not available from toxicity studies.

It is difficult at best to argue in favor of any one of these models over any other. As noted by Cornfield (22), all such models are subject to major shortcomings:

- The choice of model drastically affects the calculated dose corresponding to an acceptable risk level, several orders of magnitude in some cases.
- Observed dose response data can often be fitted equally well by any of the models, providing no obvious statistical basis for response behavior in the low-dose range.
- No firm scientific basis now exists for choosing among alternative extrapolation techniques.

Nonetheless, the use of such techniques in the case of carcinogenesis permits the evaluator to rank the substances involved in terms of their potential hazard and to make a priority assessment relative to control actions. For example, application of any of these techniques—even accepting the dose-response data at face value—would show that any cancer risk associated with saccharin is at worst several orders of magnitude less than that for aflatoxin or dimethylnitrosamine (6).

In a discussion of extrapolation methods prepared for the Office of Technology Assessment, Crump and Masterman (26) caution that the selected method should, to the extent possible, reflect the known or plausible information relative to the biological mechanisms involved. For example, these authors point out the theoretical arguments for treating "directly acting carcinogens" that have been advanced by several writers (27–29). The "directly acting carcinogens" are those for which either the agent itself or a metabolite acts at the cellular level to produce a heritable change that eventually leads to tumor formation. In such cases, and in others where the agent adds to an ongoing background process of tumor formation, the argument can be made that the dose–response function is linear at low doses (i.e., has a positive slope) (30). Because of the theoretical arguments that can be advanced for low-dose linearity of the dose–response function and the conservatism (relative to risk estimation) that is entailed, many scientists advocate extrapolation methods encompassing low-dose linearity unless it can be ruled out on the basis of knowledge of the mechanism involved.

Until we increase our present knowledge about the mechanisms involved in cancer, it will not be possible to utilize extrapolation models with certainty. However, these approaches are finding increasing use in the regulatory arena [e.g., the FDA proposal relative to animal drugs (31) that entails the one-hit model] and lend support to the valid need for relative risk assessment in food safety.

FOOD SAFETY REGULATION

Recognizing that we are only deluding ourselves when we seek the noble goal of absolute food safety—for such is not possible—what should be the basis for food safety regulation? This question is normally addressed in terms of the controversial Delaney Clause.

There are probably four options currently possible for dealing with the Delaney Clause. The first is to leave it as it is. That would satisfy relatively few and would maintain the current confusion and concern. A second approach would be to extend its coverage. This would satisfy even fewer of us and would multiply the confusion and concern, but this seems to be the direction in which we are heading. A third approach would be to simply eliminate it. That approach probably has more supporters than either of the first two approaches. However, it is not likely to happen, and even if the Delaney Clause were eliminated, the basic safety provisions of the FFD&C Act could leave us on the horns of the same regulatory dilemma. Thus we are left with the fourth approach, which is to modify the Delaney Clause in some fashion. Many different versions of this approach have been advocated.

The toxicologists have developed one collective version of a modified Delaney Clause. The Society of Toxicology Technical Committee has suggested the following wording:

> No additive shall be deemed free of risk of inducing human cancer if it is found to induce cancer when ingested by man or if it is found, after tests which are appropriate for the evaluation of the health risk of food additives, to induce cancer in animal species that metabolize the additive in a manner similar to man.

This version contains the two important aspects of metabolism and risk assessment, both of which are basic to the restoration of scientific judgment to the regulation of food safety.

Lowrance (32) has suggested that safety should be redefined as a judgment of the acceptability of risk. That suggestion has a good deal of merit, for it is only in this way that public policy will be able to keep pace with the advancing frontiers of science. The advantage of treating carcinogens from the risk assessment point of view is that it recognizes the irrefutable scientific fact that not all carcinogens are alike. Some carcinogens are much more likely to produce human cancer, and not all human cancers are alike. If risk assessment were inserted into the regulatory process, there would be a quantitative mechanism for evaluating carcinogens individually rather than as a group.

The General Food Safety Issue

In addition to the specific issue of the Delaney Clause, the broader issue of food safety has also been widely debated. For example, the regulators themselves have addressed food safety. Miller (33) has stated that a regulatory system should (*a*) establish categories of risk (i.e., unavoidable contaminants, food and color additives, spices and herbs, nutrients, etc.), (*b*) be flexible enough to permit a reasoned response, (*c*) allow, when possible, a variety of regulatory options, and (*d*) encourage the development of the safest possible food supply. However, Miller also cautions that:

> if we are to move to a New Evangelism based upon the recognition of relative risks, then it must also be an Informed Evangelism founded upon sound data which reveal the nature and extent of the hazards to which we are exposed.

A wide range of ideas have been expressed by the legislators. It is interesting to note that one of the few congressmen with scientific training (organic chemistry), Representative James G. Martin, introduced a bill, H.R. 5091, in the 96th Congress to modify food safety assessment. This bill 5091 would permit approval of a food additive if:

> based on all the data presented the benefit from the reasonable anticipated use of the food additive by the persons who can reasonably be expected to use the additive outweighs the risk to such persons from

> such use, or that the risk to humans, as extrapolated to humans from test data on animals by normal statistical methods, is so low as to be clearly acceptable in the context of overall safety and reliability of the food supply.

This bill also addresses risk assessment and the use of the maximum tolerated dose concept in evaluating animal carcinogenic studies.

In response to requirements of the Saccharin Study and Labeling Act, the National Academy of Sciences examined food safety policy and issued a report containing a number of recommendations (34). These recommendations were that a new food safety system should be established with the basic features of comprehensiveness, discrimination among risk levels, flexibility in response, education, regulatory discretion, and accountability and jurisdiction. Basically, the academy recommended that there should be a single standard for the safety of all components of food. This standard should encompass four categories of risk (high, moderate, low, and none) with flexible regulatory options for each category. (The "none" category would cover those substances showing no evidence of hazard or presumed safe on the basis of use experience.)

The range of views on the complex subject of food safety regulation are illustrated by the papers presented at the 1979 Symposium of the Institute of Food Technologists on Developing Public Policy for Food Safety (33, 35–40). Despite the divergence of views, however, all these authors espouse the importance of science and risk evaluation in food safety policy decisions.

In many aspects of life we at least subconsciously recognize that risks are involved and also make decisions about our activities based on the degree of associated risk. Hutt (41) has summarized the annual risk of death for several common activities, and some of these are illustrated in Table 7.1. The data in Table 7.1 indicate the fact that no aspect of life is free of risk. However, the risks associated with food are not as obvious as the examples cited. We have seen in previous chapters that deaths associated with foods are the exception rather than the rule. We have also seen that the more subtle toxic effects attributed to foods, although present naturally as well as because of the actions of humans, are difficult to quantitate. Nonetheless, by considering what is known and what can be determined about food risks, food safety can be put in proper perspective and meaningful priorities for attention can be assigned.

A New Food Safety Policy

The continuing efforts of government and industry to not only keep food risks at a minimum, but to further decrease those risks would be materially aided by a national food safety policy which set priorities

TABLE 7.1 Annual Risk of Death Associated with Common Activities in the United States

Type of Activity	Risk of Death Per Year Per 100,000 Population
Motor Vehicles	22.0 (1/4,550)
Home Accidents	1.2 (1/83,330)
Alcohol (cirrhosis-1974)	16.0 (1/6,250)
Air travel (one transcontinental trip per year) (1974)	0.3 (1/333,330)
Amateur Boxing (40 hours/year)	2.1 (1/47,620)
Skiing (40 hours/year)	3.0 (1/33,330)
Canoeing (40 hours/year)	40.0 (1/2,500)
Rock Climbing (40 hours/year)	100.0 (1/1,000)
Fishing (drowning)	1.0 (1/100,000)
Recreational (drowning)	1.9 (1/52,630)
Bicycling	1.0 (1/100,000)
Occupational	
Airline Pilot	30.0 (1/3,330)
Truck Driver	10.0 (1/10,000)
Fire Fighter (1971-72)	80.0 (1/1,250)
Coal Miner (1970-74)	130.0 (1/770)

[a] Based on 1975 data except as noted.

Source. P. B. Hutt, Food Drug Cosm. Law J., 33 (10), 558 (1978).

based on the risks rather than the politics and the emotion involved. Such a policy sould contain the elements of:

- Comprehensiveness
- Consistency
- Flexibility
- Peer review
- Public participation
- Education

It is critical that our national food safety policy encompass all areas of potential hazards, including those of microbiological, nutritional, environmental, natural, and food additive origin discussed in this book. To do otherwise is at best a waste of scarce scientific resources and at worst a jeopardy to our health. We must consider hazards of natural as well as man-made origin and direct our efforts in the priority areas dictated by the risks to humans. Further, we have seen in previous chapters the very large extent to which individual food-related problems can interact with one another. Thus only a comprehensive approach can be truly effective in maintaining and improving food safety.

Consistency must also be a keystone of food safety policy. This does not imply a rigid checklist or a yes–no approach, but rather a general framework permitting the treatment of individual problems on a case-by-case basis. That framework should apply whether the origin of the problem is man-made or due to nature, or both. The basis for consistent treatment must be the evaluation of risk. As noted in the preceding, such evaluation would entail identification of the type, the extent, and the onset time of adverse effects; species-to-species variability in susceptibility; differences in metabolism and kinetics among test species and humans; dose–response characteristics; and human exposure, including any particularly vulnerable subpopulations. The subsequent regulatory mechanism selected should be aimed at reducing the risks involved, again regardless of whether the origin of the problem is humans or nature (41). However, the regulatory control imposed must be consistent with the additional practicalities involved in each particular case.

Maintenance and improvement of food safety also requires flexibility in regulatory policy (and hence in the law that provides the basis for that policy). There must be sufficient latitude in the regulatory process to maintain consistency yet still permit the tailoring of regulatory controls to the individual situation. Different components of food involving the same type and level of risk should not necessarily be regulated in exactly the same way unless the risk is so great as to warrant rigid control or so minimal as to warrant little or no control. Although reduction of risk should always be the regulatory objective, the means of achieving that objective will differ and will likely involve different economic, health, and other factors depending on the food constituent involved. Unless precluded by an obviously high level of risk, which is rare in food safety, the ultimate regulatory control thus should take into account the impacts on health and on the nutritional value, the availability, the acceptability, and the consumer cost of foods. Such flexibility in regulatory policy could reduce risks across the board without unduly impacting on the food supply. In addition, it would leave open the possibility of informative labeling of certain foods or other means of conveying consumer information, thus preserving freedom of choice in the marketplace. Such labeling or other information must, however, be considerably improved over the simple and misleading kind of dire warning utilized in the case of saccharin (41). Probably the best reason for permitting, or better yet, directing, flexibility in regulatory control of food safety is the reminder of the Food Safety Council:

> The purpose of regulation should be to serve the best interests of the consumer and neither to be subservient to the producer nor to dictate to the consumer. In our current system there are many instances of arbitrary restriction on freedom of choice (42).

If we are to restore some measure of public credibility to scientists and especially regulatory scientists, peer review must play a greater role in food safety decisions. Such review in other than routine decisions could assist measurably in improving overall food safety and could greatly augment the scientific expertise available to regulatory agencies. Subsequent to an evaluation of the risks involved in a given situation, the regulatory agency(ies) involved should have the prerogative to convene an appropriate expert panel to review that evaluation. (The emphasis should be on "appropriate" and on "expert." Members should be recognized authorities in the particular areas involved and not merely self-appointed "experts.") By the same token, parties that could be affected by a subsequent regulatory decision should have the right to request and, having shown cause, be granted the formation of such an expert panel. The one caveat to the inclusion of peer review in the regulatory process is the obvious requirement of a regulatory agency to deal with emergency situations. In emergency situations requiring immediate actions, temporary controls could be invoked, to be reviewed subsequently by an expert panel and then modified as necessary. It is also important that the findings of an expert panel be made public in sufficient detail to permit public understanding of the issues involved.

Public participation has become an integral part of regulatory rulemaking. It should remain a part of any new food policy and should be incorporated into any new food law. Information gathered by a regulatory agency relative to a given food component or food problem should be made public, and comment should be invited prior to any agency action. As noted, this would include the findings of any expert panel involved in the issue. In the event of emergency, the immediate action taken should be explained to the public as soon as possible. Any subsequent modifications to the initial approach should also be made public. The information made public should include details on the types of risks and their potential extent in humans and possible regulatory options. For the options cited, the impacts should be identified, including any reduction in risks and effects on consumer costs, availability, and acceptability of foods, nutrient values, and any health-related effects. Such information must be balanced and complete, however. Negative as well as positive effects observed in animal tests, any available epidemiology or other auxiliary information, exposure data, and so on, must all be presented to present a useful explanation to the public.

The final piece in the puzzle of a new food safety policy, and in many ways the most important, is that of education. Consumers are beginning to realize that absolute food safety is impossible and are showing some willingness to weigh the risks in a given situation against what they perceive to be benefits and to act accordingly. However, few consumers have even a basic understanding of animal testing or its role in predicting human hazard. Perhaps fewer still are aware of the importance of microbiological hazards, realize that essential nutrients can be toxic, or recognize that naturally occurring foods contain toxicants. Many consumers are not aware of the fact that our life spans are increasing and that the death rates of major diseases such as heart disease and many forms of cancer are decreasing. Industry, government, and academia have collectively done a poor job of conveying to consumers the admittedly complex facets of food safety! Thus much of the public's "education" in this regard has been left to alarmists, attention seekers, and purveyors of worthless (and sometimes dangerous) health products. For public acceptance and success, any new food safety policy must encompass a mechanism for first explaining the basics of food safety to consumers. There must also be a built-in mechanism for explaining the facts of a food-related problem and the reasons for adopting a particular regulatory course of action. Accomplishing these education needs will by no means be an easy task, but with a dedication to that end incorporated into a new food safety policy, it can be done.

CONCLUSION

Although many food professionals agree on the need for a new food safety policy and even agree on many of its components, such change will be slow in arriving. Modernization of the Delaney Clause and associated food safety demonstration requirements to get past the most visible current issues could be accomplished with only minor modifications of the FFD&C Act, but even this is not likely to happen quickly. Revision of food safety law to incorporate consistent risk evaluation and permit regulatory flexibility would require considerable modification of nearly all sections of the Act as well as modifications of associated food laws. New public policy of this nature is always slow in evolving.

Despite the difficulties of achieving it, new food safety policy is critically needed and must continue to be sought after by Congress, regulatory agencies, industry, academia, professional societies, and public interest groups. All these groups must explain to the public that we cannot guarantee a risk-free society, and this pertains to food as well as other aspects of life. We must painstakingly strive to present to the public in understandable terms the whole complex story of food safety, instead of taking the easy way out—the oversimplified but alarming and confusing statement on each new food safety issue. It is only by this route that consumers will come to understand the issues and accept the fundamental changes that need to be accomplished.

REFERENCES

1. Interagency Regulatory Liaison Group, *Draft I. R. L. G. Guidelines for Selected Acute Toxicity Tests,* Environmental Protection Agency, Washington, D.C., 1979.
2. United Kingdom Department of Health and Social Security, *A Consultative Document on Guidelines for the Testing of Chemicals for Carcinogenicity,* London, 1979.
3. United Kingdom Department of Health and Social Security, *A Consultative Document on Guidelines for the Testing of Chemicals for Mutagenicity,* London, 1979.
4. Food and Drug Administration, *Fed. Reg.,* 43, 59986 (1978).
5. Select Committee on GRAS Substances, *Fed. Proc.,* 36 (11), 2519 (1977).
6. Food Safety Council, *Food Cosmet. Toxicol.,* 16, Suppl. 2 (1978).
7. Interagency Regulatory Liaison Group, *Fed. Reg.,* 44, 39858 (1979).
8. P. B. Hutt, *Food Drug Cosmet. Law J.,* 33 (10), 541 (1978).
9. IFT Expert Panel on Food Safety and Nutrition, *The Risk/Benefit Concept as Applied to Food,* Institute of Food Technologists, Chicago, 1978.
10. A. C. Kolbye, Jr., *Proceedings of the Third Annual Meeting, The Toxicology Forum,* 84, Washington, D.C., 1979.
11. National Academy of Sciences, *Principles for Evaluating Chemicals in the Environment,* Washington, D.C., 1975.

12. Office of Technology Assessment, *Cancer Testing Technology and Saccharin,* U.S. Government Printing Office, Washington, D.C., 1977.

13. I. C. Munro, *J. Environ. Pathol. Toxicol.,* 1, 183 (1977).

14. C. V. Vorhees, R. E. Butcher, R. L. Brunner, and T. J. Sobotka, *Toxicol. Appl. Pharmacol.,* 50, 267 (1979).

15. R. L. Brunner, C. V. Vorhees, L. Kinney, and R. E. Butcher, *Neurobehav. Toxicol.,* 1, 79 (1979).

16. G. M. Cramer, R. A. Ford, and R. L. Hall, *Food Cosmet. Toxicol.,* 16, 255 (1978).

17. H. R. Roberts, "Food Safety Risk Assessment," paper presented at Peer Review Seminar, Food Safety Council, Palo Alto, California, 1979.

18. N. Mantel and W. R. Bryan, *J. Nat. Cancer Inst.,* 27, 455 (1961).

19. N. Mantel, N. Bohidar, D. Brown, J. Ciminera, and J. Tukey, *Cancer Res.,* 35, 759 (1971).

20. D. Hoel, D. Gaylor, R. Kirschstein, U. Saffiotti, and M. Schneiderman, *J. Toxicol. Environ. Health,* 1, 133 (1975).

21. P. Armitage and R. Doll, "Stochastic Models for Carcinogenesis," *Proceedings of the Fourth Berkeley Symposium on Mathematical Statistics and Probability,* Berkeley, California, 1961.

22. J. Cornfield, *Science,* 198, 693 (1977).

23. J. Cornfield, F. Carlborg, and J. Van Ryzin, "Setting Tolerance on the Basis of Mathematical Treatment of Dose Response Data Extrapolated to Low Doses," *Proceedings of the First International Toxicology Congress,* Lyon, France, 1978.

24. H. Druckrey, "Quantitative Aspects of Chemical Carcinogenesis," in R. T. Truhaut, Ed., *Potential Carcinogenic Hazards from Drugs,* Springer-Verlag, Berlin, 1967, Monograph No. 7.

25. N. Chand and D. G. Hoel, "A Comparison of Models for Determining Safe Levels of Environmental Agents," in F. Proschan and R. J. Serfling, Eds., *Reliability and Biometry, Statistical Analysis of Lifelength,* Society for Industrial and Applied Mathematics, Philadelphia, Pa., 1974, pp. 681–700.

26. K. S. Crump and M. D. Masterman, "Review and Evaluation of Methods of Determining Risks from Chronic Low-Level Carcinogenic Insult," in Office of Technology Assessment, *Environmental Contaminants in Food,* U.S. Government Printing Office, Washington, D.C., 1979, pp. 154–165.

27. R. Peto, "The Carcinogenic Effects of Chronic Exposure to Very Low Levels of Toxic Substances," paper presented at NIEHS Conference on Cancer Risk Estimation, Pinehurst, North Carolina, 1976.

28. K. S. Crump, D. G. Hoel, C. H. Langley, and R. Peto, *Cancer Res.,* 36, 2973 (1976).

29. H. A. Guess, K. S. Crump, and R. Peto, *Cancer Res.,* 37, 3475 (1977).

30. G. S. Watson, *Proc. Nat. Acad. Sci.,* 74, 1341 (1977).

31. Food and Drug Administration, *Fed. Reg.,* 44, 17070 (1979).

32. W. W. Lowrance, *Of Acceptable Risk*, Kaufmann, Los Altos, California, 1976.

33. S. A. Miller, *Food Technol.*, 33 (11), 57 (1979).

34. National Academy of Sciences, *Food Safety Policy: Scientific and Societal Considerations*, Washington, D.C., 1979.

35. I.C. Munro, *Food Technol.*, 33 (11), 43 (1979).

36. R. Zeckhauser, *Food Technol.*, 33 (11), 47 (1979).

37. B. N. LaDu, Jr., *Food Technol.*, 33 (11), 53 (1979).

38. C. J. Brown, *Food Technol.*, 33 (11), 61 (1979).

39. J. M. Turner, *Food Technol.*, 33 (11), 63 (1979).

40. A. S. Clausi, *Food Technol.*, 33 (11), 65 (1979).

41. P. B. Hutt, *Food Drug Cosmet. Law J.*, 33 (10), 558 (1978).

42. Food Safety Council, *Food Technol.*, 34 (3), 79 (1980).

Author Index

Subject Index